HYDROCARBON FLUID INCLUSIONS IN PETROLIFEROUS BASINS

HYDROCARBON FLUID INCLUSIONS IN PETROLIFEROUS BASINS

VIVEKANANDAN NANDAKUMAR

J.L. JAYANTHI

ELSEVIER

Elsevier
Radarweg 29, PO Box 211, 1000 AE Amsterdam, Netherlands
The Boulevard, Langford Lane, Kidlington, Oxford OX5 1GB, United Kingdom
50 Hampshire Street, 5th Floor, Cambridge, MA 02139, United States

Library of Congress Cataloging-in-Publication Data
A catalog record for this book is available from the Library of Congress

British Library Cataloguing-in-Publication Data
A catalogue record for this book is available from the British Library

ISBN: 978-0-12-817416-6

Publisher: Candice Janco
Acquisitions Editor: Amy Shapiro
Editorial Project Manager: Billie Jean Fernandez
Production Project Manager: Bharatwaj Varatharajan
Cover Designer: Mark Rogers

Typeset by SPi Global, India

Contents

Preface

Application of fluid inclusion techniques for oil exploration activities across the globe needs to be encouraged and put into practice, considering the vital database that could be generated from drill core samples or wall cutting samples obtained during exploratory drilling. Even as renewable energy resources are increasing rapidly, oil and gas remain the primary energy source globally. Laxity in the development of new tools for oil exploration to target and discover this decreasing primary energy source could result in an energy crisis for the global community as its transitions to renewable energy resources will take many decades. Fluid inclusion techniques can help us to constrain thermal maturity of sedimentary basins where vitrinite is absent. Fluid inclusion investigations can assist in the interpretation of failed exploratory wells and can answer science questions such as the repeated well failures in petroliferous basins like the Kerala offshore in the Western Continental Shelf of India, where all the 15 exploratory wells emplaced over the years went dry. Although several laboratories are taking lead in fluid inclusion applications to basement hard rocks, only a few are researching on its applications in sedimentary rocks and their resources. This book is intended to provide a stimulus for many oil companies, laboratories and universities globally to take up research work in hydrocarbon fluid inclusions (HCFIs) in sedimentary basins.

HCFIs, similar in appearance to aqueous fluid inclusions during normal petrographic examination, fluoresce under UV-light excitation and are thus easily distinguished. They are direct samples of petroleum fluids that have either been generated in the host rocks or migrated from source rocks, or reflect petroleum in the process of migration. In essence, HCFIs are "hidden petroleum shows." These inclusions along with coeval aqueous inclusions are stores of vital geological information that can aid petroleum exploration. Parameters like oil window, a maturity index for the various strata to generate oil or gas, diagenetic and postdiagenetic temperatures, API gravity, species identification of hydrocarbons, and more, can be quantified or assessed by studying hydrocarbon fluid inclusions.

Spectroscopic techniques have been employed to determine the API gravity of petroleum oils—a commercial indicator. Discriminating HCFIs in the fluid inclusion wafers of sedimentary rocks has traditionally posed a great challenge, now overcome with special wafer preparation techniques, the key element to successful HCFIs study. The interpretation of fluorescence emission of oils in HCFIs and the determination of API gravity are the major areas of focus of this volume. Analysis of the constituents in HCFIs using Raman spectroscopy, detailed in this book, can indicate compositional differences in oils of different generations. By superimposing fluid inclusion data on the Burial History Curve, the Petroleum Charge History of a basin can be determined. This book provides comprehensive information on the fluids that have permeated or interacted with the major formations in the Mumbai and Kerala Konkan Basins of the Western Offshore of India, two representative basins in which the fluid inclusion techniques that we have developed have been tested.

HCFI analysis is significant in terms of determining the oil quality, identifying the constituents, which depends on the maturity and the determination of "oil window" of the various geological horizons in a basin. The volume introduces this nondestructive spectroscopic approach for addressing these challenges in petroleum exploration. Identifying HCFIs from a sedimentary rock sample has been a challenging task, thus, we urge readers to familiarize themselves with the methodology that has been developed to prepare fluid inclusion wafers, the first step in the process. A challenge to the use of Raman spectral studies on natural HCFIs is the presence of fluorescence emissions from minerals and aromatic compounds in HCFIs that leads to masking of Raman signals. Selection of the optimum excitation wavelength to obtain Raman signals is a further challenge. To overcome these hurdles, the special wafer preparation techniques, along with the use of fluorescence quenchers, that have been employed by us are also introduced in this book.

Rapid estimation of API gravity (APIG) from fluid inclusions in the drilling processes itself could form a significant step in managing crude oil extraction in any oil field. We report a methodology to deduce the APIG of oil from HCFIs in drill/cutting samples (single mineral grain) of well cuttings from lithologies (reservoir, or carrier beds) encountered during drilling that have the capability to act as oil-producing zones. We are introducing a procedure for capturing the fluorescence emission spectra of HCFIs trapped in minerals such as feldspars and quartz contained in specially prepared sample "wafers." This book elaborates on fluorescence emission data of pure petroleum oils of known APIG and proposes an empirical graphical tool for predicting the APIG of oils in micron-sized HCFIs through this noninvasive, nondestructive procedure that capitalizes on microscopy-based fluorescence techniques. This book also examines how best to detect the Raman signals from natural hydrocarbon-bearing fluid inclusions using an excitation wavelength of 785 nm with suitable optical parameters and special wafer preparation techniques to negate the background fluorescence.

Dr. V. Nandakumar
Dr. J.L. Jayanthi

Acknowledgments

The book "*Hydrocarbon fluid inclusions in petroliferous basins*" is based on our expertise in the research on fluid inclusions in sedimentary basins, a major scientific program of the National Centre for Earth Science Studies. Dr. V. Nandakumar and Dr. J.L. Jayanthi gratefully acknowledge the financial support of the Ministry of Earth Sciences, Government of India to carry out research on the Palaeofluids in sedimentary basins. This study would not have been possible without the support of Dr. Shailesh Nayak (Former Secretary, Ministry of Earth Sciences, Government of India), and Dr. M. Rajeevan (Secretary, Ministry of Earth Sciences, Government of India). The support, encouragement, and coordination extended by Dr. M. Baba and Dr. N.P. Kurian (Former Directors, Centre for Earth Science Studies and NCESS) are remarkable. The Directors of NCESS Dr. T.N. Prakash, Dr. V.M. Tiwari and Dr. M. Shamsuddin are acknowledged for their timely help and support during this hydrocarbon fluid inclusion study at NCESS. Prof. Somnath Das Gupta, Chairman, Research Advisory Committee of NCESS is thanked for his ardent support and encouragement throughout the scientific program. Dr. (Mrs.) Parvinder Maini, Scientist-G, MoES is thanked for her support for the initiation and execution of this program. Mr. P. Radhakrishnan, Section Officer, Ministry of Earth Sciences, Govt. of India is thanked for his encouragement and advice.

Prof. M. Santosh, Foreign Expert and Professor of Geology, School of Earth Science & Resources, University of Geosciences, Beijing, China, the patriarch of fluid inclusion studies in India has always been a guiding star to V. Nandakumar. The idea of having such a book on hydrocarbon fluid inclusions emerged during a discussion of V. Nandakumar with Prof. M. Santosh. Prof. K.P. Thrivikramji, Former Head of the Department of Geology, Kariavattom, University of Kerala, a visionary and a mentor guided V. Nandakumar to undertake challenging research topics in petroleum geology. The support extended by Prof. M.K. Panigrahi, Department of Geology & Geophysics, Indian Institute of Technology, Kharagpur, and Prof. Medury Sastri, Head R&D, Gemmological Institute of India & Former Head, Spectroscopy Division, BARC, Mumbai during the program has been invaluable.

Deeply and sincerely cherishing the contributions of Dr. Jan Kihle, Senior Geologist, IFE, Norway for his invaluable support and guidance for this research to Dr. Nandakumar. Prof. Robert H. Goldstein, Professor of Geology, University of Kansas, USA, is specially thanked his scientific interactions during the PACROFI conferences.

The Oil & Natural Gas Corporation (ONGC, Govt. of India) is thanked for logistic support by providing core and cutting samples of exploratory wells from the Western offshore, India. The active participation and coordination of Mr. H. Upadhyay (Former GM Geology, KDMIPE, Dehradun), Mr. D.K. Pande (Former Director-Exploration, ONGC), Mr. P.K. Bhowmick (Former Head, KDMIPE), Dr. R.K. Shukla (Former HoI, KDMIPE, Dehradun), Mr. U.C. Pradhan (Former GM Geology, KDMIPE, Dehradun) and Dr. Dawae (Former GM Geology, RGL, ONGC, Panvel) are acknowledged. Dr. S.K. Biswas (Former ED and HoI, KDMIPE, Dehradun) is thanked for his active encouragement and consultation during the development of the research program.

Prof. Simon L. Harley, Professor in Lower Crustal Processes, Department of Geology and Geophysics, University of Edinburgh, UK had been a constant encouragement to Dr. Nandakumar for all kinds of scientific endeavors. Dr. Nandakumar is having a long standing scientific collaboration and friendship with Prof. S.L. Harley. We are grateful to Dr. Subhash Narayanan (Scientist G and Group Head (Rtd), NCESS) for his suggestions and timely advice and Jayanthi is thankful to him for his research guidance leading to PhD. Prof. Sudershanakumar (Late) and Prof. Sobhana Warrier (University College, Thiruvananthapuram) are also acknowledged. Prof. Sobhana Warrier introduced V. Nandakumar to higher level of physics and had been a teacher, a friend, and a mentor. Mr. B.D. Mukharjee of Leica Microsystems is acknowledged and M/s Agilent Technologies Ltd is thanked for the GCMS analysis. Our colleague Mr. Nishant Neelayi (Scientific Assistant, NCESS) deserves special mention for his technical support. Mr. K. Eldhose (Technical Assistant, NCESS), S.S. Anoop, Mr. Shivapriya (Scientific Assistant, NCESS), and T.K. Silpa are thanked for their help in preparing the fluid inclusion wafers. Dr. D. Padmalal, Scientist-G and Group Head of NCESS is specially thanked for his encouragement and support.

The authors are very much thankful to Ms. Amy Shapiro (Editor) and Ms. Billie Jean Fernadaz (Editorial Project Manager) of Elsevier, USA for their timely help, support and communications throughout the book project and Mr. Bharatwaj Varatharajan (Production Project Manager) is thanked for his support during production time.

Dr. V. Nandakumar

I am extremely grateful to my family for making me available as a scientist to undertake challenges in fluid inclusion research. I thank my wife Dr. Bindu Valiyakkil Balakrishnan (Gynecologist, Corniche Hospital, Abu Dhabi, UAE) for her love, affection, and constant support for all my endeavors. I thank my daughter Greeshma B. Nandakumar and my son N. Nandakishore for their love and understanding. I'm deeply indebted to my mother Mrs. M. Sarvamma (Former Commissioner, Travancore Devaswom Board, Govt. of Kerala) for her love, care, the firm support and the guidance throughout my career as a scientist. My father Mr. N. Vivekanandan (Late) kindled the scientific spirit in me to grow as a scientist, and a good human being. Dr. Baiju Madhavan and Dr. Cibi Mukundan (Kettering, UK) had always been a constant encouragement throughout my career. My friend late Mr. K.V. Krishnakumar (Indian Air Force) is remembered for his love, affection, and support throughout my school, college days, and in my career. My childhood friends (Stella Marys and St. Chrysostom convent, Nellimood, Thiruvananthapuram) have been incredibly supportive in such endeavors.

Dr. Jayanthi J.L.

I give all the glory and honour to my Lord and Saviour Jesus Christ, 'The Light of the World', who is always with me and made me a scientist. I am deeply indebted to my husband Pr. Joseph Nadhanael (Jose) for his unconditional love, encouragement, and prayers for me. I am extremely thankful for his wholehearted support, understanding and guidance in my career without which this would not be possible for a mother of two small kids. I thank my kids, Jeremiah Nadhanael Joseph and Jemimah Nadhanael Joseph for their love towards me. I express my deep sense of gratitude to my parents and in-laws for their love, care, support, and prayers. I thank my parents Mr. Jayaraj H and Mrs. Lalitha Jayaraj and in-laws, Mr. Aliyas Kallupurackal and Mrs. Marykutty Aliyas for all their love and support. It is my pleasure to acknowledge my colleagues at NCESS, Dr. Tiju I. Varghese (Project Scientist), Mr. Nishant Neelayi (Scientific Assistant), Dr. Nilanjana Sorcar (Scientist C), Mr. Shivapriya (Scientific Assistant), Mr. Arunlal (Research Fellow), Ms. Silpa Thankan (Inspire Fellow) and Mrs. Sreejambika (Office Assistant) for their support in getting some figures/references for the completion of this book.

CHAPTER 1

Introduction to fluid inclusions

V. Nandakumar[a] and J.L. Jayanthi[b]
[a]Scientist - G, National Facility for Geofluids Research and Raman Analysis, National Centre for Earth Science Studies (NCESS), Thiruvananthapuram, Kerala, India
[b]Project Scientist - C, National Facility for Geofluids Research and Raman Analysis, National Centre for Earth Science Studies (NCESS), Thiruvananthapuram, Kerala, India

1.1 Entrapped fluids: Fluid inclusions

When a crystal grows in the presence of a fluid phase, some of the fluid may get trapped in the imperfections in the growing crystals to form fluid inclusions. The trapped fluid may be liquid, vapor, or supercritical fluid and the fluid composition may include pure water, brines of various salinity, gas or gas-bearing liquids, petroleum, silicate, sulfides, carbon melts, etc. Thus, fluid inclusions represent the trapped portions of gas, liquid, or melts from which the crystal had grown and could be used to establish the environment in which a rock or mineral might be formed. In the literature, the term fluid inclusion has been used for inclusions that are trapped as fluid and remains as fluid at surface temperature (Roedder, 1984). In the early 19th century, eminent British scientists Sir Humphrey Davy and Sir David Brewster published fascinating accounts of very large fluid inclusions in quartz. But it was the pioneering work of Henry Clifton Sorby in the mid-1800s following the rapid development of optical microscopy that elevated the status of fluid inclusions from simple objects of curiosity to that of considerable scientific merit and importance.

After crystallization, the minerals of almost all terrestrial sedimentary, metamorphic, and igneous rocks get fractured one or more times and these fractures could have been healed in the presence of liquid or gaseous fluids. During these processes of crystal growth and fracture healing, small quantities of the surrounding fluid medium get trapped as fluid inclusion in the host crystal. The subject of fluid inclusions in sedimentary systems must include all of those systems forming at the Earth's surface and those that extend more deeply below the surface until conditions leave the digenetic-catagenetic realm probably at temperatures below about 140 °C. This would include evaporates formed from evaporation of saline surface and groundwater, and products of physical and chemical reactions in the subaerial realm where

Hydrocarbon Fluid Inclusions in Petroliferous Basins
https://doi.org/10.1016/B978-0-12-817416-6.00006-X

a multitude of fluid exists including oil, gas, and aqueous fluids of various origin.

Geologically, fluid inclusions are small voids that contain a variety of liquids, which are often found in natural minerals and rocks. They are regarded as small sealed vials, often less than 10 μm in size that host fossil fluids that exist when the minerals grew or healed after fracture formation (Roedder, 1984; Goldstein and Reynolds, 1994). The composition of fluid inclusions varies considerably, and may comprise liquid, solid, and/or gaseous phases, depending on the fluid source and pressure–temperature conditions experienced. These phases commonly include water, dissolved gases, and salts; in extreme cases, inclusions may host daughter minerals, high-pressure vapor phases, and complex organic mixtures. Among these, the inclusions, which contain hydrocarbon fluids, which originated from petroleum that once migrated through the rocks before becoming trapped, are of special interest to the petroleum industry (Goldstein, 2001; Munz, 2001). Fig. 1.1 shows some aqueous and hydrocarbon-bearing fluid inclusions (HCFIs) from the Mumbai offshore basin, India.

1.2 Significance of geofluids

Fluids play a vital role in virtually all crustal and mantle processes. The circulation of fluids has important effects on the transport of chemical constituents and heat and are the principal contributors in the formation of hydrothermal ore deposits. The mechanisms by which crustal rocks deform are strongly influenced by the presence of water, as well as by the pore-fluid pressure. It has also been suggested that fluids play a very significant role in earthquakes. On a broader scale, pore-fluid pressure influences the mechanical processes that control rock deformation in accretionary wedges associated with subduction zones. The volume of fluids carried to deeper levels in subduction zones appears to influence the rate and depth of melting and determines the site of volcanism in the overlying plate (The Role of Fluids in Crustal Processes, 1990). Water and other geofluids play an important role in the geochemical and archeological evolution of the Earth and other planetary bodies in the solar system. These fluids are responsible for the formation of hydrothermal mineral deposits, affect eruption behavior in volcanic systems and the geophysical properties of the mantle, and significantly affect the way in which rocks deform and fracture. Water is required for life to develop and survive and the search for life beyond Earth is naturally a search for water in the solar system and beyond (Bodnar, 2005). The minerals of

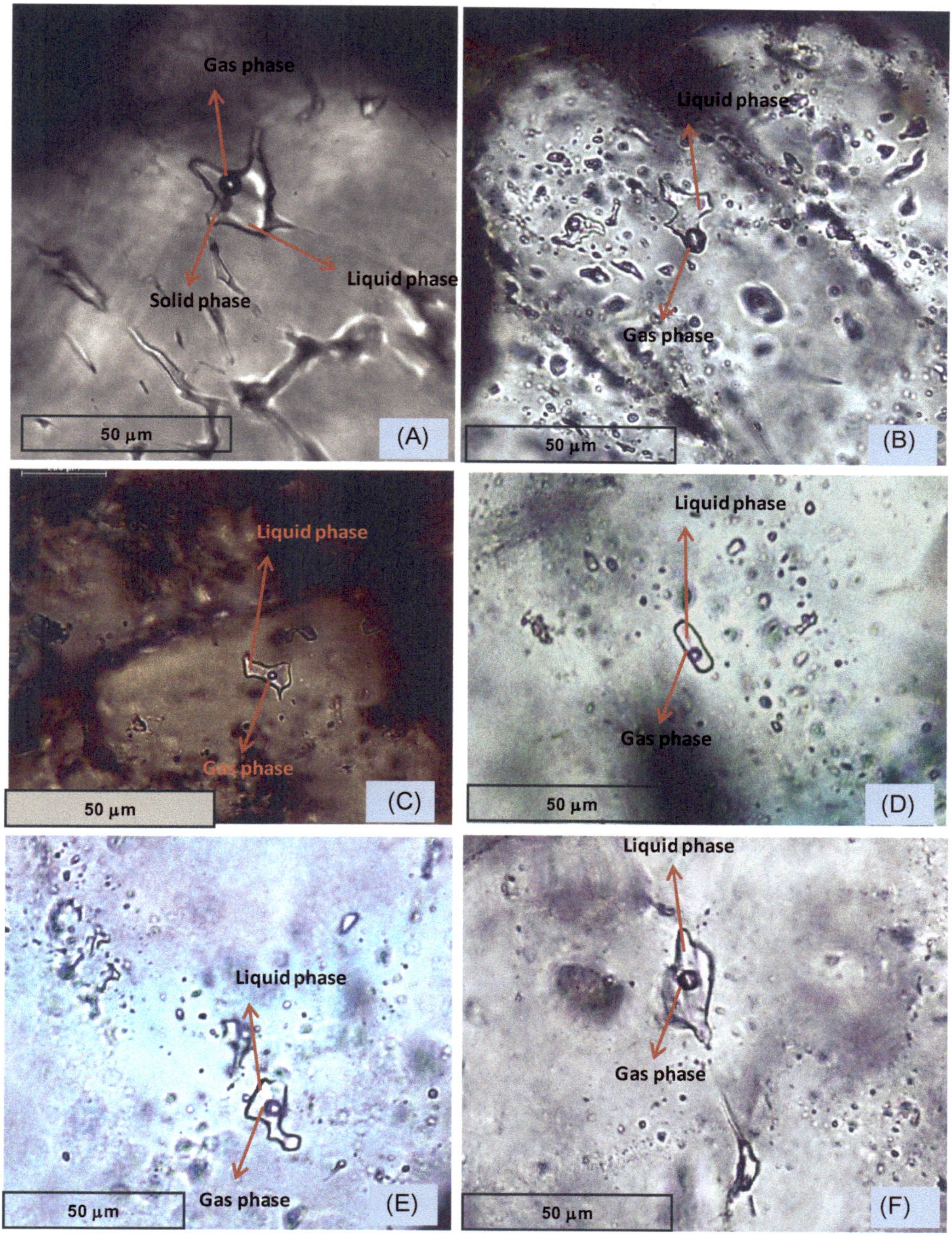

Fig. 1.1 Examples of multiphase (A), biphase aqueous (C, D, F) and hydrocarbon-bearing fluid inclusions (B, E) from Mumbai offshore basin, India.

many meteoritic and lunar samples and of terrestrial igneous rocks have grown from fluid silicate melts. All crystals in all terrestrial and extraterrestrial samples have grown from fluids with an exception of those crystals in metamorphic samples that have grown in the solid state. New crystals in many sedimentary and some metamorphic rocks and in almost all ore deposits are formed from aqueous fluid containing various solutes.

Aqueous fluids play a key role in mass transfer processes in the Earth, including the generation of magmas in the Earth's mantle above subduction zones, the release of fluids from crystalline magmas, the production and migration of fluids during mountain building, the formation of hydrothermal ore deposits, and the interaction of fluids released from the deep Earth with the hydrosphere and atmosphere. As the only topmost part of the Earth's crust is accessible to direct sampling of fluids (i.e., discharge from geothermal systems and deep drill holes), most of the information we have about deep Earth fluids comes from studying palaeo-fluids that are preserved as fluid inclusion in minerals.

Important fluids in the sedimentary realm include atmospheric gases, freshwater of meteoric origin, lake water, seawater, mixed water, evaporated water, formation waters deep in basins, oil, and natural gas. Preserving a record of the distribution and composition of these fluids from the past should contribute significantly to the studies of paleoclimate and global-change research, essential for improving understanding of diagenetic-catagenetic systems and can provide useful information in petroleum geology.

1.3 Fluids of the sedimentary realm

Freshwater of meteoric origin is the most important fluid associated with the sedimentary realm and includes freshwater precipitating on sediment and soil surfaces as rain water, freshwater from lakes and ground water, and water of meteoric derivations, compositionally modified through rock water interaction and evaporation. Seawater is another fluid that is common near the surface of the earth. Today, this fluid is relatively uniform in composition and maintains salinity near 35 ppt. But in the past, the total salinity might have varied, isotopic composition varied, and ratios of major, minor, and trace ions may have been different from what they are today. The chemistry of seawater may have changed so significantly through time that even the major carbonate minerals precipitating from it have changed through time giving rise to ancient seas dominated by aragonite/high-Mg calcite precipitation and seas dominated by calcite precipitation. This variation in seawater chemistry is further supported by secular variation in the composition of evaporate minerals, with seawater of one age giving rise to Mg-rich bitterness and seawater of another age giving rise to K-rich bitterness. Understanding secular variations in the composition of seawater is important as a paleoclimate indicator, for understanding the evolution of organisms, and is also important for understanding the paleoecology of ancient environments, tectonic controls on global cycling of chemical constituents in nature, another digenetic effect of seawater composition in carbonate sediments (Goldstein and Reynolds, 1994). Seawater that has been modified through evaporation or meteoric dilution falls into another class of fluids important in sedimentary systems. Brackish waters, created by dilution of seawater with meteoric water, are important digenetic fluids, which may be responsible for dissolution of carbonate minerals and precipitation of calcite and dolomitization (Land, 1973; Goldstein and Reynolds, 1994). Seawater that has been modified through evaporation is a diagenetically reactive fluid that may be important in precipitating calcite, dolomite, and evaporates.

As one moves more deeply into the sedimentary sections, a wide variety of subsurface fluids may be encountered, along with the temperature and pressure increase normally encountered along with the geothermal gradient. Aqueous fluids ultimately may have been derived from meteoric and marine fluids, but typically, these have been altered so extensively through processes of evaporation at the surface, evaporate dissolution, or other type of rock-water interaction that they have achieved compositions significantly different from their parent fluids. These basinal fluids typically are very salty, may achieve concentrations well above those of marine fluids, and contain ion ratios that reflect their evolution. Many of these fluids are quite important for precipitation of digenetic minerals such as feldspar, quartz, anhydrite, calcite, and dolomite. In addition to aqueous systems, the subsurface may contain petroleum and various compositions of natural gas derived from organic matter.

1.4 Fluid inclusions in sedimentary and diagenetic systems

Detrital quartz grains from sediments often contain abundant fluid inclusions but these are usually unrelated to the fluids developed during burial and compaction of the sediment. Fluid inclusions related to diagenetic or low-grade metamorphic fluid processes are best preserved in the larger veins, vugs, geodes, and concretions sometimes present in these rocks. Diagenetic quartz and carbonate overgrowths and cements in medium to coarse grain sediments should, in theory, also contain fluid inclusions. Fluid dynamics in sedimentary basins is of tremendous interest, both from a scientific and an economic point of view. Integration between fluid inclusion and present-day fluid data provides the time aspect necessary for the reconstruction of fluid flow paths, and can be used for mapping fluid dynamics both on a regional basin scale and on the local scale of petroleum reservoirs. Aqueous fluids dominate in sedimentary systems, and actively participate in diagenetic processes (Goldstein, 2001).

Diagenetic reactions are important in controlling porosity and permeability in oil and gas reservoirs. Diagenesis of sedimentary rocks and carbonates can be closely linked to origin of the fluid responsible for digenetic alteration. Diagenesis can even be linked to the surface environment. Most sediments are deposited by marine fluids at the Earth's surface temperatures. Common diagenetic phases known to precipitate from seawater include fine grained cements of Mg-calcite or aragonite compositions and coarser Mg-calcite and aragonite cements with a variety of morphologies. Low-Mg calcite and dolomite are also known from these systems. When seawater is mixed with freshwater of meteoric origin, there is some evidence for dissolution of carbonate minerals as well as evidence for precipitation low-Mg calcite in some settings and replacement of carbonate minerals with dolomite. At and below the water table, meteoric fluids may mix with other meteoric fluids of differing CO_2 content, encouraging dissolution of carbonate minerals. Dissolution of unstable carbonate minerals and out gassing of CO_2 encourages precipitation of calcite in this setting. In all of the above settings, minor silicate digenesis is also possible and is greatly dependent on the composition of the pore fluids and their modification by chemical components provided through interaction with the sediment.

As one goes deeper into the sedimentary section into a burial setting, or as warm fluids from deeper in a basin are injected towards the surface, many diagenetic reactions are driven by changing temperature. Among these is dissolution of minerals as well as precipitation of feldspar, quartz, and carbonates. Therefore, a sequence of diagenetic phases in ancient rock, determining the diagenetic environment of its precipitation is determined by two or three basic characteristic of the fluid that precipitated it—salinity, temperature, and pressure. These three parameters are those most easily determined from fluid inclusion analysis of the right suit of fluid inclusions, and the analysis of fluid inclusions in diagenetic minerals is probably the most unambiguous method for determining diagenetic history. Most fluids in the diagenetic realm are incredibly complex in composition. One of the best examples of this is seawater, a highly concentrated fluid (about 3.5 wt% NaCl) that has major amounts of Na^+, Mg^{2+}, Ca^{2+}, K^+, Sr^{2+}, Cl^-, $SO4^{2-}$, HCO_3^-, Br^-, F^-, and B and minor amounts of many other elements. In addition, the composition of organic and inorganic gases and hydrocarbon liquids is also quite complex.

Seawater: Seawater is one of the most important fluids in the sedimentary realm. Many diagenetic systems can clearly be tied to the marine sediment–water interface. Such ties are crosscutting stratigraphic relationships that either exclude the possibility of the fluids other than those derived from seawater, or that directly link diagenesis to a marine process such as deposition of marine internal sediment or submarine erosion.

Freshwater: Relatively pure H_2O is common in sedimentary systems, especially near the surface and in shallow parts of a basin. In sedimentary systems that have not been subject to hydrothermal activity, freshwater exists in the one-phase field, at a pressure above the vapor–liquid field boundary or boiling curve. It is not until these fluids are trapped or re-equilibrated in fluid inclusions at high temperature, and then cooled, that they reach the two-phase field boundary. The fresh-water system is best identified by its behavior during microthermometric cooling of fluid inclusions. The melting temperature of ice for these fluid inclusions is 0 °C. At room temperature, fluid inclusions with this composition will have an aqueous liquid phase and may or may not have a bubble or a gas phase.

Methane: Relatively pure methane is common near the surface where it results from biogenic degradation of organic matter (fermentation or CO_2 reduction), and deeper in the subsurface where it results from the thermogenic cracking of kerogen. In sedimentary systems, a single methane phase exists, above the −82.1 °C critical point. At room temperature, methane fluid inclusions commonly are identified by their dark and single-phase appearance. Methane inclusions could be identified by cooling because they commonly separate into distinct gas and liquid phase below the critical point and melt at the triple point for methane at −182.5 °C. At room temperature, crushing studies show that these inclusions exist at high pressure.

H_2O-NaCl: Aqueous solutions dominated by NaCl are common in many natural systems of the sedimentary realm. In fact, this system is so common and such a good approximation for even more complex salt solutions, that when the composition is not well known, scientists commonly interpret their fluid inclusion salinities as if they were from the H_2O-NaCl systems calculating salinities in wt% NaCl equivalent.

H_2O-NaCl-$CaCl_2$: Aqueous solutions dominated by NaCl and $CaCl_2$ are well-known deep in sedimentary basins where brines have undergone significant rock-water interaction. Although brine composition deep in basin can be quite complex and may have many more components, this class of brines serves as a good model for the composition of many sedimentary brines.

H_2O-NaCl-CH_4: Saltwater solutions containing dissolved methane are important, particularly in the subsurface of sedimentary basins. In general, the solubility of methane into brine is relatively low, so it is common for methane to reach saturation with respect to aqueous solutions. With increasing depth in a basin (high temperature and high pressure) methane's solubility increases. Increase in salinity may also decrease the solubility of methane.

Petroleum: Petroleum in fluid inclusions is best identified by its tendency to fluoresce when excited with ultraviolet (UV) light. Some petroleum fluid inclusions do not fluoresce brightly enough to be detected with standard microscopic methods; these include some condensates (gas-rich inclusions with a thin rim of oil) and some oil inclusions that have lost their volatiles. Although most oil inclusions are colorless in transmitted light, some may be identified by yellowish or brownish color.

1.5 Occurrence and classification of fluid inclusions

Most inclusions are smaller than 0.1 mm (100 μm) and are observed with the aid of a suitable optical microscope. The usual size range encountered during microscope examination is between about 2 and 20 μm. Very small inclusions are much more abundant than larger ones.

The 10 minerals in which inclusions are most commonly reported are listed below:

1. Quartz—SiO_2
2. Fluorite—CaF_2
3. Halite—NaCl
4. Calcite—$CaCO_3$
5. Apatite—$Ca_5(PO_4)_3$ (OH, F, Cl)
6. Dolomite—$CaMg(CO_3)_2$
7. Sphalerite—(Zn, Fe) S
8. Barite—$BaSo_4$
9. Topaz—$Al_2SiO_4(F,OH)_2$
10. Cassiterite—SnO_2

Apart from halite, these approximate to the most abundant transparent ore and gangue minerals associated with hydrothermal deposits. A feature common to all these minerals is that they are transparent and lightly colored, which is the most fundamental prerequisite for any optical study of inclusions.

This is not to say that opaque ore minerals are devoid of inclusions. Galena may contain extremely well formed cavities often visible on fresh cleavage surface. Inclusions in soft, easily cleaved minerals such as barite and calcite are highly susceptible to leakage and necking down and are therefore often regarded as of little value.

The overall abundance and distribution of inclusions in a single crystal depend partially on the primary growth conditions and partially on the post-crystallization history of the sample. Inclusions tend to occur in cluster and planar or curviplanar groupings, which are often crystallographically controlled. Often, the early growth of a mineral is characterized by a higher abundance of inclusions than during later growth. This is clearly seen in certain types of quartz crystals that have a milky appearance near the base, but clear termination.

1.5.1 Genetic classification

Petrographic analysis helps in classifying the fluid inclusions based on their origin and it was summarized by Roedder (1979, 1984). Studies on the origin of the inclusions creating a paragenesis will help to postulate their relative timing of entrapment, temperature, pressure, and fluid composition.

Terms that are normally used in classifying the fluid inclusions in terms of origin are primary, secondary, and pseudosecondary. (1) The term primary normally is used to refer to inclusions that are entrapped during and as a direct result of crystal growth. (2) Secondary fluid inclusions are those trapped after crystal growth is complete, normally when fractures or deformation features cut across all growth zones of a particular mineral phase and heal-to-trap inclusions. The implication of this healing is that it is by a process that does not require introduction of new ions to fill the microfracture, but employs ions redistributed from the fracture surface that have dissolved and reprecipitated to achieve a state of lower surface free energy. (3) Pseudosecondary fluid inclusions are those entrapped before crystal growth is complete, but not necessarily as a direct result of crystal growth. They are similar in origin to secondary fluid inclusions in that they normally are trapped along microfractures or other deformation features that have healed,

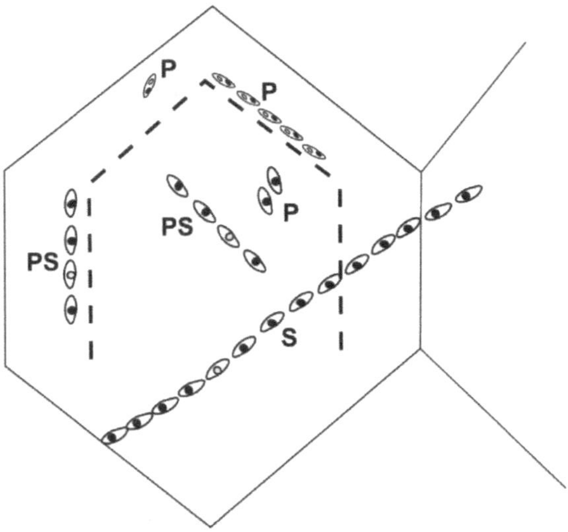

Fig. 1.2 Schematic diagram of halite crystal showing primary (P), pseudosecondary (PS) and secondary (S) inclusions. The dashed line denotes growth zoning. Inclusions along the growth planes are denoted as primary (P). The S trail, extends to the surface of the crystal to another crystal.

and their entrapment is followed by further crystal growth. Some features of the primary, secondary, and pseudosecondary inclusions are discussed here.

(i) *Primary (P) inclusions*: Primary fluid inclusions are formed on sealed imperfections on crystal surfaces during precipitation and contain the fluid present at the moment of sealing. Texturally, they tend to be solitary or isolated, and are best identified by being trapped parallel to a growth zone or crystal face (Fig. 1.2). When analyzed, these inclusions yield information on the conditions of formation or crystallization of the host mineral. All primary inclusions are surrounded by host mineral deposited at about the same time that the fluid was trapped.

(ii) *Secondary (S) inclusions*: Secondary inclusions are those that form by any process after the crystallization of the bulk of the host is essentially complete. Inclusions are incorporated into the host mineral during a later process after the crystal has been formed and therefore secondary fluid inclusions form after crystal growth. They often occur along healed fractures (i.e., fractures that develop after the formation of the host mineral, and trap fluid available then) and may cut across growth zones (Fig. 1.2). They also occur in planes or along cleavage directions or twin planes. Texturally, secondary inclusions can be recognized by their occurrence in trails or clusters that often cut across grain boundaries. The generally accepted mechanism for forming secondary inclusion involves the development of post-crystallization fractures initiated during mechanical or thermal stress. Late fluids get entrapped to form the characteristic trails of secondary inclusions that typically crosscut earlier generation fluid inclusion trails before the fracture gets healed.

(iii) *Pseudosecondary (PS) inclusions*: Psuedosecondary inclusions are developed in a way similar to secondary inclusions; the only difference is that fracturing and healing take place before crystal growth has terminated. So pseudosecondary fluid inclusions form when a crystal fractures during its growth and/or fluids are trapped in such fractures and heals before the crystallization is complete. These inclusions will therefore occur along trails that end abruptly against grain boundaries, or one of the growth zones or may abruptly end inside the crystals as a scratch (Fig. 1.2).

Primary or pseudosecondary inclusions reflect the character of the fluids present during growth, whereas secondary inclusions represent later fluids possibly introduced late in the system. Often it is practical to identify "possible primary inclusions" rather than make an unambiguous identification. Secondary inclusions are much easier to recognize because they typically occur as flattened cavities in planar groups cross cutting many crystal boundaries. This contrasts with planes of PS inclusions, which characteristically terminate abruptly within the crystal.

In samples that have suffered multiple stages of fracturing and re-healing, many different generations of secondary inclusions may be present. It is rarely possible to distinguish early and late generations by visual inspection alone, but with the aid of a through thermometric study, it is possible to build up a comprehensive picture of the overall fluid inclusion paragenesis.

This threefold classification of inclusion origin bears no implications as to whether the inclusions have leaked and refilled, or re-equilibrated in some other way. There may be another situation like the origin that cannot be identified called "indeterminable origin." It is simply a classification of the origin of the fluid inclusion vacuole; re-equilibration, the process of alteration of composition or density of the fluid in the vacuole, must be considered separately (Bodnar, 2003a,b). This classification represents petrographically based relative interpretations about timing of the formation of the inclusion vacuole and does not refer to the actual timing of entrapment of the fluid within the inclusion vacuole.

Classification of inclusion origin is a petrographic interpretation and not based on any analytical data. The only criterion that is applicable in identifying fluid inclusions as primary is petrographic evidence for a relationship to growth of the crystal. This evidence can be as follows: One of the most important is the shape and orientation of the inclusions relative to growth of the crystal. If inclusions have a particular shape and orientation consistent with entrapment during growth of the crystal, then this can be used as strong evidence for a primary origin. Such inclusions commonly are elongated in the direction of growth of the crystal and have flat bases and tapering tips. Entrapment on the leeward (later growth) side of an obvious defect (such as a fractured surface) after subsequent crystal growth is another useful criterion. Sometimes, primary inclusions can develop due to the obstruction in the growth of the parent crystal and can include solid phases (accidental solid) and fluid phases (immiscible gas or liquid). Distribution of a visible concentration of inclusions along a single growth zone is used commonly as evidence for a primary origin. Similarly, if fluid inclusion size, shape, orientation, or concentration vary by concentric growth zone, then the inclusions are best assigned a primary origin. Likewise, if fluid inclusions are concentrated in a particular growth sector or associated with the position of particular growth faces, then this is used commonly as evidence for primary origin. Concentric and sector growth zoning can be observed in transmitted light by observing variation in inclusion density or mineral color, but it is most easily identified by observing the compositional variation in the crystal using BSE, CL, or EDS/WDS element mapping.

Secondary fluid inclusions form after crystal growth is complete and is a process of healing of microcracks and deformation features. Some classifications of secondary fluid inclusions are based on origin and transgranular versus intragranular occurrence (Simmons and Richter, 1976; Kranz, 1983; Van den Kerkhof and Hein, 2001). To identify fluid inclusions as secondary, they must appear to occur in arrays that cut across all growth zones of a crystal. Fluid inclusions trapped by the healing of microfractures typically occur in planar arrays or along curved surfaces that cut across growth zonation. Secondary fluid inclusions are trapped along cleavage or twin planes, and such crystallographically controlled planes of secondary fluid inclusions are easily distinguishable from primary fluid inclusions by the tendency for secondary planes to cross one another and the tendency for primary fluid inclusions to mimic crystal terminations.

Pseudosecondary fluid inclusions are identical to secondary fluid inclusions since they have the same origin except that they are trapped before mineral growth is complete. The criteria discussed for the identification of secondary inclusions can be applied to pseudosecondary fluid inclusions except that the fractures or deformation features do not cut across all growth zones of the mineral. For pseudosecondary fluid inclusions, planar arrays of fluid inclusions end abruptly at a growth zone boundary. They are subsequently overgrown by later mineral growth lacking fluid inclusions trapped along the same fracture. Recognition of pseudosecondary inclusions requires that multiple planar arrays of inclusions terminate against a growth zone. A single planar array of inclusions terminating against a growth zone is not sufficient to classify inclusions as pseudosecondary, because all fractures end somewhere, whether secondary or pseudosecondary.

Secondary inclusions—cross-cutting relationships: Determining the timing of entrapment of multiple planar arrays of secondary inclusions is among the most important and difficult petrographic techniques in fluid inclusion work. As many samples from the sedimentary environment are dominated by secondary inclusions, and as secondary inclusions have the potential for recording nearly every stage in the P-T-X history of a geologic setting, establishing the relative timing of entrapment of various secondary inclusions is essential. The technique employed in establishing relative timing (Touret, 1981) is a basic cross-cutting relationship. To establish this relationship, two planar arrays (inclusions along healed fractures) must lie along cross-cutting paths. Each planar array should contain inclusions of distinctive size, shape, or orientation so that the inclusions of one array can be distinguished petrographically from inclusions of the other array.

1.5.2 Compositional classification

Fluid inclusion composition varies considerably, depending on the fluid source and temperatures and may be host to aqueous or hydrocarbon liquids, solids, and gaseous phases (Figs. 1.1 and 1.3). Aqueous fluid inclusions comprise mainly water with dissolved salts or daughter minerals along with minor dissolved inorganic, organic, and trace noble gases whereas hydrocarbon-bearing fluid inclusions (HCFIs) generally comprise liquid hydrocarbons, low carbon-number gases, and occasionally dark solid phases. Elaborate classification schemes have been proposed by many people based on the differing proportions of solids, liquids and vapor in the inclusions. Such schemes are certainly easier to use than paragenetic classification because subjectivity is kept to a minimum point. It is often convenient though to group inclusions according to the major phases at room temperature and are as follows:

Monophase liquid (L) inclusions: Inclusions that are entirely filled with liquid—usually aqueous.

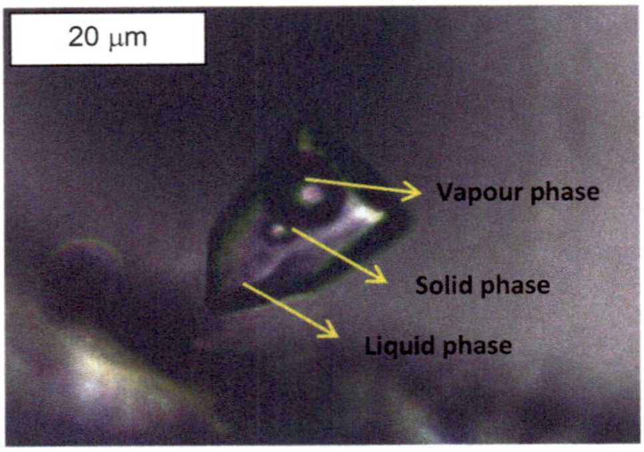

Fig. 1.3 A multiphase fluid inclusion from KK-basin at 3115–3120 m depth with sandstone claystone lithology.

Monophase vapor (V) inclusions: Inclusions that are entirely filled with a low-density vapor phase (usually a mixture of CO_2, CH_4, H_2O, and N_2) without any visible liquid.

Biphase Liquid-rich, (L± V) inclusions: The liquid phase dominates but a small vapor bubble is always present and occupies up to a minimum of about 40%–50% of the total volume. Biphase liquid-rich inclusions are shown in Fig.1.4A and B.

bi-phase Vapor-rich, (V±L) inclusions: (A vapor phase dominates and occupies more than half of the inclusion value. However, a thin rim of liquid is still visible).

Multiphase Solid (S± L± V) and multisolid (S± L± V) inclusions: Both are similar in containing one or more solid crystalline phases (daughter minerals) as an essential component in addition to liquid and vapor. If the solid occupy greater than 50%, the inclusion is referred as multisolid rather than multiphase; Multiphase inclusions are shown in Figs. 1.3, 1.4D and 1.1A.

Immiscible liquid (L_1 ±L_2 ±V) inclusions: These are characterized by the presence of two immiscible liquids, one being aqueous and the other CO_2_ rich less commonly (oil). The CO_2-rich phase may contain substantial amount of CH_4 or N_2. A CO_2-rich vapor bubble may also be present, depending on the overall density of the CO_2-rich phase. In three-phase inclusions of this kind, the aqueous phase always surrounds the CO_2 -rich liquid, which in turn encloses the CO_2-rich vapor bubble.

Fluid inclusion petrography

Fluid inclusion petrography is done with the transmitted-light microscope. Different combinations of phases may be observable within fluid

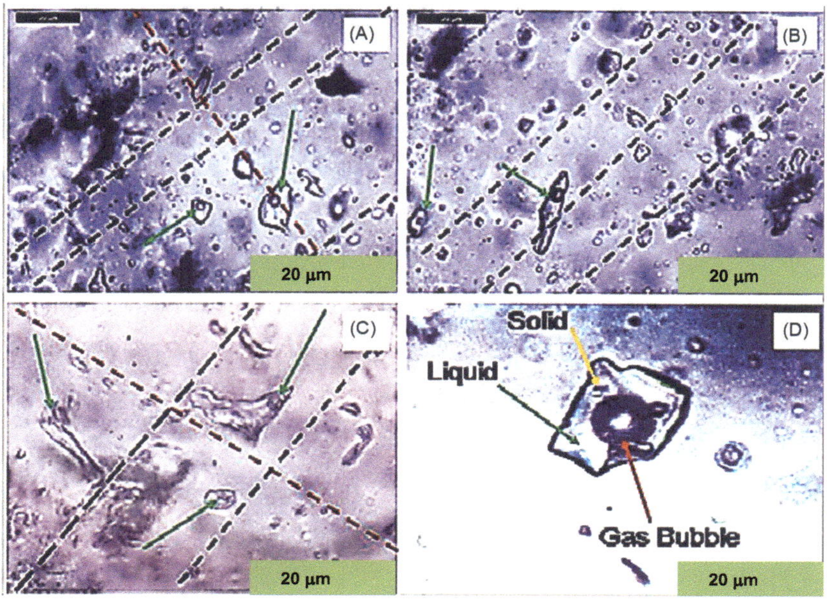

Fig. 1.4 Trails of fluid inclusions in quartz (from the Ratnagiri offshore, Mumbai basin, India). Biphase fluid inclusions are shown in A, B, and C. Multiphase fluid inclusion shown in D.

inclusions. Most of the inclusions may contain a single liquid or gaseous phase. In inclusions with more than one fluid phase, the wetting properties and density control the distribution of the phases, normally with the lowest density fluid floating to the center or top of the inclusion and the higher density phase appearing in the exterior part of the inclusion. Usually, two fluid phases are present. In aqueous inclusions, this normally consists of a dominant liquid phase and a small vapor or gas bubble. The bubble normally appears quite dark compared to the surrounding liquid because of the difference in refractive index between it and the surrounding liquid, and because of its common spherical shape. The bubble may be embedded itself in a corner of the inclusion, may have floated to the top of the inclusion, or may be dancing around the inclusion in pseudo-Brownian motion. In general, fluid inclusion is viewed as a two-dimensional representation of a three-dimensional object. Therefore, the area of the inclusion occupied by the bubble appears greater than the volume accurately occupied by it.

Solid phases in fluid inclusions

Solid phases can also be present in fluid inclusions. This include aqueous inclusions containing crystals that are precipitated during cooling, solid hydrocarbons that are precipitated from oil, solids accidentally entrapped in inclusions as they were formed, and melt inclusions containing mostly glass or crystalline solids. The solid phases are accidentally trapped, may have settled or precipitated on the surface of a growing crystal and then are incorporated as part of a fluid inclusion when the host crystal grew around it. Alternatively, daughter solids are solid phases actually precipitated from the inclusion fluid after entrapment when it is cooled.

Daughter solids are common in aqueous inclusions. The appearance of the outer margin of fluid inclusions is controlled by inclusion shape and the difference in refractive index between the mineral and the fluid. The discernible nature of fluid inclusions from the host mineral depends on refractive index difference between them (Burruss, 1981). Van den Kerkhof and Hein (2001) have compiled a useful table of the optical properties of inclusion fluids, enclosing minerals, and common solid phases found within fluid inclusions. Usually, the inclusion shape and difference in refractive index control how bright or dark the inclusion appears in transmitted light. Inclusions with flat upper and lower surfaces transmit light well and appear bright. Negative crystals with irregular and globular inclusions have a focusing and defocusing effect on the light, which could lead to irregular illumination of the inclusion interior. The aqueous and hydrocarbon liquid inclusions typically appear bright and gas-filled inclusions typically appear dark (Goldstein et al., 1991).

Aqueous inclusions

Aqueous inclusions are water-rich inclusions consisting of a single liquid phase, when trapped at low temperatures. Those trapped or re-equilibrated at higher temperatures consist of a liquid phase containing a gas bubble. The liquid phase typically has a bright appearance in transmitted light. The gas phase normally appears dark. When high-temperature (boiling) liquids are entrapped, inclusions will have a large bubble of water vapor with minimal liquid phase (Diamond, 2003). Carbon dioxide-rich aqueous fluid inclusions are present in metamorphic rocks and granitoid vein systems. Carbon dioxide-rich aqueous inclusions contain multiple fluid phases. In three-phase inclusions, the outer part of the inclusion is occupied by a single phase of aqueous liquid and a slightly darker area of liquid carbon dioxide and inside it a bubble of CO_2 gas, commonly in pseudo-Brownian motion. Two-phase inclusions consist of an outer area of carbon dioxide liquid and an inner area with a gas bubble, or an outer area with aqueous liquid and an inner area with carbon dioxide existing as a single fluid phase.

Hydrocarbon/oil inclusions

Oil, oil-aqueous, condensate, and bitumen inclusions containing solid and liquid hydrocarbons are common in diagenetic systems. Under transmitted light, aqueous inclusions and HCFIs cannot be distinguished. But under UV illumination, HCFIs will fluoresce and can be distinguished. Some oils in inclusions have a yellowish or brownish color in transmitted light. Fluid inclusions containing liquid oil also may contain a small amount of brownish bitumen-like solid. The solids either were entrapped as accidental solids or formed as daughter solids. Most commonly, oil inclusions consist of two fluid phases, a dominant liquid phase and a small gas phase. As in aqueous inclusions, the gas bubble typically appears dark. In oil inclusions containing a gas bubble, the bubble is more likely to be in pseudo-Brownian motion than in aqueous inclusions with the same volume ratio of gas to liquid. When heated during microthermometry, these inclusions typically homogenize to the liquid phase. In contrast, some two-phase inclusions are dominated by the gas phase, with only a thin rim of liquid oil around the exterior of the inclusion. The liquid oil phase is equivalent to a condensate liquid and when heated during microthermometry, these inclusions homogenize to the gas phase. Single-phase inclusions are also common. Many of these consist of liquid oil and condense out of the gas phase during cooling.

Most hydrocarbon liquids have a refractive index higher than the aqueous liquid. Oil-aqueous inclusions are found with three fluid phases, an aqueous liquid occupying the exterior of the inclusion, liquid oil inside it, and a bubble of gas inside the oil. Single-phase inclusions are also common. Many of these consist of liquid oil and are condensed out of the gas phase during cooling. Single-phase gas inclusions appear to contain only a single fluid phase that have a relatively dark appearance due to their high relief, which commonly discriminates them from aqueous liquid inclusions. Such dark inclusions might have a variety of compositions consisting of methane, carbon dioxide, nitrogen, or a variety of other compounds (Van den Kerkhof and Thiéry, 2001). The fluid simply exists under pressure–temperature conditions above the homogenization temperature or critical point and it may have homogenized to liquid, to gas, or through critical behavior (Diamond, 2003). Inclusions such as these may be common in sedimentary rocks, where they are dominated by methane. Petrographic analyses of fluid inclusions are detailed in Chapter 5.

1.6 The nature of petroleum fluids

Petroleum fluids are characterized by their complexity. A classification scheme can be based on chemical or physical properties and depends commonly on the application, whether it is in geology, geochemistry, or reservoir engineering. Genetic characterization of source and maturation of petroleum fluids has been the focus of research in organic geochemistry for decades.

The main components of petroleum are hydrocarbons, nitrogen-sulfur-oxygen-bearing (NSO) compounds and inorganic gases, such as N_2, CO_2, and H_2S. The hydrocarbons in petroleum can be subdivided according to their structure into: (1) paraffins, which consist of normal and branched alkanes. Alkanes are often called saturated hydrocarbons, since all atoms are connected with single bonds; (2) naphthenes or cycloalkanes, which consist of compounds with cyclic structures. The naphthenes are also saturated compounds, which only contain single bonds; and (3) aromatics are compounds which contain cyclic structures with double bonds. The simplest aromatic compound is benzene, C_6H_6, consisting of a ring structure with six carbon atoms. The simplest hydrocarbon compound is methane, CH_4. Methane has a critical point at low temperature and will only exist as a supercritical fluid or gas under geological conditions. As the number of carbon atoms in the molecule increases, the possible ways of combining the atoms increase. Compounds that have the same formula but different structures are called isomers. The number of isomers increases rapidly with increasing number of carbon atoms. Increase in the number of carbon atoms in the molecule also leads to increase in the critical temperature and decrease in the critical pressure.

1.7 Hydrocarbon-bearing fluid inclusions

Hydrocarbon bearing fluid inclusions (HCFIs) occur in sedimentary rocks such as sandstones and limestones and in cements, particularly in minerals, e.g., quartz, calcite, and feldspar, and contain micron-scale samples of trapped hydrocarbon oil. In addition to hydrocarbon fluids, at room temperature, HCFIs can also contain traces of other fluids such as H_2O, gasses such as CO_2 and CH_4, and solids that were either entrapped along with/or precipitated from the liquid phase at later times. The accurate determination of the composition of the trapped oil in fluid inclusions can yield (together with microthermometric studies) fundamental information about the history of oil formation and oil migration in sedimentary basins, which is important to both the petroleum production and exploration industries and to the geological community.

HCFIs may occur both in reservoirs and in migration pathways, however the most commonly documented are associated with reservoirs, and are trapped in diagenetic cements, overgrowths, or secondary fractures in quartz, feldspar, and calcite. HCFIs may develop at any time from the onset of reservoir filling to the present day, with trapping most likely during reservoir filling rather than later stages when the water is displaced. This may be attributed to quartz cementation being a slow, temperature-mediated process and the fact that the micron-sized oil droplets sticking to the quartz grains inhibit quartz growth. The trapping of oil within quartz cement may take millions of years whereas healing of secondary fractures is much more rapid, and may preserve the later stages of reservoir filling. The abundance of HCFI in clastic sediments may be correlated with porosity and permeability, whereas in chalk or limestone reservoirs the HCFI may be more irregular favoring cemented fractures, recrystallized fossils, or coarse-grained cement (Munz, 2001, George et al., 2007; Burruss et al., 1983, Blamey and Ryder, 2007). Some secondary fluid inclusion trails observed during our studies on fluid inclusions from the Mumbai offshore basin, India, are shown in Figs. 1.4–1.6.

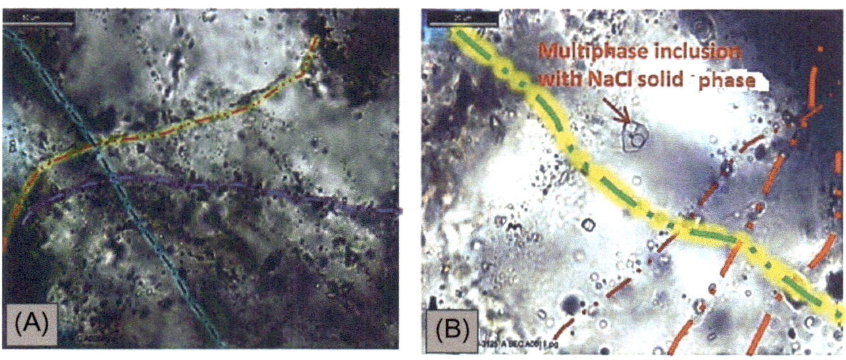

Fig. 1.5 Fluid inclusion trails (A and B) observed in the Mumbai offshore basin, India, (B) shows multiphase inclusions with NaCl solids.

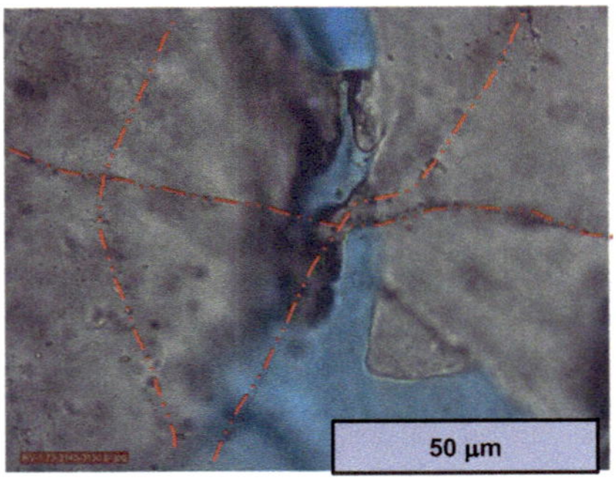

Fig. 1.6 Fluid inclusion trails cross-cutting the mineral grain boundary (Mumbai offshore basin, India).

1.8 Significance of fluid inclusion study

Some of the attributes of basin development like pressure, temperature, oil window, burial history, etc. can be determined using fluid inclusions in sedimentary and diagenetic minerals. Important fluids in the sedimentary realm include atmospheric gases, freshwater of meteoric origin, lake water, seawater, mixed water, evaporated water, and formation waters deep in basins, oil, and natural gas. Preserving a record of the distribution and composition of these fluids from the past should contribute significantly to studies of paleoclimate, global-climate change research, diagenetic systems, and provides useful information in petroleum geology.

In studies of global climate change, fluid inclusions can be used as sensitive indicators of the paleo-temperature of surface environments. Fluid inclusions also preserve microsamples of ancient seawater and atmosphere, the analysis of which could figure prominently into discussions of past changes in the chemistry of the atmosphere and oceans. In petroleum geology, fluid inclusions have proven to be useful indicators of migration pathways of hydrocarbons; they can delineate the evolution of the chemistry of hydrocarbons; and they remain important in understanding the thermal history of basins and relating fluid migration events to evolution of reservoir systems. In studies of diagenesis, fluid inclusions can be the most definitive record. Most diagenetic systems are closely linked to temperature and salinity of the fluid. Thus, fluid inclusions are sensitive indicators of diagenetic environments (Goldstein, 2001).

HCFIs generally occur in diagenetic cements or grains and contain complex mixtures of mainly organic compounds depending on their source/s.

Accurate analysis of the entrapped hydrocarbons in HCFIs can yield vital information about the history, evolution, and migration of petroleum fluids and is thus crucial data for the petroleum exploration industry. Studying HCFIs is advantageous because the trapped fluids are representative of the actual hydrocarbon fluids that existed when the inclusions were sealed in the mineral. This sealing-in process preserves the petroleum fluid, thus isolating it from subsequent infiltration of petroleum fluids and events in oil reservoirs such as loss of charge, water washing, or biodegradation. It also preserves the fluid from contamination during the drilling processes used to extract samples from the ground (George et al., 2007).

Fluid inclusions, despite their small size, are highly valuable to understand many geological processes. When these cavities within the rock were sealed, they trapped the original fluid at these fossil pressure-volume-temperature (PVT) conditions. This PVT data can be used for modeling fluid phase behavior and give an indication of the "oil window" at which oil formation occurs (Munz, 2001). Geologists can learn much about the processes, fluid compositions, temperatures, and pressure conditions in geologic systems from fluid inclusions that form at a particular time, such as during ore formation in base-metal or gold deposits, or from the migration of hydrocarbon fluids in petroleum basins. Not all fluid inclusions present within a sample may represent the key event under study and therefore an experienced fluid inclusion analyst is needed to examine the paragenetic relationships between fluid inclusion assemblages, mineral growth, or fracture healing. By so doing, the geologist is able to study the migration of fluids, including petroleum fluids, and understand what processes led to the migration or trapping of petroleum.

One of the earliest and most understanding examples of the usefulness of inclusions is found in the study of ore deposition. The ore-forming fluid for a deposit was assumed to have been rich only in those constituents now present, even in proportion to their abundance in the deposit. However, most ore fluids contained, in addition to the ore elements, deposited large amounts of volatile constituents and soluble salts that passed through the deposit leaving almost no trace except the fluid inclusion and hence much has been learned about the process of ore deposition from a study of these inclusions. The most direct approach to the study of ancient crustal fluids is through the analysis of fluid inclusions (Munz, 2001). Fluid inclusions provide the geologist with what can only be described as unique samples of the total spectrum of fluids that have interacted with and developed in the earth crust and upper mantle throughout the geological time (Nandakumar and

Jayanthi, 2016). Historically, they are the most important contributors towards understanding the character, origin, evolution of hydrothermal ore-forming fluids, and ultimately ore genesis. Fluid inclusion studies have acquired prime importance during the last decades as the research on the role of fluid phases in sedimentary, diagenetic, metamorphic, and magmatic processes have shed light on the following:

Temperature: The use of fluid inclusions for deciphering the temperature of past geologic events was first proposed by Sorby (1858). This use of fluid inclusion for geothermometry is a result of differential shrinkage of the host mineral and the inclusion fluid on cooling from the temperature of trapping to that of observation.

Pressure: Data obtained from fluid inclusions can provide information on the pressure of the environment at the time of trapping. The measured pressure range varies from near atmospheric to many kilobars.

Density: If the composition and density of each of the phases now present in a fluid inclusion (liquid, gas, or crystal) can be determined along with their individual volumes, the total average density of the material in the inclusion can be calculated. The circulation of fluids in the Earth's crust is driven by density differences. An important objective of most fluid inclusion studies is to calculate fluid densities and isochores, and to estimate the pressure and temperature conditions of trapping, either by combining information from solids or from coexisting fluid phases. For petroleum inclusions, it may also be of great interest to estimate fluid properties under pressure and temperature conditions other than the trapping conditions.

Composition: With the rapid development of modern analytical techniques, it is now possible to obtain an extraordinary amount of information on the composition of fluid inclusions. Microthermometry and Raman spectroscopic analysis detailed in the later chapters of this book discuss the compositional identification of fluid inclusions. Composition of the fluids can be inferred by measuring the temperature of first melting/the eutectic melting temperature (Goldstein and Reynolds, 1994).

Geothermometry and Geobarometry: Fluid inclusions have been used as geothermobarometers and they provide an important guide to pressure–temperature (PT) conditions during mineral formation. Fluid inclusions are one of the only geobarometers available for the sedimentary realm. Because pressure can be translated to burial depth or water depth, there is a variety of environmental information and burial history information that can be gleaned from fluid inclusions (Goldstein, 2001). Procedures for interpreting the degree to which aqueous fluid inclusion assemblages (FIA—it is a group of fluid inclusions that were entrapped at the same time and defined as "the most finely discriminated, petrographically associated group of inclusions" or "groups of inclusions that can be defined by petrographic means as the most finely discriminated events of fluid inclusion entrapment." The concept of fluid inclusion assemblage (FIA) was proposed by Goldstein and Reynolds (1994), and Goldstein (2003) defines FIA as the most finely discriminated fluid inclusion trapping event that can be identified with petrographic studies. The FIAs that are best for interpreting temperature of entrapment consist of variably sized and shaped aqueous inclusions in which 90% of the homogenization temperatures are within 10–15 °C. These simple observations rule out problems with thermal re-equilibration, and thus, each homogenization temperature measured, in such an FIA, is a record of the minimum entrapment temperature of the fluid inclusion. The degree to which these homogenization temperatures underestimate true entrapment temperature depends on the amount of gas dissolved in aqueous inclusions, the compositional control on the isochoric slope, and the pressure of entrapment. These factors control the 'pressure correction' that corrects homogenization temperature to true entrapment temperature.

Patterns of fluid flow: Secondary fluid inclusions can shed light on fluid migration pathways and fossil hydrogeological systems. If one assumes that the abundance of secondary inclusion in rocks is dependent on the fluid flux through the rock, then it is possible to outline areas where fluid activity has been most pronounced. These areas may be important from the mineral exploration viewpoint in delineating zones where hydrothermal activity was at a maximum and therefore deposits are most likely to occur. Fluid inclusions provide one of the best records of the history of petroleum migration through rocks. When HCFIs are placed in a paragenetic framework, the relative timing of petroleum migration can be determined. The temperature information from HCFIs can be used to place the timing of oil migration; the compositions of HCFIs can be used to trace the migration history of oils in a basin (Goldstein and Reynolds, 1994). Thus, a microscopic "oil show" in a HCFI may carry much importance.

Fluid inclusion studies in mineral exploration: The primary and secondary inclusions in ore and gangue minerals can be used to "fingerprint" certain types of ore-forming fluids, to characterize particular ore mineral assemblages and to define areas where these fluids are most likely to concentrate. It is also possible to predict whether the determined PT state of the fluid favors ore deposition. The use of fluid inclusions in mineral exploration has received varying degrees of attention over the years. The more direct use of fluid inclusions in exploration mainly rely on defining an empirical relationship between some inclusion characteristic and mineralization. Methods for using fluid inclusions to assist target selection on a regional scale or for more localized definition of likely zones of focussed fluid flow or ore shoots can be subdivided into three categories (Wilkinson, 2001; Randive et al., 2014) such as the occurrence or relative abundance of a specific inclusion type, systematic variations in microthermometric properties, and the systematic variations in other properties like decrepitating behavior and inclusion chemistry.

Fluid inclusion studies in oil exploration and development: It should be possible to identify migration pathways and even possible reservoir rocks, using data on the abundance and distribution of hydrocarbon or gas-rich inclusions in diagenetic minerals from sedimentary rocks (Burruss, 1981). Inclusion data could make a significant contribution to studies concerned with the depth of burial during diagenesis and temperature–time profiles for sedimentary basins (Hazeldine et al., 1984). Such studies could prove to be of value for oil exploration by helping to establish whether a sedimentary basin had been heated beyond the maturation point of the oil. In recent years, fluid inclusion work has become an important tool in oil exploration and development. Not only have fluid inclusions figured prominently in understanding the diagenetic controls on porosity evolution and thermal history, but also fluid inclusion work has been important in evaluating the history and pathways of hydrocarbon migration. Petroleum fluoresces with UV epi-illumination, and its color of fluorescence is tied to maturity and API gravity (Lumb, 1978; McLimans, 1987; Tsui, 1990), providing an important record of the composition of petroleum fluid inclusions. Furthermore, careful petrographic analysis of fluid inclusions within the context of diagenetic history has also proven useful. Individual events of petroleum migration can be recognized by fluid inclusion trails crosscutting the grain boundaries and cements. A detailed study of these relationships, together with homogenization temperature measurements, is quite useful in understanding reservoir systematics (Pottorf et al., 1997; Brennan and Goldstein, 1998). Furthermore, the extraction and analysis of HCFIs provides useful information for determining hydrocarbon migration history and for evaluating the composition of hydrocarbons (Burruss, 1987; Karlsen et al., 1993; Hall et al., 1996; Smith, 1997). In recent years, attempts have been made to interpret the *PVT* relationships of oil in fluid inclusions to better evaluate the temperature and pressure of entrapment. MacLeod et al. determined the volume of gas and liquid in fluid inclusions to supplement modeling of the *PVT* relationships (Macleod et al., 1996). Although this appears to be useful, information that may help in the interpretation of oil-filled fluid inclusions, an analysis of the errors involved in the volume determinations, and assumptions about composition would be a useful addition.

Fluid inclusions and diagenetic systems: Most diagenetic systems are defined by the temperature, pressure, and salinity of the fluids active in them. Therefore, fluid inclusions are one of the best techniques for constraining the diagenetic history of sedimentary rocks since these are the samples of fluids responsible for diagenesis. Fluid inclusions are a direct record of diagenetic systems and have been successfully applied to all diagenetic systems (Goldstein and Reynold, 1994). Such systems include low-temperature and high-temperature ones that have never been heated and buried, and include many low temperature and high-temperature systems that have been buried and heated beyond their conditions of entrapment. Fluid inclusion microthermometry coupled with petrography can be employed for understanding the problems in diagenesis; in providing useful temperature, pressure, and fluid chemistry data on ancient diagenetic systems that is not possible through any other ways.

Fluid inclusions and global change: The future change in Earth's surface environment is one of the most important areas of analysis. One of the only ways of predicting the future is in understanding the past, and fluid inclusions will figure prominently in this type of research. Fluid inclusions of seawater in marine carbonate minerals are unique samples that may provide us with a record of past change in composition of the seawater-atmosphere system (Johnson and Goldstein, 1993). Fluid inclusions in marine evaporates can also provide a record of past change in the composition of the seawater-atmosphere system (Goldstein, 2001).

Other applications: There are a number of other areas where fluid inclusion studies can be applied to practical problems. One obvious though poorly documented area is the use of fluid inclusions as tracers for defining the provenance of minerals for sedimentological purposes. Another important application is in *gem*ology and gem testing. Using simple microscope observations, it is possible to distinguish between certain synthetic and natural gemstones and even on occasions to identify the actual deposit from which a particular gemstone was derived. The influence of fluid inclusions in determining the overall "strength" of a rock is also poorly understood yet. It is seen that the "rock bursts" or face failures in certain deep salt mines are influenced by the compressed gas inclusions held under high pressure within the salt. Whether fluid inclusions can be important factors related to the problem of rock bursts in deeper hard rock mines can be worthy of future investigation. Another possible use of fluid inclusions is in the determination of the timing of hydrothermal events aiding to date hydrothermal mineral deposits.

1.9 Importance of hydrocarbon-bearing fluid inclusions in petroleum exploration

Aqueous fluid inclusions comprise mainly water with dissolved salts or daughter minerals along with minor dissolved inorganic, organic, and trace noble gases whereas HCFIs generally comprise liquid hydrocarbons, low carbon-number gases, and occasionally dark solid phases (Fig. 1.7).

The presence of petroleum inclusions in cements or in healed fractures demonstrates that oil was present at the time the cement formed or a fracture healed. This simple observation aids in understanding the relative time of oil generation and migration in the context of the diagenetic and tectonic history of the rock. The rapid expansion of work on petroleum inclusions is due to the simple fact that petroleum inclusions are "hidden petroleum shows" (Lisk et al., 2002). The petrographic setting of HCFIs combined with a burial history reconstruction can place tight limits on the time of oil migration (Burruss et al., 1985). When this approach is coupled with the current state-of-the-art methods of basin analysis, including thermochronology from fission track analysis, microthermometry of petroleum and aqueous inclusions, and modeling of the PVT properties of petroleum, we can obtain a detailed picture of the time, temperature, and pressure of petroleum migration (Burruss, 2003).

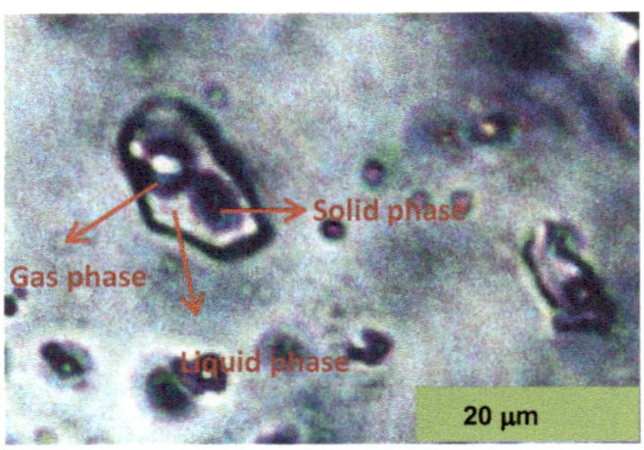

Fig. 1.7 HCFIs showing three phases: gas phase, liquid oil, and solid bitumen at a depth of 3495–3500 m in a dry well from the Mumbai offshore basin, India.

The hydrocarbon fluids in HCFIs may have highly complex compositions but are principally liquid with variable amounts of gas (light alkanes, CO_2, N_2, and H_2S), paraffins, napthalenes, aromatics, resins, and sometimes waxes. The apparent HCFI colour depends on a combination of the wavelengths absorbed by the various chromophores present in the trapped hydrocarbon liquid and the associated mineral environment. Consequently, their colour, when viewed under a microscope with plain white light illumination, may vary considerably from transparent to yellow, dark brown, or black. The mineralogical properties, which can also influence the perceived colour, include birefringence, polarization, reflectivity, and variations in refractive index.

HCFIs provide fewer phase changes but they do fluoresce under UV or visible light illumination, which can provide another source of information about fluid composition. The petroleum composition of HCFIs is typically obtained by crushing a small amount of the HCFI-containing material and extracting the petroleum fluid for analysis by gas chromatography and mass spectrometry. The key disadvantages of the crush method are the destruction of the sample, probable mixing of fluids from different source inclusions, and potential contamination (George et al., 2001; Parnell et al., 2001). Therefore, the oil composition data derived from these bulk analyses should be carefully considered as mixing of heterogeneous fluid inclusion populations may have occurred.

HCFIs are therefore a desirable goal for those studying petroleum migration and is best achieved via optical methods. Optical methods have several advantages that include: nondestructive-noncontact analysis, high degree of sensitivity, ability to undertake micron-scale analysis of single inclusions, high-end technology and methodologies, high diagnostic potential, and relatively simple instrumentation (Guilhaumou et al., 1990; Piranon and Pradier, 1992). Of particular interest to the petroleum industry are inclusions that contain hydrocarbon fluids, which originated from petroleum that once migrated through the rocks before becoming trapped. These hydrocarbon-bearing fluid inclusions (HCFIs) are useful for learning about the processes, fluid compositions, temperatures, and pressure conditions in geologic systems such as the migration of hydrocarbon fluids in petroleum basins and petroleum charge history.

The chapters of this book detail the petrographic, microthermometric, and optical spectroscopic approaches to characterize HCFIs for exploring its potential to be an adjuvant tool in the petroleum exploration industry. Two well samples from two typical offshore basins in India have been discussed for the purpose.

References

Blamey, N.J.F., Ryder, A.G., 2007. Hydrocarbon fluid inclusion fluorescence: A review. In: Geddes, C.D. (Ed.), Reviews in Fluorescence. Springer, New York, pp. 299–334.

Bodnar, R.J., 2003a. Introduction to fluid inclusions. In: Samson, I., Anderson, A., Marshall, D. (Eds.), Fluid Inclusions: Analysis and Interpretation. Mineral. Assoc. Canada, Short Course, Vol. 32, pp. 1–8.

Bodnar, R.J., 2003b. Interpretation of data from aqueous-electrolyte fluid inclusions. In: Samson, I., Anderson, A., Marshall, D. (Eds.), Fluid Inclusions: Analysis and Interpretation. Mineralogical Association of Canada: Short Couse Series 32, Vancouver, pp. 81–100.

Bodnar, R.J., 2005. Fluids in planetary systems. Elements 1, 9–12. https://doi.org/10.2113/gselements.1.1.9.

Brennan, S.T., Goldstein, R.H., 1998. Fluid and thermal history of an exhumed petroleum reservoir. PACROFI Program and abstracts. Las Vegas (unpaginated).

Burruss, R.C., 1981. Hydrocarbon fluid inclusions in studies of sedimentary diagenesis. In: Hollister, L.S., Crawford, M.L. (Eds.), Short Course in Fluid Inclusions: Application to Petrology. Mineral. Assoc. Canada, Short Course Handbook, Vol. 6, pp. 138–156.

Burruss, R.C., 1987. Crushing-cell, capillary column gas chromatography of petroleum fluid inclusions: method and application to petroleum source rocks, reservoirs, and low temperature hydrothermal ores. In: American Current Research on Fluid Inclusions. Socorro, NM, Abstracts (unpaginated).

Burruss, R.C., 2003. Petroleum fluid inclusions, an introduction. In: Fluid Inclusions: Analysis and Interpretation. Vol. 2003. Mineralogical Association of Canada: Vancouver, pp. 159–169.

Burruss, R.C., Cercone, K.R., Harris, P.M., 1983. Fluid inclusion petrography and tectonic burial history of the Al Ali no. 2 well; evidence for the timing of diagenesis and oil migration, northern Oman Foredeep. Geology 11 (10), 567–570.

Burruss, R.C., Cercone, K.R., HARRIS, P.M., 1985. Timing of hydrocarbon migration: Evidence from fluid inclusions in calcite cements, tectonics, and burial history. In: Schneidermann, N., Harris, P.M. (Eds.), Carbonate Cements. Vol. 26. SEPM Special Publ, pp. 277–289.

Diamond, L., 2003. Introduction to gas bearing fluid inclusions. In: Samson, I., Anderson, A., Marshall, D. (Eds.), Fluid Inclusions: Analysis and Interpretation. Mineral Assoc. Can. Short Course Ser, Vol. 32, pp. 101–158.

George, S.C., Ruble, T.E., Dutkiewicz, A., Eadington, P.E., 2001. Assessing the maturity of oil trapped in fluid inclusions using molecular geochemistry data and visually determined fluorescence colours. Appl. Geochem. 16 (4), 451–473.

George, S.C., Volk, H., Ahmed, M., 2007. Geochemical analysis techniques and geological applications of oil-bearing fluid inclusions, with some Australian case studies. J. Petrol. Sci. Eng. 57 (1–2), 119–138.

Goldstein, R.H., 2001. Fluid inclusions in sedimentary and diagenetic systems. Lithos 55, 159–193.

Goldstein, R.H., 2003. Petrographic analysis of fluid inclusions. In: Samson, I., Anderson, A., Marshall, D. (Eds.), Fluid Inclusions: Analysis and Interpretation. Mineralogical Association of Canada, Short Course. 32, pp. 9–53.

Goldstein, R.H., Reynolds, T.J., 1994. Systematics of Fluid Inclusions in Diagenetic Minerals. (SEPM (Society for Sedimentary Geology) short course 31, USA ISBN #1-56576-00805).

Goldstein, R.H., Stephens, B.P., Lehrmann, D.J., 1991. Fluid inclusions elucidate conditions of dolomitization. In: Eocene of Enewetak Atoll and Mid-Cretaceous Valles Platform of Mexico. Dolomieu Conference on Carbonate Platforms and Dolomitization: Abstracts, Ortisei, Italy, pp. 92–93.

Guilhaumou, N., Szydlowskii, N., Pradier, B., 1990. Characterization of hydrocarbon fluid inclusions by infra-red and fluorescence micro spectrometry. Mineral. Mag. 54 (375), 311–324.
Hazeldine, R.S., Samson, I.M., Cornford, C., 1984. Dating diagenesis in a petroleum basin, a new fluid inclusion method. Nature 307, 354–357.
Hall, D.L., Sterner, S.M., Shentwu, W., Bigge, M.A., 1996. Applying Fluid Inclusions to Petroleum Exploration andProduction. Fluid Inclusion Technologies, Inc. USA.
Johnson, W.J., Goldstein, R.H., 1993. Cambrian seawater preserved as inclusions in marine low magnesium calcite cement. Nature 362, 335–337.
Karlsen, D.A., Nedkvitne, T., Larter, S.R., Bjørlykke, K., 1993. Hydrocarbon composition of authigenic inclusions, application to elucidation of petroleum reservoir filling history. Geochemica et Cosmochimica Acta 57, 3641–3659.
Kranz, R.L., 1983. Microcracks in rocks: A review. In: Friedman, M., Toksoz, M.N. (Eds.), Tectonophysics. Vol. 100, pp. 449–480 (1–3).
Land, L.S., 1973. Contemporaneous dolomitization of Middle Pleistocene reefs by meteoric water, North Jamaica. Bull. Mar. Sci. 23, 64–92.
Lisk, M., O'Brien, G.W., Eadington, P.J., 2002. Quantitative evaluation of the oil-leg potential in the Oliver gas field, Timer Sea, Australia. Am. Assoc. Pet. Geol. Bull. 86, 1531–1542.
Lumb, M.D. (Ed.), 1978. Organic luminescence. In: Luminescence Spectroscopy. Academic Press, New York, pp. 93–148.
Macleod, G., Larter, S.R., Aplin, A.C., Pedersen, K.S., Booth, T.A., 1996. Determination of the effective composition of single petroleum inclusions using confocal scanning laser microscopy and PVT simulation. In: Brown, P.E., Hagemann, S.G. (Eds.), Biennial Pan-American Conference on Research on Fluid Inclusions (PACROFI VI) Madison Wisconsin, USA, pp. 81–82.
McLimans, R.K., 1987. The application of fluid inclusions to migration of oil and diagenesis in petroleum reservoirs. Appl. Geochem. 2 (5–6), 585–603.
Munz, I.A., 2001. Petroleum inclusions in sedimentary basins: systematics, analytical methods and applications. Lithos 55 (1–4), 195–212.
Nandakumar, V., Jayanthi, J.L., 2016. Hydrocarbon fluid inclusions, API gravity of oil, signature fluorescence emissions and emission ratios: an example from Mumbai offshore, India. Energy Fuel 2016 (30), 3776–3782.
National Research Council, 1990. The Role of Fluids in Crustal Processes (Studies in Geophysics). National Academy Press, Washington DC.
Parnell, J., Middleton, D., Honghan, C., Hall, D., 2001. The use of integrated fluid inclusion studies in constraining oil charge history and reservoir compartmentation: examples from the Jeanne d'Arc Basin, offshore. Newfoundland. Mar. Pet. Geol. 18 (5), 535–549.
Piranon, J., Pradier, B., 1992. Ultraviolet-fluorescence alteration of hydrocarbon fluid inclusions. Org. Geochem. 18, 501–509.
Pottorf, R.J., Gray, G.G., Kozar, M.G., Fitchen, W.M., Richardson, M., 1997. Paleothermometry techniques applied to burial history and hydrocarbon migration analyses, Tampico-Misantla Basin, Mexico. American Association of Petroleum Geologists Abstracts with Programs, p. A94.
Randive, K., Hari, K.R., Dora, M.L., Maple, D.B., 2014. Study of fluid inclusions: methods, techniques and applications. Gond. Geol. Mag. 29 (1 and 2), 19–28.
Roedder, E., 1979. Fluid inclusions as samples of ore fluids. In: Barnes, H.L. (Ed.), Geochemistry of Hydrothermal Ore Deposits, second ed. Wiley, New York, pp. 684–737.
Roedder, E., 1984. Fluid inclusions. Rev. Mineral. 12, 1–644.
Simmons, G., Richter, D., 1976. Microcracks in rocks. In: Strens RGJ(Ed) the Physics and Chemistry of Minerals and Rocks. Wiley, Toronto, pp. 105–137.

Smith, M.P., 1997. Fluid inclusion well logs: petroleum migration, seals, and proximity to pay. ECROFI XIV Abstracts, Nancy, France, p. 310.
Sorby, H.C., 1858. On the Microscopical, structure of crystals, indicating the origin of minerals and rocks. Q. J. Geol. Soc. 14, 453–500. https://doi.org/10.1144/GSL.JGS.1858.014.01-02.44.
Touret, J., 1981. Fluid inclusions in high grade metamorphic rocks. In: Hollister, L.S., Crawford, M.L. (Eds.), Mineralogical Association of Canada Short Course Handbook. Vol. 6, pp. 182–208.
Tsui, T.F., 1990. Characterizing fluid inclusion oils via UV fluorescence microspectrophotometry—a method for projecting oil quality and constraining oil migration history (abstract). AAPG Bull. 74, 781.
Van den Kerkhof, A., Hein, U.F., 2001. Fluid inclusion petrography. Lithos 55 (1–4), 27–47.
Van den Kerkhof, A., Thiéry, R., 2001. Carbonic inclusions. Lithos 55 (1–4), 49–68.
Wilkinson, J.J., 2001. Fluid inclusions in hydrothermal ore deposits. Lithos 55, 229–272.

Further reading

Aplin, A.C., Macleod, G., Larter, S.R., Pedersen, K.S., Sorensen, H., Booth, T., 1999. Combined use of confocal laser scanning microscopy and PVT simulation for estimating the composition and physical properties of petroleum in fluid inclusions. Mar. Pet. Geol. 16 (1999), 97–110.
Hunt, J.M., 1979. Petroleum Geochemistry and Geology. W.H. Freeman and Company, San Francisco, p. 617.
Rosasco, G.J., Roedder, E., 1979. Application of a new Raman microprobe spectrometer to nondestructive analysis of sulfate and other ions in individual phases in fluid in-clusions in minerals. Geochim. Cosmochim. Acta 43, 1907–1915.

CHAPTER 2

Nondestructive analytical techniques for fluid inclusions

V. Nandakumar[a] and J.L. Jayanthi[b]

[a]Scientist - G, National Facility for Geofluids Research and Raman Analysis, National Centre for Earth Science Studies (NCESS), Thiruvananthapuram, Kerala, India

[b]Project Scientist - C, National Facility for Geofluids Research and Raman Analysis, National Centre for Earth Science Studies (NCESS), Thiruvananthapuram, Kerala, India

2.1 Fluid inclusion analysis: Basics

Fluid migration across the Earth's strata is a complex process. When hydrocarbons and water move through various minerals and cements in rock layers, trace quantities are entrapped within micron-scale isolated cavities. These fluid inclusions form during diagenetic and catagenetic events whereby cement is added to open pore space, or during which compacted or tectonic fractures are created and sealed. Despite the small size, fluid inclusions contain a wealth of useful information that retains certain characteristics of the original fluid, such as hydrocarbon composition, bulk density, brine salinity, and API gravity. By studying these fluids, we get an opportunity to study the past by collecting fundamental data that cannot be generated by any other means. These data are often critical for an accurate and complete assessment of the processes that govern petroleum systems and can indicate the economic potential of reservoirs. An essential part of almost every application of fluid inclusion research is the petrography. Thus, most research papers using fluid inclusions should have a petrography section, presenting the overall paragenesis, evidence for timing of entrapment of fluid inclusions, fluid inclusion assemblages, phases observed, and their distribution. Fluid inclusion petrography and microthermometry are applicable to lithologies and reservoirs of any age—provided suitable fluid inclusion populations are found. Fluid inclusion petrography is performed using doubly polished sections/wafers of rocks under transmitted plane-polarized light and epi-illumination with a high-intensity UV source. Petroleum and aqueous inclusion populations are identified along with relevant variables, such as fluorescence color, distribution, and abundance. Thus, migration pathways can often be distinguished from paleo accumulations based on the visual

Hydrocarbon Fluid Inclusions in Petroliferous Basins
https://doi.org/10.1016/B978-0-12-817416-6.00004-6

distribution of liquid petroleum inclusions. Fluid inclusion characteristics often reflect the maturity and effectiveness of local organic matter for generating petroleum fluids.

Hydrocarbon-bearing fluid inclusions (HCFIs) occur in minerals such as quartz, feldspar, calcite, cements between mineral grains, healed microfractures and contain micron scale samples of trapped hydrocarbon oil. In addition to hydrocarbon fluids, at room temperature, minerals can also contain traces of other fluids such as H_2O, gases such as CO_2 and CH_4, and solids that were either entrapped along with or precipitated from the liquid phase. The detailed analysis of fluid inclusions can yield fundamental information about the history of oil formation in sedimentary basins (along with microthermometric studies), which is important to both the petroleum production and exploration industries and to the geological community (McLimans, 1987; Munz, 2001; Burruss, 2003). Destructive analysis of HCFIs can be done either as bulk sample analysis or as a single inclusion analysis. The bulk fluid inclusion analysis is typically done by crushing rock samples and extracting the fluids trapped in inclusions for chromatographic analysis. The oil composition data obtained in this way may suffer from a variety of problems including mixing of fluids from multiple nonhomogenous HCFI populations and contamination from material within the rock sample itself or lab contamination. Therefore, analysis of single inclusions is significant. It requires opening of HCFIs (laser ablation or high-energy beam ions), but the very low quantities of oil typically contained in HCFI ($\sim$10 μm) restrict analysis with gas chromatography and/or mass spectroscopy to rather large inclusions or groups of inclusions (Blamey and Ryder, 2007). Munz (2001) has detailed the destructive and nondestructive analytical methods for fluid inclusion analysis that comprise nondestructive techniques for characterization of various properties of single inclusions and destructive bulk techniques for characterization of composition.

Optical microspectroscopy, on the other hand, offers good diagnostic potential for nondestructive HCFI analysis. The most common spectroscopic methods are based on infrared absorption, Raman, and fluorescence spectroscopy. UV fluorescence is commonly used for identification of HCFIs and the high sensitivity of fluorescence and a wide range of available techniques makes it a potentially viable tool for HCFI analysis. The fluorescence of petroleum products that originates from the presence of a polycyclic aromatic fraction is directly governed by the chemical composition and physical properties of the oil. In general, heavy oils (low-API gravity) tend to have relatively broad, more shifted to the red and less intense fluorescence bands, lower quantum yields, and shorter fluorescence lifetimes as compared to light oils (high-API gravity). These effects are produced by higher concentrations of fluorophores and quenchers in heavy oil, which yield higher rates of energy transfer and quenching. The use of fluorescence color is widely used as a qualitative guide for assessing the maturity of crude oil in bulk and in HCFI (George et al., 2001), but this technique is intrinsically prone to error (Hagemann and Hollerbach, 1986; Oxtoby, 2002; George et al., 2002). The techniques based on measurement of fluorescence intensity and spectral distribution may be affected by instrumental factors such as excitation source intensity variation, detector spectral sensitivity variation and also by the physical properties of the sample such as opacity, geometry, turbidity, scattering properties, and photo bleaching, but are manageable with careful handling of the optics and sample. Time-resolved-based fluorescence techniques are also not largely free of these constraints, but they can contain information that is lost in time-averaging processes inherent to steady-state methods. These techniques have also been used to characterize various petroleum products through the last many years (Przyjalgowski et al., 2005). The most commonly used factor in time-resolved studies of petroleum oils is fluorescence lifetime, which is also governed by the complex interplay between energy transfer and quenching, with heavy oils having typically shorter lifetimes than light oils. The fluorescence lifetime is emission/excitation wavelength dependent since each emission/excitation wavelength represents different populations of fluorophores.

Fluid inclusions are small samples of pore fluid crystallographically trapped in rocks during diagenesis, catagenesis, or fracture healing processes which contain composition and density information that can be translated to temperature, pressure, and composition. These data are useful for understanding petroleum migration, diagenesis and developing basin models. Towards this fluid inclusion, wafers (transparent, polished slabs of rock material) are prepared and studied optically with a petrographic microscope. Samples are viewed under transmitted plane-polarized white light as well as under reflected ultraviolet or blue-violet illumination. Aromatic species within natural oils and condensate inclusions render them fluorescent under UV light. Hence, aqueous inclusions, nonfluorescent gas, and fluorescent condensates and oils can be identified along with their relationship to each other, diagenetic features (such as physical and chemical compaction), and the rock matrix that can be resolved.

After fluids are trapped at a certain pressure and temperature in a mineral forming at some depth in the Earth's crust or mantle, that mineral will most likely undergo a stage of cooling and decompression. It is believed that at the time of trapping, the inclusion represents a homogenous system (single phase) sometimes if a saturated fluid contains a suspension of solid particles that may be a heterogeneous system also. After trapping and on cooling, differential contraction between the fluid and host mineral results in the appearance of a vapor bubble. During this time, the fluid may give two immiscible liquids or can give some solute components to form daughter minerals, and therefore at room temperature many inclusions are a multiphase system. There is no change in mass or volume of the fluids happening after entrapment; therefore, it is assumed that the inclusions behave as constant volume-constant composition system (Shepherd et al., 1985). This means that the fluid inclusions follow a path of constant density. Lines of constant density on a P–T diagram are known as isochores. Fluids of different compositions will have isochores with different slopes in the P–T space.

To study phase transitions in fluid inclusions, fluid inclusion wafers are broken into small chips that are mounted on a special heating and freezing stage. These chips are then observed while the stage is either heated or cooled, and the phase changes which take place during a heating or cooling cycle are carefully recorded along with the temperatures at which such changes occur.

2.2 Destructive and nondestructive analytical methods for fluid inclusions

HCFIs are small encapsulations of oil and gas that offer an invaluable opportunity to get a better understanding on the evolution of petroleum systems. Observations on paleofluid compositions supplement insights on the fluids that are responsible for the cumulative processes throughout the basin history. A wide range of approaches have been used to extract geochemical information from petroleum inclusions and these techniques can be grouped into optical, spectrographic, and thermometric nondestructive methods, or destructive chemical analyses of bulk samples or individual inclusions. Typically, optical methods documenting the distribution and visual properties of petroleum inclusions are used to provide petrographic context for subsequent specialized geochemical analyses of petroleum inclusions. Additional nondestructive techniques such as Raman spectroscopy can then be applied to provide some further insights into the composition of the trapped fluids that are detailed in Chapter 7. The complex nature of petroleum generally requires direct access to the fluid for a more complete understanding of the geochemical aspects.

A variety of destructive techniques have been developed, initially to analyze bulk samples released by mechanical crushing and more recently through ablation type techniques that allow the composition of individual inclusions to be characterized. Screening geochemical techniques that utilize mechanical crushing of bulk samples to analyze petroleum inclusions using mass spectrometry without prior chromatographic separation have become routine analyses. Other geochemical techniques more geared towards detailed molecular information such as biomarkers utilize chromatographic separation prior to mass spectrometry. Evaluation of the isotopic composition of petroleum inclusions is also possible for both bulk samples and compound specific analyses. The use of lasers to open individual inclusions allows the released contents to be analyzed by thermal extraction-gas chromatography–mass spectrometry (GC–MS), or mass spectrometric mapping of minerals using time-of-flight Secondary Ion Mass Spectrometry (ToF-SIMS), a surface-sensitive analytical method that uses ion beams to ablate into minerals. The continued evolution of techniques to analyze the incredibly small volume of hydrocarbons trapped within fluid inclusions has progressed to a point where there is little that can be done to evaluate a live oil or gas sample that cannot be achieved for a fluid inclusion sample.

2.2.1 Destructive fluid inclusion analytical techniques

To assess the content of individual fluid inclusions

Two techniques that can provide direct access to the content of HCFIs are (i) laser ablation of trapping minerals followed by thermal extraction and GC–MS (Greenwood et al., 1998; Volk et al., 2010; Zhang et al., 2012), and (ii) ion milling followed by time-of-flight Secondary Ion Mass Spectrometry (ToF-SIMS) (Siljeström et al., 2010, 2013).

i. *Laser ablation GCMS*

The use of inductively coupled plasma-mass spectrometry (ICPMS) methods allows inorganic geochemical properties of single inclusions to be determined (Heinrich et al., 2003; Horn et al., 2006; Sindern, 2017), and several microprobes for laser ablation-ICPMS are commercially available. However, ICP-MS is not suitable for the analysis of organic molecules. For the analysis of HCFIs, laser ablation systems coupled with GC–MS detection have been developed in some research laboratories (Hode et al., 2006; Greenwood et al., 1998). At CSIRO, Greenwood et al. (1998) used a laser micropyrolysis system to analyze 10–100 random oil inclusions using thermal decrepitation induced by an infrared Nd:YAG laser beam with a wavelength of 1064-nm coupled online to a GC–MS system. The geochemical signature obtained from the groups of inclusions analyzed from synthetic and natural samples showed data similar to those obtained by online crushing of the bulk samples, but also showed some molecular weight discrimination and some pyrolysis artifacts induced by the laser. Hode et al. (2006) extracted fluid inclusion oils in a vacuum chamber using an infrared Er:YAG laser with a wavelength of 2940 nm, followed by condensation of this oil onto a cold finger. They then dissolved the condensed oil in hexane and analyzed it off-line by GCMS. These two studies were not targeting any individual inclusion, since the content of inclusions is liberated by thermal decrepitation after heating the trapped fluid, rather than by ablating the mineral enclosing the inclusion.

Volk et al. (2010) demonstrated that it is possible to analyze the content of a single petroleum inclusion by interfacing a femtosecond laser to a GC–MS instrument. Clean ablation of the host quartz mineral, followed by the opening of a single inclusion, allowed thermal extraction and analysis of the content by GC–MS. No cyclic biomarkers could be detected due to the high-molecular-weight discrimination of the extracted petroleum. However, the low-molecular-weight hydrocarbon distribution recorded from a single inclusion was similar to the low-molecular-weight hydrocarbon distribution of the same sample analyzed by online bulk crushing, and simple biomarkers such as branched isoprenoids were detected. Zhang et al. (2012) achieved the opening and analysis of the contents of individual fluid inclusions using a 193-nm excimer laser. Their system used a newly designed GC inlet system that enabled the detection of C4–C30 hydrocarbons, extending the molecular weight significantly beyond the C21 limit of previous online laser-based hydrocarbon analysis systems for oil inclusions. Hydrocarbon-based parameters derived from their study indicated that the blue fluorescing inclusions are of higher thermal maturity than the yellow fluorescing inclusions (Zhang et al., 2012; Pan et al., 2017).

ii. Time-of-Flight Secondary Ion Mass Spectrometry (ToF-SIMS)

Time-of-flight secondary ion mass spectrometry (ToF-SIMS) enables analysis of the composition and spatial distribution of molecules and chemical structures on surfaces (Thiel and Sjövall, 2011). In geosciences, it has been used to map out biomarkers at microscopic levels (Thiel et al., 2007) or even within single cells (Leefmann et al., 2013). Siljeström et al. (2009, 2010, 2013) demonstrated how ToF-SIMS can be used to detect biomarkers such as hopanes and steranes (the two most commonly used biomarker classes in the application of organic geochemistry to petroleum exploration) in nonfractionated crude oils, without extraction and chemical preparation. They validated their method by spiking crude oils with authentic standards and by monitoring specific mass ranges at high resolution. Their tests on detection limits suggested that it should be possible to detect hopanes and steranes in single fluid inclusions containing petroleum with similar biomarker concentration for inclusions as small as 10 μm^2 in diameter. Analysis of four individual oil bearing fluid inclusions found in a 1430-Ma sandstone from the Roper Superbasin in Northern Australia reported the presence of n-alkanes/branched alkanes, monocyclic alkanes, bicyclic alkanes, aromatic hydrocarbons, and tetracyclic and pentacyclic hydrocarbons, which that were in agreement with those obtained from bulk crushing of inclusions trapped in the same samples (Dutkiewicz et al., 2003; Volk et al., 2003).

Molecular and isotopic signature of groups of petroleum inclusions

A prerequisite for analyzing the molecular composition and isotopic signature of petroleum inclusions is minimizing contamination as the volume of oil released from petroleum inclusions is very small. To this end, minerals containing petroleum inclusions need to be isolated and rigorously cleaned. Mineral separation to obtain particular mineral separates (often quartz, without any clays) in which petroleum inclusions are hosted is usually advisable. Details of the separation methods and cleaning protocols employed at CSIRO are given in George et al. (2012) and Jones and Macleod (2000).

Cleaning of petroleum inclusion samples prior to bulk crushing generally involves disaggregation using gentle hand crushing, rinsing in water (if the host minerals are not water-soluble salts), followed by digestion in H_2O_2 and acids such as HCl, HNO_3, H_2SO_4, Aqua Regia, and chromic acid (if the host minerals are not carbonates). Various polar organic solvents (e.g., methanol, furan, chloroform, and dichloromethane) can also be used to help clean the mineral surfaces and, indeed, are the only choice when attempting to analyze petroleum inclusions in carbonates (George et al., 1998; Shariatinia et al., 2013). In samples where individual grains or cements containing petroleum inclusions cannot be isolated, sand-sized rock fragments can also be used (Dutkiewicz et al., 2003, 2004; Volk et al., 2003).

Off-line crushing of crystals, sand grains, or small rock fragments (the crush-and-leach technique) is the most widely applied approach to obtain petroleum inclusion extracts that can be analyzed by gas chromatography with flame-ionization detection (FID-GC) (Jensenius and Burruss, 1990), or the more widely applied and diagnostic approach of GC–MS (Karlsen et al., 1993; George et al., 1997; Volk et al., 2002, 2005; Killops et al., 2009; Shariatinia et al., 2013). Crushing can be performed using a pestle and mortar or in crushers developed for this purpose (Volk and George, 2019). Petroleum inclusion samples from oil migration pathways (George et al., 2004) have been successfully analyzed using this technique, in combination with batch crushing (George et al., 2007). Petroleum inclusions in carbonate veins can be analyzed by this approach (Peters and George, 2018). Prior to GC–MS analysis, fractionation of fluid inclusion oil using high-performance liquid chromatography (Pang et al., 1998) has been undertaken, however, in the vast majority of studies using the off-line crushing technique. Liquid chromatographic separation before GC–MS analysis is not applicable, due to the very small amounts of inclusion oil dissolved in the solvent.

Polar compounds present in petroleum inclusions have been analyzed by GC–MS and are enriched in petroleum inclusion oils relative to corresponding crude oils (Peters et al., 2018). The phenolic, carbazolic, and benzocarbazolic compounds are only minor constituents of the petroleum inclusion oils, whereas they appear much more abundant in the crude oils. Noah et al. (2018) demonstrated that polar compounds in fluid inclusion oils can be analyzed using Fourier-transform ion cyclotron resonance mass spectrometry (FT-ICRIRC-MS). Many of the polar compounds in petroleum inclusion oils have never been studied in detail and indeed remain largely unidentified and uncharacterized, thus representing a promising field for future analytical developments. For large petroleum inclusion extracts, it is possible to obtain isotopic information on individual hydrocarbons. A key advantage of the off-line crushing technique is that the petroleum inclusion extracts obtained by this approach can be analyzed with methods similar to those applied to conventional oil or rock extract fractions, facilitating more direct comparison and correlation. Detailed distributions of aliphatic biomarkers within petroleum inclusions can be obtained from off-line crushing, which is very useful as these are the most widely applied group of compounds for oil-source correlations. The off-line crushing technique has three main disadvantages: (i) it is prone to contaminant overprinting that can be problematic if not recognized, (ii) as a bulk method, the detailed geochemical information obtained represents an average of all oil inclusions present in the aliquot of minerals crushed, and (iii) gases and light hydrocarbons are lost during the preparation of the fluid inclusion extract.

Difficulties with some of the problems with off-line analysis can be overcome by techniques that use online crushing or that yield information on individual inclusions. The composition of dissolved or free gases in individual fluid inclusions can be inferred from Raman and FT-IR spectroscopic methods (Dubessy et al., 2001; Pironon et al., 2001; Frezzotti et al., 2012; Chen et al., 2016b; Toboła, 2018), but there are significant uncertainties in calibrating these analyses to conventional GC data, and no information on the isotopic composition of gas species can be obtained. For this reason, a range of bulk analysis techniques of fluid inclusions have been developed, some of which were reviewed in Salvi and Williams-Jones (2003). Compared to the crush-and-leach extraction approach, crushing or thermally decrepitating samples with petroleum inclusions in an online system allow information from gases and light hydrocarbons to be obtained. Murray (1957) published the composition of petroleum trapped in a single quartz crystal. Detailed molecular and isotopic information from fluid inclusion gases requires the bulk crushing of samples containing fluid inclusions. Potter and Longstaffe (2007) provide details for an online gas chromatography continuous-flow isotope-ratio mass spectrometry (GC/CirMS) method for extraction, separation, and measurement of fluid inclusion volatiles. The higher hydrocarbons tend to have a reduced intensity compared to molecular signatures obtained by off-line crush-and-leach techniques.

2.2.1.1 Direct MS detection of petroleum inclusions

Detailed molecular information from petroleum inclusions by mass spectrometry typically requires initial separation using gas chromatography (GC). A further screening technique, called fluid inclusion stratigraphy (FIS) by fluid inclusion technologies (FIT, acquired by Schlumberger in 2015), is a patented technique aimed at the analysis of organic and inorganic volatiles in fluid inclusions from cuttings, core, or outcrop samples. Quadrupole mass spectrometers attached to an automated high-vacuum sample introduction system (Hall et al., 1997) provide information on low-molecular-weight hydrocarbons as well as CO_2, H_2S, and other organic compounds. FIS has also been applied to fine-grained rocks for analysis of unconventional plays, where optical observation of fluid inclusions in thin sections is much harder due to the small grain size. Fluid inclusion volatile (FIV) is a proprietary technique that represents another screening tool that uses direct mass spectrometric detection without chromatographic separation to analyze volatiles. The key components used to identify hydrocarbon migration are methane, ethane, C3+ gases, paraffins, naphthenes, and ratios between these compounds. This technique was used in combination with other geochemical methods to develop an improved understanding of down dip-oil potential in a mature exploration play (Pottorf et al., 2008).

2.2.2 Nondestructive fluid inclusion analytical techniques

Petrography and establishing a paragenetic sequence in the context of fluid inclusion entrapment are fundamental components of fluid inclusion study. Fluid inclusions containing petroleum are common in sedimentary sequences that have undergone diagenetic transformations (Munz, 2001). They are usually found within diagenetic phases such as quartz and feldspar overgrowths, carbonate cements, annealed microfractures through authigenic and detrital minerals, and vein-filling minerals.

Goldstein and Reynolds (1994) introduced the concept of the fluid inclusion assemblage as a group of fluid inclusions that were all trapped at the same time. Each fluid inclusion assemblage is therefore a fluid event that can be used to represent a discrete event in the geological history. The fluid inclusion assemblages method works well for aqueous inclusions (Goldstein, 2001), but Eadington and Kempton (2008) showed that similar approaches to recognizing oil inclusion assemblages can be more difficult. Bourdet et al. (2014) suggested that HCFI assemblages in a single petroleum reservoir might have a wide range of fluorescence colors, suggesting that the reservoir was filled with a number of different hydrocarbon charges, but trapped at about the same time in basin history. Some studies also incorporate coeval aqueous inclusions into HCFI assemblages (Bourdet et al., 2012). The first step of a fluid inclusion study is to carefully inspect the doubly polished section to determine whether petroleum inclusions are present and, if so, to determine their location in a paragenetic sense to construct relative timing. A transmitted-light microscope with a polarizer and a UV lamp makes a petrographic equipment. However, additional methods may be required to gain further insights into textural and paragenetic relationships of minerals that host fluid inclusions. Scanning electron microscopy in back-scattered electron (SEM-BSE) imaging mode, cathodoluminescence (SEM-CL), energy dispersive X-ray spectroscopy (SEM-EDS), and wavelength-dispersive X-ray spectroscopy (WDS) element mapping may need to be used. Practical aspects about preparing samples for fluid inclusion petrography, the tools used to study textural relationships, and the classifications of fluid inclusion assemblages are provided in Goldstein (2003).

The most common way to detect HCFI is screening of thin sections by epifluorescence microscopy under UV light (e.g., Burruss, 1981; McLimans, 1987; Stasiuk and Snowdon, 1997; Duggan et al., 2001; Chen et al., 2017; Zhao et al., 2017). Because of their bright fluorescence emission colors, UV fluorescence microscopy is an effective way of differentiating aqueous from petroleum fluids. However, hydrocarbon gas inclusions remain undetected using fluorescence colors. Since aqueous and petroleum inclusions are described at room temperature, their size, shape, color under transmitted and fluorescent light, and the proportions of liquid, vapor, and solid phases in the inclusions should be observed carefully. Even though relative phase volumes can be estimated using conventional microscopes, more accurate determinations of liquid to vapor ratios are required for PVT modeling of oil-bearing fluid inclusions and a confocal microscope used for this purpose (Pironon et al., 1998; Aplin et al., 1999; Bourdet and Pironon, 2008; Bourdet et al., 2010; Ping et al., 2013, 2017). Fluid inclusions can be classified into primary, secondary, pseudosecondary, and indeterminable inclusions according to criteria described in numerous excellent review papers and textbooks (e.g., Roedder, 1984; Goldstein and Reynolds, 1994; Van den Kerkhof and Hein, 2001; Goldstein, 2003).

2.3 Microthermometry

Microthermometric measurements using a heating-freezing stage are used to examine fluid inclusions in more detail. As sealed viles, fluid inclusions can provide insight into pressure-volume-temperature (PVT) conditions of the geological history. Phase changes are observed as a function of temperature (Shepherd et al., 1985), providing insights on fluid composition and PVT properties under which the fluid inclusions were trapped. Homogenization temperature (Th) recorded on HCFIs/petroleum inclusions and coeval aqueous fluid inclusions allows inference about trapping temperatures where Th gives a minimum temperature of trapping. The actual trapping temperature will lie somewhere along the isochore above the two-phase region of the PT diagram, and may be considerably higher than Th, especially for HCFIs with relatively flat isochores reflecting less dense and more compressible fluids. The isochores of fluid inclusions containing heavy oils will be more vertical than those of gas-condensate inclusions, which contain fluids that are more compressible. Hence, it requires a greater correction to infer true trapping temperatures from Th at a given pressure. (See the figure of the P–T phase diagram showing immiscibility curves and isochores of HCFIs shown in Chapter 5.) Isochores in HCFIs are having a more variable nature compared to isochores from aqueous inclusions; therefore, it is generally more practical to infer trapping temperatures of HCFIs from coeval aqueous inclusions rather than from HCFIs themselves. But this requires coeval aqueous inclusions to be carefully identified.

It is good to cross-check trapping temperatures inferred from coeval aqueous inclusions against trapping temperatures of HCFIs that are obtained using Th. Th of oil inclusions are generally lower than those of coeval aqueous fluid inclusions. Therefore, Munz (2001) recommended measuring Th of HCFIs first in order to avoid stretching, leakage, and decrepitation. First and Last ice melting temperatures can also be determined from microthermometric observations.

2.4 PVTX modeling

PVTX modeling of petroleum inclusions has significantly increased in the last 21 years. Critical inputs for these models are the determination of Th and accurate liquid–vapor (L/V) ratios of HCFIs. The liquid–vapor ratios for regularly shaped inclusions can be roughly estimated using a conventional epifluorescence microscope. But sufficiently accurate determinations of L/V ratios require the use of a Confocal Laser Scanning Microscope (CLSM), where the emission light is filtered by a confocal pinhole (Pironon et al., 1998; Aplin et al., 1999). The laser causes fluorescence in the liquid, but not in the vapor of the inclusion, enabling a clear distinction between the two phases. A series of layered images stepped along the z-direction is used to calculate the total volume and the volumetric ratio between liquid and vapor. To calculate the phase envelope and isochore, a composition that matches the physical properties of the trapped petroleum is needed, requiring information on the fluid composition and the average molecular weight of the trapped petroleum. Different phase equilibrium models and programs are derived as part of this study for determining the tapping pressure, temperature, molar volume (Aplin et al., 1999; Bakker, 2003; Thiery et al., 2000;, Grimmer et al., 2003; Bourdet et al., 2008, 2010; Ping et al., 2012), etc.

2.5 Fluorescence methods

Fluorescence microscopy is also widely used to study petroleum inclusions (e.g., Burruss, 1991; Kihle, 1995; Stasiuk and Snowdon, 1997; Blamey and Ryder, 2009; Baba et al., 2019), and the fluorescence properties of oils can be related to the nature of these fluids (Hagemann and Hollerbach, 1986; Barwise and Hay, 1996). Blamey and Ryder (2009) have given a review on the fluorescence of HCFIs and the spectroscopic methods employed for their analysis. Some studies show HCFI's fluorescence color to infer the thermal maturity of trapped paleo-fluids (Burruss et al., 1985; McLimans, 1987; Bodnar, 1990; Stasiuk and Snowdon, 1997; Parnell et al., 2001; Parnell, 2010; Conliffe et al., 2017; Matapour and Karlsen, 2017) with the assumption that inclusion oils with higher API gravities fluoresce in the blue end of the visible spectrum, whereas inclusion oils of lower API gravity show a shift of fluorescence colors towards the red region. George et al. (2001) demonstrated that the extension of this relationship to the use of the fluorescence colors of oil inclusions as a thermal maturity guide is not always justified, while the relationship of API gravity and the fluorescence properties of crude oils was discussed by a few authors (Hagemann and Hollerbach, 1986; Barwise and Hay, 1996).

While much of the fluorescence in oils is due to aromatic hydrocarbons, polar compounds are also known to have fluorescence properties. Polar compounds are preferentially adsorbed onto charged mineral surfaces (e.g., Clementz, 1976; Horstad et al., 1990), and fluid inclusion oils trapped in minerals generally contain a higher proportion of polar compounds than associated reservoir oils (Karlsen et al., 1993; Macleod et al., 1993; Nedkvitne et al., 1993; Bhullar et al., 1999). This was confirmed in a study that separated fluid inclusion oils using high-performance liquid chromatography, with some exceptions (Pang et al., 1998). Work on synthetic fluid inclusions also provides strong evidence for preferential trapping of polar compounds in inclusions (e.g., Stasiuk and Snowdon, 1997; Chen et al., 2015). Naturally occurring petroleum inclusions may also exhibit marked differences in fluorescence properties and L/V ratios, even in cases where PVTX relationships demonstrate that they all belong to the same fluid inclusion assemblage (e.g., Bourdet et al., 2012). To avoid subjective descriptions of fluorescence colors, the intensity of visible light wavelengths from fluorescing petroleum inclusions can be recorded and plotted as Chromaticity Index (CIE) parameters. The standard observer chromaticity diagram (Deaton, 1987) is used in many petroleum inclusion studies (e.g., McLimans, 1987; Blanchet et al., 2003; Schubert et al., 2007; Bourdet et al., 2012, 2014; Evans et al., 2014; Ferrill et al., 2014; Szabo et al., 2016).

Bourdet and Eadington (2012) reported on the relationships between crude and inclusion oil fluorescence, including relationships to polarity fractions (saturates-aromatics-resins-asphaltenes or SARA), API gravity of stock tank oils, and fluorescence and Fourier-Transform infrared (FTIR) spectroscopy results from petroleum in trapped fluid inclusions and stock tank crude oils. They recommended ways to distinguish fluorescence properties largely driven by thermal maturity from those where phase-controlled molecular fractionation and trapping mechanisms of inclusion oil exert the strongest control on fluorescence color. Their interpretational framework makes complementary use of fluorescence spectra, FTIR spectroscopy results, liquid–vapor relationships in inclusion and their dependency on temperature combined with PVT modeling. UV spectroscopy can also be used to assess the content of adsorbed and included oil in reservoir sandstones. Such methods may help identify zones of paleo- or present-day oil saturation, often associated with the presence of oil inclusions, which can be measured using the Grains containing Oil Inclusions (GOI) petrographic method (Lisk and Eadington, 1994). CSIRO subsequently developed the methods quantitative grains with fluorescence (QGF) (Liu and Eadington, 2005; Liu et al., 2003, 2007, 2017) and quantitative grains with fluorescence on solvent extractable hydrocarbons (QGF-E) (Liu and Eadington, 2005; Liu et al., 2007; Guo et al., 2016). A detailed description of the UV spectroscopic approaches that have been developed over the years can be found in Liu et al. (2017). Fluorescence spectroscopic analysis on HCFIs using a 405-nm diode laser and its applications are discussed in Chapter 5.

2.6 Raman and Fourier-transform infrared (FT-IR) spectroscopy

Fluorescence spectroscopic studies on fluid inclusions are accomplished by Raman spectroscopy applied on water inclusions or gas inclusions with low to no fluorescence, as well as by FTIR spectroscopy applied to HCFIs (e.g., Barres et al., 1987; Videtich et al., 1988; Pironon and Pradier, 1992; Kihle, 1995; Bourdet et al., 2014; Chen et al., 2016a; Zhao et al., 2019). Both techniques are able to detect transitions between vibrational energy levels, providing inference about the constituents of trapped fluids. Raman spectroscopy uses monochromatic lasers to cause Raman shifts, typically reported as wave numbers. The method is very useful to characterize fluids with small molecules, such as N_2, CO_2, H_2S, and CH_4. FTIR spectroscopy measures the adsorption of incident light of a polychromatic excitation source in the 1100–2000 nm range, and the functional groups that can be detected are -CH, -CH_2, and -CH_3 radicals, mainly contained in aliphatic structures, CH_4, CO_2, and liquid water (Bourdet et al., 2014). Quantitative estimates of the -CH_2:-CH_3 abundance based on peak areas at diagnostic wave numbers have been used to infer the average alkane chain length of trapped petroleum (Pironon and Barrés, 1990).

The application of Raman spectroscopy is much more difficult in oil inclusions due to their fluorescence, which is orders of magnitude stronger than Raman scattering. Long-wavelength laser beams can help reduce this interference. However, Pironon et al. (1991) found that even when using 1064-nm laser excitation in the near-infrared range (NIR FT-Raman), most of the petroleum inclusions that are fluorescent during visible Raman microspectroscopy (514 nm excitation) are still fluorescent in the NIR range. Bourdet et al. (2011) applied UV-Raman using short wavelength excitation from a frequency doubled argon ion laser at 244 nm, below the fluorescence maximum of benzene where UV interference was reduced, but it was not possible to distinguish low-molecular weight from high-molecular-weight aromatic compounds using this method. The applications of Raman spectroscopy for the characterization of HCFIs are discussed in Chapter 7.

In order to avoid difficulties such as the mixing up of inclusion contents and contamination in destructive analytical techniques, we emphasize the use of nondestructive analytical techniques such as petrography, microthermometry, fluorescence spectroscopy, and Raman spectroscopy for discussing the analysis of HCFIs in the coming chapters.

2.6.1 Geological information from fluid inclusions

(a) *Fluid composition*: Fluid composition is achieved indirectly by measuring the first melting temperature (known as the eutectic melting T or T_{FM}). After freezing the inclusion, it is heated slowly while being carefully observed under the microscope. The temperature of first melting of solid (ice) is recorded (and is usually marked by the first movement of the bubble). Comparing this temperature to eutectic melting points on published phase diagrams for binary and ternary systems would allow us to predict the composition of the fluid. Laser Raman microspectroscopy can also determine fluid composition. The inclusions can be physically opened/crushed, and the fluids can be chemically analyzed for getting the fluid composition.

(b) *Salinity of the fluids*: The HF stage is cooled with the help of liquid N_2 during which phase changes in the inclusion are carefully monitored. After the liquid in the inclusion solidifies completely, the stage is heated slowly while the inclusion is observed. The temperature at which the last piece of solid melts (T_{LM}) is then recorded. This will correspond to the freezing temperature of the inclusion. Knowing the freezing points of pure H_2O and CO_2, the recorded freezing point

(T_{LM}) is put into an equation of state (PV = nRT), and the freezing point depression for the system is then directly related to the amount of impurities in this system, providing us with some information on the concentration of salts in this fluid/salinity of the fluid. Therefore, salinity is the measure of amounts of solutes in aqueous solution, including electrolytes (e.g., NaCl, $CaCl_2$) and nonelectrolytes (e.g., CO_2, H_2S). Often, the concentrations of the individual solutes are not known in multicomponent fluid inclusions, but phase transitions can be measured that are sensitive to the overall salinity [e.g., Tm(Ice), Tm(Hydrohalite), Tm(Halite), Tm(Clathrate)]. In these cases, the salinity is conventionally reported in NaCl (or $CaCl_2$) equivalents, e.g., "NaCl-equivalent = 10%" means that the inclusion shows phase transitions that are consistent with a mass fraction of 10% NaCl in the aqueous phase. Salinity can be calculated by the Equation:

$$1.78 \times T_{LM} - 0.0442 \times (T_{LM})^2 + 0.00557 \times (T_{LM})^3.$$

where T_{LM} is the final/last melting temperature of a frozen solid form of fluid phases in an inclusion.

(c) *Fluid density*: The density of the fluid can be calculated from knowledge of the T_{LM} and T_h. Putting these values into equations of state for fluids of known composition (known system) helps us to determine the density of the fluid and calculate isochores.

(d) *Temperatures of entrapment*: Microthermometric techniques rely on heating two—phase inclusions on the stage until they homogenize (change to one phase). The temperature at which homogenization takes place can therefore be considered a minimum temperature for fluid entrapment. If the pressure of entrapment (or mineral formation) is known by some independent means (phase relations or geobarometry), then the temperature of entrapment can be immediately determined by the intersection of the isochore passing through the homogenization temperature with this pressure. The difference between the homogenization temperature and the temperature of entrapment is known as the pressure correction.

(e) *P–T history of the sample*: The textural relations exhibited by a particular fluid inclusion assemblage often give clues as to the P–T history of the sample. Pressure–temperature (PT) conditions of trapping can be calculated from the isochores (lines of constant density). In many studies, microthermometric data of coeval aqueous inclusions are used instead to constrain trapping conditions of petroleum inclusions or are used to verify the trapping temperatures of petroleum inclusions inferred from the microthermometry of petroleum inclusions combined with PVT modeling. This is because aqueous fluids have lower compressibility, hence steeper isochoric pressure–temperature (P–T) gradients, resulting in smaller differences between homogenization temperatures (Th) and trapping temperatures (Tt).

2.6.2 Petrographic analysis of fluid inclusions

Petrography is regarded as the most important prerequisite of any fluid inclusion study. Petrography is the study of rocks in thin sections by means of a petrologic microscope (i.e., an instrument that employs polarized light that vibrates in a single plane). Petrography is primarily concerned with the systematic classification and precise description of rocks; the most basic objective of any fluid inclusion study is to determine the relative timing of entrapment of fluid inclusions. Fluid inclusions may be entrapped during or after mineral growth. Obviously, those inclusions trapped during mineral growth are those most likely to be representative of the conditions of mineral growth, and those trapped after mineral growth may record later conditions. Most geologic samples, however, consist of an assemblage of various minerals that have grown and recrystallized at different times, and each crystal may consist of many growth zones, recording many events of mineral growth spread out over geologic time. When minerals are fractured, deformed, or recrystallized, fluid inclusions are likely to be trapped after mineral precipitation. Individual samples may record multiple events of fracturing and subsequent fluid inclusion entrapment. Clearly, if the entrapment of fluid inclusions can be put in a sequence, then a detailed history can be interpreted that will help the researcher in interpreting fairly a geologic history of a set of samples. Thus, for most studies, the researcher must establish crosscutting and superpositional relationships for mineral paragenesis and for fluid inclusion entrapment. Petrography is the easiest method for evaluating how many fluid phases were present at the time of fluid inclusion entrapment.

2.6.3 HCFI sample preparation

Goldstein (2003) proposes a procedure for fluid inclusion studies with emphasis on the criteria of selecting fluid inclusions for detailed petrography, microthermometry, and spectroscopic analysis. An overview of descriptive and genetic classifications of fluid inclusions in single crystals and in massive rocks is given with the intention of further differentiating the commonly used terms "primary" and "secondary" fluid inclusions. Petrographic microscopy of a rock sample is the first and at the same time an essential step of any fluid inclusion study. A proper interpretation of fluid inclusions can be made only when textural relationships between fluid inclusions and the host mineral are examined. Primary fluid inclusions are trapped during crystal growth, whereas secondary fluid inclusions can be trapped at any time after the growth of the host crystal. Fluid inclusions in microcracks that formed during crystal growth and ceased to grow before the crystal growth is completed are referred to as "pseudosecondary."

The most important aspect of fluid inclusion studies in sedimentary rocks is the petrographic analysis of samples. The first step is typically the gathering of core or cutting samples obtained from exploratory drilling by oil companies. Then the samples are prepared to prevent alteration of the fluid inclusions during heating because of problems with thermal re-equilibration and metastability. Therefore, fluid inclusion studies must avoid samples that have been heated in ways other than normal geologic heating in sedimentary basins. This means thin sections and their rock chips, which have been heated for mounting, are not appropriate for a fluid inclusion study. Outcrop samples that have been heated by fires are also inappropriate. Core and plugs that have been subjected to porosity and permeability analysis typically have been heated as well and hence must be avoided. When sampling a core, it is important to understand that they have not been subjected to whole-core analysis for porosity and permeability, because of heating. During preparation, it is important to use the best section making tools possible to get the samples doubly polished. However, all mounting techniques must be done at room temperature to avoid any heating of the samples. Furthermore, cutting of samples must be done to avoid fracturing and heating, so most labs use low-speed cutting saws that avoid damage to the sample. Finally, all preparation must be done to avoid damage; and one must be quite gentle in preparing the samples, never exerting too much pressure and never allowing samples to overheat due to friction.

Petrographic work alone can lead to significant interpretations about the environment of entrapment and the postentrapment history of fluid inclusions (Goldstein, 1993). The first part of this approach requires isolation of FIAs; 10–15 numbers of fluid inclusions that were initially entrapped under the same set of conditions can be identified. Once an FIA has been isolated, it must be ascertained that the liquid phase of inclusions is aqueous and not hydrocarbon in composition. This is accomplished by studying samples using UV epifluorescence microscopy. If inclusions fluoresce, then they are composed of oil and if they lack fluorescence, then they are most likely to be aqueous in composition. Some oil inclusions show variable vapor/liquid ratios and are likely the result of heterogeneous trapping; most oil inclusions have very similar vapor/liquid ratios. Petrography is the easiest method for evaluating how many fluid phases were present at the time of fluid inclusion entrapment and under what conditions of pressure and temperature the inclusions were entrapped (Goldstein, 1993, 2003). Goldstein (2001, 2003) detailed the fluid inclusions in the sedimentary system and its petrography, and Burruss (2003) discussed an introduction to hydrocarbon fluid inclusions.

Wafer preparation procedure for fluid inclusion studies from sedimentary rock samples requires a different approach than that used for samples from metamorphic/igneous rocks. Sedimentary rocks are inconsistent rocks and are friable. Therefore, in order to make it into a robust rock, a mounting media is used. The cutting and core samples obtained were crushed, washed, dried, and sieved using a 2 mm sieve. The sieved samples are then impregnated with a suitable epoxy resin-hardener mixture, doped with a fluorescent-quenching dye (waxoline blue), and then made into a doubly polished wafer for fluid inclusion studies (Fig. 2.1). While preparing a fluid inclusion wafer, the preparation procedure should be a slow process and it should be temperature controlled. Such specially prepared wafers are used for petrographic, microthermometric, and spectroscopic studies. Therefore, we are neither disturbing nor destroying the fluid inclusions within the grains/cements and our methodology is nondestructive and optical.

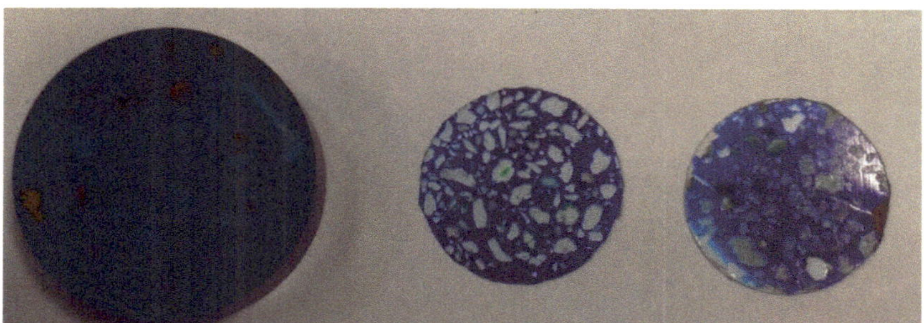

Fig. 2.1 Images of some blocks and wafers.

2.6.4 Petrography of HCFIs

The hydrocarbon fluids in HCFIs may have complex compositions but are principally liquid with variable amounts of gas (light alkanes, CO_2, N_2, H_2S), paraffins, naphthalene, aromatics, resins, and sometimes waxes. The apparent HCFI color depends on a combination of the wavelengths absorbed by the various chromophores present in the trapped hydrocarbon liquid and the associated optical parameters of the host mineral. Consequently, their color, when viewed under a microscope with plain white light illumination, may vary considerably from transparent to yellow, dark brown, or black. The mineralogical properties which can also influence the perceived color include birefringence, polarization, reflectivity, and variations in refractive index.

The specially prepared wafers as discussed above could then be microscopically examined for detailed petrographic studies. Then the wafers are microscopically viewed to detect the presence of hydrocarbons under UV-light. While analyzing the HCFIS from a Mumbai offshore basin, India, HCFls occurring in secondary trails showed good fluorescence under UV illumination and are mostly biphase inclusions (i.e., 80%–85% liquid and 20%–15% gas). The dark spot in the inclusion is the vapor bubble, which does not fluoresce under UV excitation (Fig. 2.2).

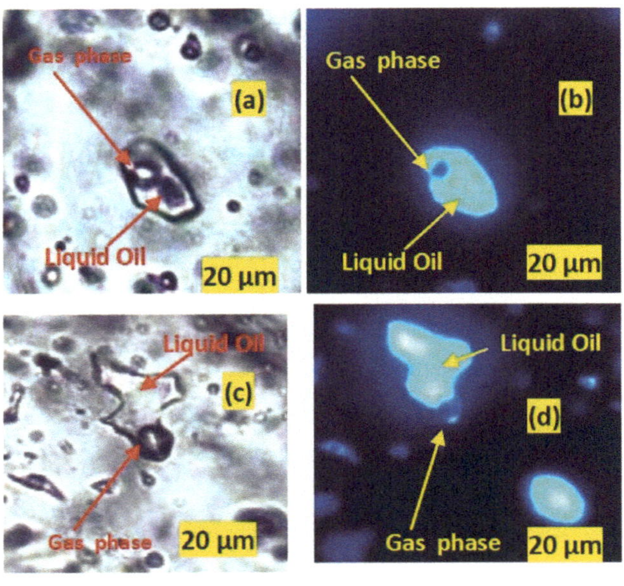

Fig. 2.2 Paired - transmitted light (a & c) and UV (b & d) photomicrographs of fluorescing biphase oil inclusions.

2.7 Microthermometric analysis of fluid inclusions

One of the key approaches in the fluid inclusion study is microthermometry (Roedder, 1984; Shepherd et al., 1985). The determination of temperatures of phase changes within fluid inclusions during the heating and cooling of samples is termed microthermometry. The technique is invaluable for estimating the temperatures at which minerals form, the thermal history a rock has experienced, and the compositions of the fluids that interacted with a rock in its history.

The sample containing the fluid inclusion is placed in a chamber that can be either heated or cooled and then observed using a microscope while the temperature is varied. The temperatures at which phase changes occur are recorded, thus providing information about homogenization temperature, brine salinity (for aqueous inclusions), or other data. HCFIs provide fewer phase changes but they do fluoresce under UV or visible light illumination, which can provide another source of information about fluid composition. The petroleum composition of HCFI is typically obtained by crushing a small amount of the HCFI containing material and extracting the petroleum fluid for analysis by gas chromatography (George et al., 2001) and mass spectrometry (Parnell et al., 2001). The key disadvantages of the crushing method are the destruction of the sample, probable mixing of fluids from inclusions of different sources/generations, and potential contamination. Therefore, the oil composition data derived from these bulk analyses have to be carefully considered as the mixing of heterogeneous fluid inclusion populations may have occurred. The nondestructive quantitative analysis of individual HCFIs is therefore a desirable goal for studying petroleum migration and is best achieved via optical methods. Optical methods have several general advantages that include nondestructive/noncontact analysis, reasonable sensitivity, ability to undertake micron-scale analysis of single inclusions, mature technology and methodologies, high diagnostic potential, and relatively simple instrumentation (Guilhaumou et al., 1990; Pironon and Pradier, 1992).

Fluid inclusion microthermometry is performed through advanced quantification of aqueous and petroleum fluid inclusions using a specially designed temperature-controlled chamber attached to a petrographic microscope (Fig. 2.3). Phase changes and other observations within individual fluid inclusions are recorded and compared with appropriate phase diagrams or calibration curves to derive data such as temperature, salinity, and API gravity. Temperatures can be related to petroleum emplacement, cementation

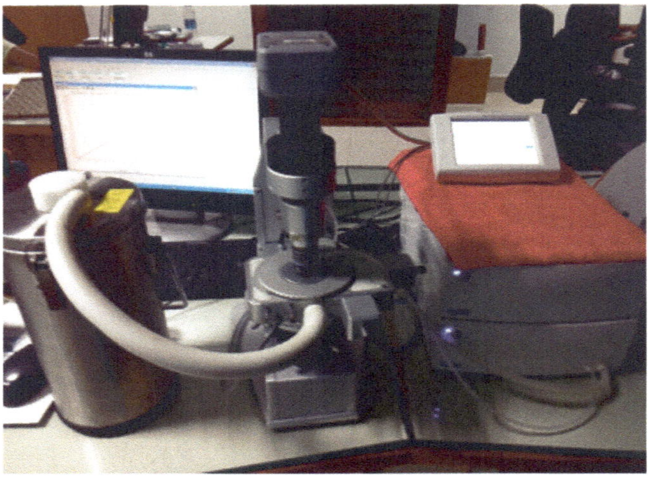

Fig. 2.3 Imaging station and the heating-freezing stage used for fluid inclusion analysis.

events, or maximum thermal exposure. The proximity to bubble point or dew point at trapping can be evaluated, and API gravity can be estimated. Salinities are used to infer fluid sources (which can be related to regional plumbing systems) and also to evaluate the composition ofirreducible water within reservoirs for calculations of water saturation.

One of the key methodologies in the fluid inclusion study is microthermometry. The sample containing the fluid inclusions are placed in a chamber that can be either heated or cooled and then observed using a microscope while the temperature is varied. The temperatures at which phase changes occur are recorded, thus providing information about homogenization temperature, brine salinity (for aqueous inclusions), or other data.

2.7.1 Heating-freezing stages

At least three types of dual-purpose heating-freezing stage specifically designed for thermometric analysis are commercially available. Many laboratories continue to employ Leitz 350, ReichartThermovar, or Mettler FP 53 melting-point stages. These well-tried instruments are useful up to +350 °C, but at low temperatures their operations are limited to −30 °C to −50 °C. The Chaixmeca, Linkam TH 600, and Reynolds stages meet specifications (−180 to +600 °C) and are intended for use with most makes of standard transmitted-light microscopes. With the Linkam stage, the choice of automatic and manual controls ensures a wide range of operating conditions and requires minimum effort to use. Though the Reynolds stage is manually operated, it is just as easy to use and employs a novel form of control, which gives fine temperature adjustment. But the present control units supplied with the Chaixmeca stage offer a poor combination of manual and semiautomatic functions. When operating above +200 °C, there is little advantage in being able to measure temperature differences of 0.1 °C. However, at sub-ambient temperatures where small differences in temperature often correspond to large differences in composition, a resolution of 0.1 °C is desirable.

Chaixmeca stage: The Chaixmeca stage comprises of a metal box, which carries a resistance heater, coolant, and platinum resistance thermometer. A silica condenser lens is designed to focus light directly into the inclusion. The top section consists of a demountable glass plate and a 1.8-mm-thick quartz window. The optimum sample size is considered to be 0.5 cm in diameter and 1 mm in thickness. The temperature of the block is changed by either energizing the resistance heater or passing a mixture of gaseous and liquid N_2 (boiling point at −196 °C) through the cooling coils. One of the main drawbacks is the need to remove the quartz window and to fit a plastic anticondenser sleeve over the lens when freezing inclusions. On heating, reverse adjustments are needed. This stage was widely used for fluid inclusion work about a decade ago.

Reynolds stage: The Reynolds stage is different from Chaixmeca and Linkam and operates by passing preheated or precooled gas over and around the sample. Having low thermal mass, the stage is fast to respond and does not require cooling coils to protect the objective lenses. The cell accepts sample wafers up to 20 mm in diameter and permits inclusions to be studied across the full field of view without repositioning the sample.

Linkam TH 600 stage: This stage has a very low thermal mass and is intended to be a fast-response device. The sample rests directly on a small silver block supported at the center of a double-walled anodized aluminum thermal cell. The upper and lower demountable sections of the cell carry double-glazed thin glass windows, and when assembled totally isolate the sample and thermal block from the external environment. At high temperatures, the double wall of the outer cell serves as a water jacket. The Linkam heating and freezing stage is similar to that of the Chaixmeca, except that the flow of coolant gas (N_2) is held constant and the temperature is changed by controlling power to the resistance heater. The temperature control unit is fully automatic and can be programmed to produce a wide range of heating/

freezing rates or switched to manual mode if decided. Samples up to 2 cm diameter can be accommodated in the cell and by repositioning the sample, inclusion throughout the entire specimen can be studied. Because the method of heat exchange is wholly by thermal conduction, good contact between the sample and the silver block is essential. For microthermometric analysis, we have placed the sample on a Linkam TH 600 stage that is attached to a Linkam imaging station (Fig. 2.3) working with Linksys32 software.

2.7.2 Important observations from heating-freezing experiment

Trapping temperature (Tt): The temperature of trapping (formation) of a fluid inclusion and Homogenization temperature (Th) are two important observations in fluid inclusion microthermometry (Fig. 2.4). There is normally a difference between the two. If we are dealing with aqueous FIA that is undersaturated with respect to a gas phase, then Th is significantly lower

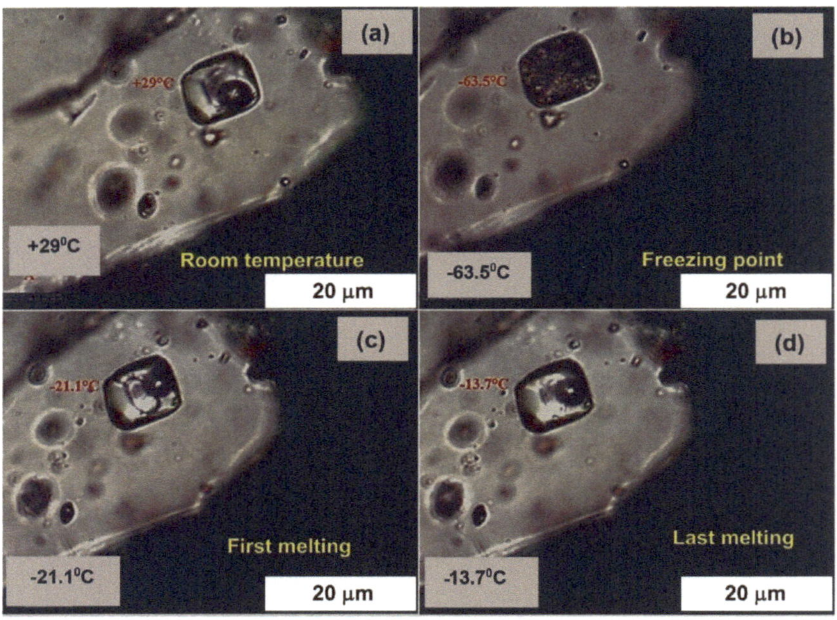

Fig. 2.4 Phase changes in fluid inclusions during heating-freezing runs, where the phase changes in a biphase fluid inclusion are shown; fluid inclusion appearance (a) at room temperature (+29 °C), (b) completely freeze out at freezing point (−63.5 °C), (c) first melts at −21.1 °C and (d) last/final melting at −13.7 °C.

than Tt. To get to the Tt, one needs to apply a pressure correction. For oil FIAs, that difference is even greater, with the Th much lower than the Tt. On the other hand, if there is evidence that we are saturated with respect to a gas phase, such as methane, then the Th can be equal to the Tt. That evidence is typically represented by some FIAs having a variable liquid–vapor ratio, caused by heterogeneous entrapment from separate gas bubbles at the time of entrapment. Many aqueous fluids in sedimentary systems have some dissolved methane but were undersaturated with respect to methane at the time of fluid inclusion entrapment. Th of these inclusions are still below Tt, but they are closer, requiring pressure corrections to get the Tt. The aqueous inclusions trapped with oil inclusions should normally be close to Tt, because there is likely methane around and close to saturation in the aqueous phase. Some workers opined that the presence of oil with the aqueous inclusions negates the need for a pressure correction on the aqueous inclusions and some disagree with this. We will see both approaches. The PVT data can be plotted using isochores for both oil and aqueous inclusions.

Homogenization temperature (Th): The temperature at which a fluid inclusion transforms from a multiphase/ two phase (heterogeneous) to a one-phase (homogeneous) state (Fig. 2.5). Homogenization may occur in several different modes, via several different phase transitions. (1) homogenization to the liquid state via a bubble-point transition, $Th(L+V \rightarrow L)$;

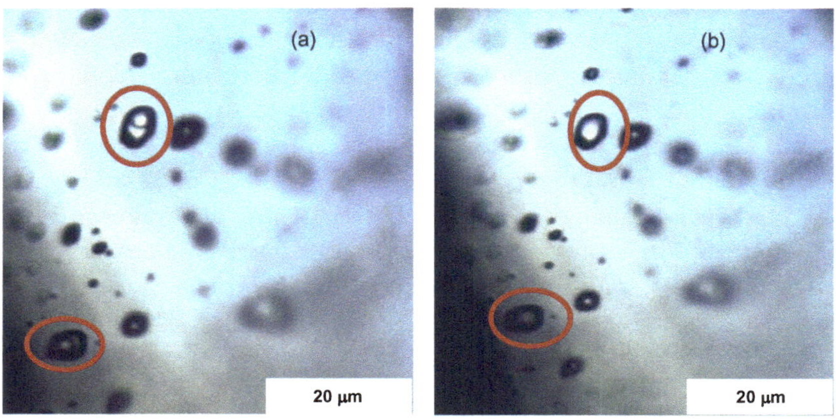

Fig. 2.5 Phase changes in a biphase fluid inclusion from the KK-basin (at 4565–4570 m depth, in a secondary trail) during microthermometric heating runs (a) room temperature (b) homogenizes when temperature (Th) reached at 83 °C, which falls within the oil window.

(2) homogenization to the vapor state via a dew-point transition, Th(L +V → V); (3) homogenization via a critical transition to a supercritical fluid: Th(L + V → supercritical fluid) or Th(critical) by fading of L-V meniscus; (4) Solid phase (S) homogenization to the liquid via a liquid (melting, dissolution) transition, Th(S + L → L) and homogenization via a vaporous (sublimation) transition, Th(S + V → V).

Eutectic temperature: The temperature changes observed in fluid inclusion studies are classified as Eutectic temperature (Te), initial/first melting temperature (T_{FM}), and final/last melting temperature (T_{LM}). Eutectic temperature is the minimum temperature of liquid stability in a specified system. For any system considered at a specified, fixed pressure, the eutectic is a unique, characteristic temperature that is associated with a unique, characteristic mixture of the components (the eutectic composition). When heating saline aqueous fluid inclusions from low temperature, melting first occurs at the eutectic temperature. However, for bulk composition of the fluid inclusion, only a small amount of liquid forms at the eutectic temperature and so it remains invisible. Such inclusions must be heated well above their eutectic temperature before enough liquid has formed to be visible. The temperature at which liquid is first observed upon heating may be termed the apparent eutectic temperature. In gas inclusions and in gas-bearing aqueous inclusions, the first liquid produced does not necessarily form at the characteristic system eutectic, but at a higher temperature dependent on the bulk composition and density of the inclusion. In such inclusions and whenever it is not clear which system is involved the temperature at which liquid is first observed upon progressive heating is termed as the initial melting temperature (Diamond, 2003).

Initial/First melting temperature (T_{FM}): The temperature at which liquid is first observed to form upon progressive heating of the frozen fluid formed due to the supercooling of the inclusion during microthermometry operations (Fig. 2.4). In gas-bearing inclusions, this temperature does not necessarily correspond to the eutectic temperature or the apparent eutectic temperature.

Final/Last melting temperature (T_{LM}): The temperature at which the frozen solid is observed to melt (or dissolve or dissociate) completely upon progressive heating of a fluid inclusion (Fig. 2.4). A considerable temperature span is generally observed between the initial and final melting temperatures of a given solid.

2.8 Optical spectroscopic methods for HCFI analysis

Hydrocarbon-bearing fluid inclusions (HCFIs) generally occur in diagenetic cements or grains and contain complex mixtures of mainly organic compounds depending on their source/s. Accurate analysis of the chemical composition of the entrapped hydrocarbons in HCFI can yield vital information about the history, evolution, and migration of petroleum fluids, and is thus crucial data for the petroleum exploration industry. Studying HCFIs is advantageous because the trapped fluids are representative of the actual hydrocarbon fluids that existed when the inclusions were sealed in the mineral. This sealing-in process preserves the petroleum fluid, thus isolating it from subsequent infiltration of other petroleum fluids and events in oil reservoirs such as loss of charge, water washing, or biodegradation (Blamey and Ryder, 2007).

The accurate characterization of the fluids entrapped in inclusions is very challenging. HCFI samples are "very priced" (usually obtained from drilling), and thus a noncontact, nondestructive, analytical method is appropriate. The small size of HCFIs necessitates the use of microscopy-based techniques. Fluorescence spectroscopy offers the best combination of high sensitivity, diagnostic potential, and relatively uncomplicated instrumentation. Optical-spectroscopic methods provide valuable data on the composition and properties of petroleum inclusions and entrapped fluids and are representative of the whole hydrocarbon fluids that once existed. Most importantly, optical procedures (a nondestructive/noncontact analysis) are of reasonable sensitivity facilitating analysis of individual micron-size inclusions (Blamey and Ryder, 2007; Guilhaumou et al., 1990; Przyjalgowski et al., 2005). The study of HCFIs through optical methods precludes mixing, and therefore by using nondestructive optical methods, more precise control on discriminating individual inclusions can be achieved.

2.8.1 Microscope-based florescence/luminescence spectroscopy

Fluorescence emission occurs as a result of a radiative electronic transition in which an electron jumps from a higher energy state to a lower one, the difference in energy being released as a photon. An electron must be first excited into higher energy states by means of UV or visible light irradiation. After excitation, the nuclei adjust their positions to the new excited situation, so that the interatomic distances equal the equilibrium distance belonging to the excited state. This process is called relaxation. During relaxation, usually there is no emission. The system can return to the ground state spontaneously under emission of radiation from the lowest level of the excited state. The emission occurs at a lower energy than the absorption due to the relaxation process. The energy difference between the maximum of the lowest excitation band and that of the emission band is called the Stokes shift. The luminescence is usually characterized by its quantum yield and lifetime.

Florescence spectroscopy is the measurement and analysis of various features that are related to the luminescence quantum yield and lifetime. The quantum yield is the ratio of the number of photons emitted to the number of photons absorbed. High quantum yield fluorescent centers display the brightest emission (Panczer et al., 2012). The lifetime is related to the average time that the fluorescence center spends in the excited state prior to its return to the ground state. It is defined as the time required for the fluorescence intensity to drop to 1/e of its original value. The fluorescence quantum yield and lifetime are related to a number of factors such as temperature, concentration, etc. that can increase or decrease the energy losses. The intensity of fluorescence centers is a function of their concentration, absorbing power at the excitation wavelength, and their quantum yield of the emission wavelength. A fluorescence emission spectrum represents the fluorescence intensity measured over a range of emission wavelengths at a fixed excitation wavelength. The fluorescence excitation spectrum is a plot of the fluorescence intensity at a particular emission wavelength for a range of excitation wavelengths. Steady-state spectroscopy is the methodology followed by most of the fluid inclusion spectroscopists. Steady-state or continuous wave (CW) fluorescence is a process where the excitation sources are continuously working at a constant intensity over a period necessary to perform the measurement. The result is an emission spectrum as discussed in Chapter 5.

2.8.2 Fluorescence of crude oils/HCFIs

The use of fluorescence for the analysis of crude oil is being practiced for the past 60 years, particularly for mud logging where UV light is used to detect the presence of oil in drilling mud and in core and cutting samples. The fluorescence of crude petroleum oils emanates from the aromatic hydrocarbon fraction and the fluorescence emission is strongly influenced by the chemical composition (e.g., fluorophore and quencher concentrations) and physical characteristics (e.g., viscosity and optical density) of the oil. Unfortunately, crude petroleum oils encompass a very wide range of physical and chemical characteristics, making the fluorescence analysis of crude petroleum oils a bit strenuous.

Crude oils vary in appearance from black tars to clear liquids, indicating a complex absorption profile. The fluorescence emission is due to the presence of a multitude of aromatic hydrocarbons in varying concentrations. The nature of the emission is governed by the complex interplay between reabsorption, energy transfer, and quenching caused by the high concentrations of fluorophores and quenchers in petroleum oils. The complexity of crude oils usually prevents the resolution of any specific chemical component in terms of fluorescence emission parameters. Factors such as the specific chemical composition (concentration of fluorophores and quenching species) and physical parameters (such as viscosity and optical density) control emission properties such as emission and excitation intensity, emission and excitation wavelength, and fluorescence lifetime. Generally, light oils (high-API gravity) have relatively narrow, intense fluorescence emission bands with small Stokes' shifts. In contrast, heavy oils (low-API gravity) have emissions that tend to be broader, weaker, and red shifted. The differences in emissions are attributed to a higher concentration of fluorophores and quenchers present in the heavier oils, thus leading to an increased rate of energy transfer and quenching to produce broader, weaker, red-shifted emissions (Downare and Mullins, 1995; Ralston et al., 1996, Ryder, 2005; Blamey and Ryder, 2007). Details of the measurement of fluorescence emission from HCFIs and further analysis are described in Chapter 6.

2.8.3 Laser Raman microspectroscopy

Laser Raman Microspectroscopy is a nondestructive technique which analyzes the vibration of molecular bonds when exposed to laser light and is important in recording the volatile content of fluids and identification of solid phases. Lasers have opened new fields of investigation in science and technology. It has given us a versatile gadget for the study of interaction of light and matter. The powerful beam of laser has become an important tool for spectroscopic analysis.

In 1928, Prof. CV Raman, an Indian Nobel laureate, discovered a phenomenon, known as the Raman Effect, by which molecular structures of different substances can be investigated by passing monochromatic light through them. He found that when light passes through a transparent substance, it is scattered and emerges with a change of frequency caused due to the vibration of molecules in the substance. This produces additional lines (known as Raman lines) in the scattered light spectrum. The use of lasers has enabled the recording of Raman lines within seconds, which otherwise would require long exposure times or a few hours using ordinary light sources. The analysis of Raman lines could bring out fundamental properties of the substances such as its vibrational characteristics that lead to the compositional identification.

When radiation passes through a transparent medium, the species present scatter a fraction of the beam in all directions. In 1928, Prof. CV Raman observed that the wavelength of a small fraction of the radiation scattered by certain molecules differs from the incident beam, and the shifts in wavelength depend on the chemical structure of the molecules responsible for scattering (Raman, 1928). Raman scattering is a weak effect (10^{-3}–10^{-6} of the irradiated light), and thus a monochromatic laser beam is normally used to study it.

The scattered radiation is of three types, namely, Rayleigh, Stokes, and anti-Stokes (Fig. 2.6). Energy exchange between the photon and molecule leads to inelastic scattering. The strongest scattering is the Rayleigh scatter, and the wavelength of Rayleigh radiation is exactly the same as that of the excitation source and is significantly more intense than the Stokes or anti-Stokes radiation. The Raman peaks appearing at lower energies relative to the excitation energy look like the Stokes shift found in fluorescence, and thus they are called the Stokes lines. Shifts towards higher energies are termed as anti-Stokes and are generally much weaker than Stokes lines. Fluorescence may interfere with the observation of Stokes shift but not always with anti-Stokes, which therefore may be more useful despite the lower intensity. In Raman scattering, the scattered photon has different energy (frequency, wavelength) than the incident photon: Stokes lines are those in which the photon has lost energy to the molecule, and anti-Stokes lines are those in which the photon has gained energy from the molecule. Since molecular energy levels are quantized, this produces discrete lines from which we can gain information on the molecule itself.

The magnitude of the Raman shift is independent of the wavelength of excitation and the shift pattern is independent of the type of laser (Panczer et al., 2012). The wavelength of the Rayleigh radiation is exactly the same as the wavelength of the exciting radiation. The interaction of the incident light with the vibration modes in the sample causes losses and gains of energy in the incident beam called Stoke and anti-Stokes scattering respectively. The relative intensities of the two processes depend on the population of

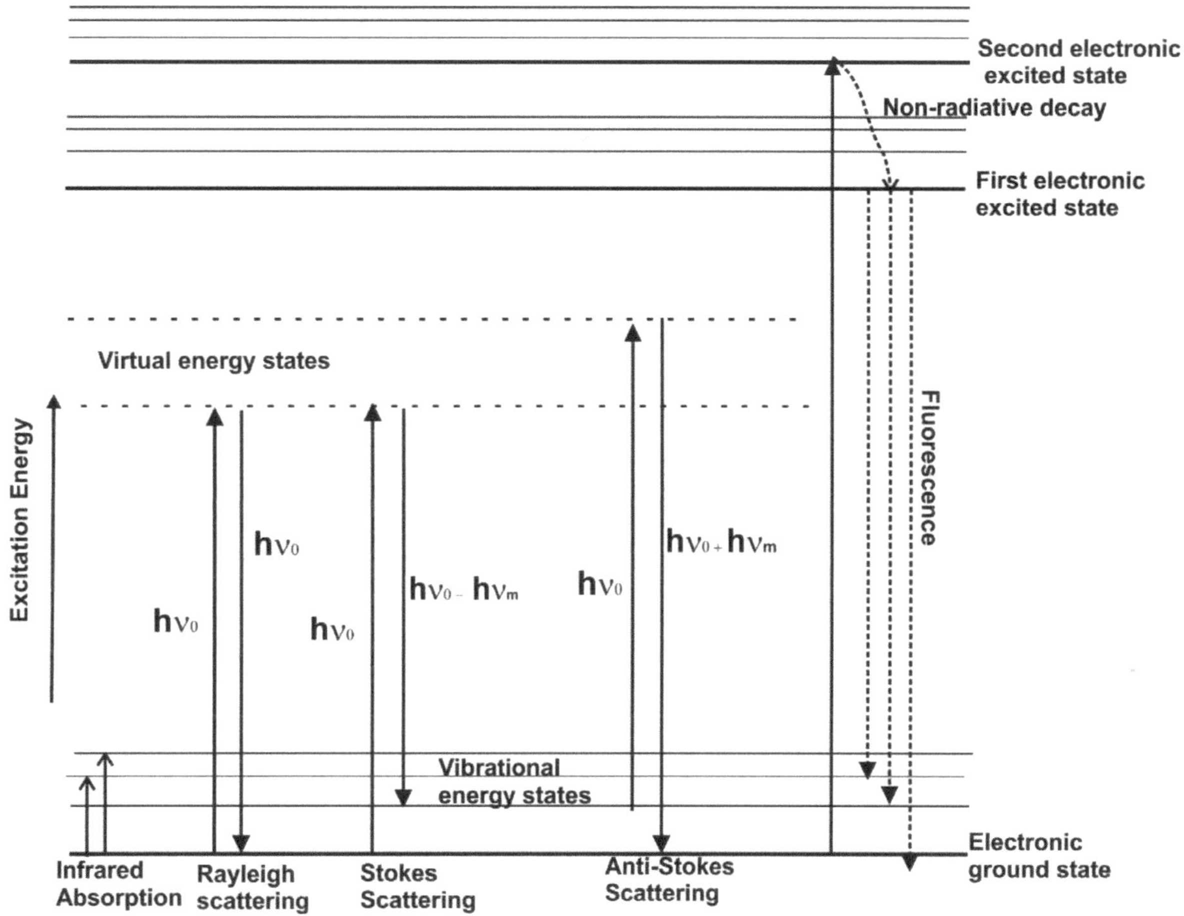

Fig. 2.6 Jablonski energy level diagram showing Fluorescence and vibrational scattering events. *Modified after Panczer G, De Ligny D, Mendoza C, Gaft M, Seydoux-Guillaume A-M and Wang X, Raman and Fluorescence, EMU Notes in Mineralogy, Vol. 12 (2012), (Chapter 2, 61–82).*

the various states of the molecule. The energy loss process is the most probable because there will always be more molecules in the lower state than in the higher states and only the more intense Stokes side of the spectrum is usually analyzed. A Stokes Raman spectrum is a plot of intensity of the scattering vs the energy loss, expressed in wave numbers relative to the source (Raman shift), i.e., the changes in wave number compared to the incident light. The peaks in the spectrum correspond to the energies of the vibrational modes of the different species in the sample. Quantification of a mixture of species is possible if appropriate scattering efficiencies are known for different peaks.

A straightforward way to explain the Raman scattering of light is by a quantum mechanical model, which considers the interaction of photons with molecules in terms of energy-transfer mechanisms (Colthup et al., 1975; Karr, 1975). A molecule has different vibrational energy levels, the ground state $n = 0$, and the excited states $n = 1$, $n = 2$, $n = 3$, etc., which are separated by a quantum of energy $\Delta E = h\nu_m$, where h is Planck's constant and ν_m is the frequency of the molecular vibration. The incident visible light ($\lambda = 400$–750 nm) with energy ν_0 induces transitions to virtual vibrational energy levels in molecules. A virtual level is not an actual energy level of the molecule. It is generated when light photons interact with the molecule, raising its energy. This virtual level is unstable, and light is instantaneously released as scattered radiation. Returning to the initial state occurs by emitting light of frequency ν_0, $\nu_0 - \nu_m$, and $\nu_0 + \nu_m$. The Rayleigh or elastic scattering occurs when the transition starts and finishes at the same vibrational energy level without loss of energy (i.e., no frequency change; ν_0). Inelastic scattering (Raman effect) induces a change to lower ($\nu_0 - \nu_m$) and higher ($\nu_0 + \nu_m$) frequencies in scattered light, which are known as Stokes and anti-Stokes lines, with ν_m representing a fundamental rotational, vibrational, or lattice frequency of the molecule. Rayleigh scattering can account for most of the light scattered by molecules, the Raman effect being extremely weak—in the order of some 10^{-6}–10^{-8} of incident photons—and variable, as the intensity of the Raman scattering is proportional to the fourth power of the frequency of the incident light.

Raman microspectrometers

Raman spectra from micrometer-sized objects within transparent samples can be obtained by coupling a research grade microscope to a Raman spectrometer. The first Raman microspectrometers had a photomultiplier as a detector. This type of monochannel scanning spectra can be extremely slow, and therefore the analysis of a single fluid inclusion may take several hours. The Microdil-28 of Dilor was the first commercial Raman microspectrometer with a multichannel detector. The conventional Raman microspectrometers, using excitation in the visible range of the spectrum and a multichannel detector, are probably the ideal instruments for the study of most fluid inclusions. They produce excellent spectra in a very short time. Only a limited number of species in fluid inclusions can be analyzed quantitatively by Raman spectroscopy, namely the polyatomic gas species and very few polynuclear species in solution. Raman analysis has consequently been mainly successful for gaseous components in fluid inclusions and for supercritical fluid inclusions.

Fluorescence masking of Raman signal

When a substance absorbed light or other electromagnetic radiations, the emission of light may be in the form of fluorescence. In most cases, the emitted light has a longer wavelength and therefore lower energy than the absorbed radiation. The weak Raman scattering can be completely masked by fluorescence, which is several orders of magnitude stronger. Three features of the doubly polished rock wafer may cause fluorescence: the surface, the matrix mineral, and/or the fluid inclusions. Surface fluorescence may be due to incompletely dissolved remnants (even optically invisible films) of epoxy and thermoplastic resins used in the preparation of the sample. Several minerals are well known for their fluorescence, including not only fluorite, but also calcite and plagioclase feldspar, sometimes even quartz. Cracks and fractures filled with a fluorescent mounting medium may prevent analysis altogether. Fluid inclusions are generally fluorescent if they contain cyclic or aromatic hydrocarbons or fluorescent daughter minerals.

One possible way to reject fluorescence is to use a pulsed laser and time resolved detection to allow the Raman photons to be discriminated from the fluorescence background (Fleger et al., 2009; Martyshkin et al., 2004). Another possibility to avoid fluorescence is to select the excitation in which the Raman signal exists and fluorescence is absent (Barbilla et al., 1999). The principle of bleaching can be applied to organic bearing samples where an irradiation of the sample by a laser beam with intense power for a few minutes can decompose organic fluorescent species. Fluorescence can also be avoided by tuning the laser excitation wavelength far away from the electronic transitions to the NIR region. This is possible with Fourier-transformed Raman spectroscopy.

2.8.3.1 Laser Raman spectroscopy

Laser Raman Spectroscopy is a sensitive, nondestructive analytical technique that can be used to determine molecular species in fluids. Modern Raman Spectrometers that are coupled to research grade microscopes are compact (benchtop size) and user-friendly. Laser light focused with a microscope objective illuminates a diffraction-limited spot, covers an area within 1–2 μm in diameter with a 100 × magnification. High quality Raman spectra can be acquired in seconds to minutes that previously required tens of minutes to hours.

Raman Spectroscopy is based on the inelastic scattering of incident photons (light) by the vibrations of molecular bonds or the lattice motions of crystalline materials. This inelastic scattering process shifts the frequency of a small fraction (about 10^{-8}) of the incident photons by the frequency of the vibration of each type of bond in a molecule. To observe these frequency-shifted photons, an intense monochromatic light source is needed to illuminate the sample (lasers are ideal). Very efficient monochromatic filters are then used to separate and measure the small fraction of the Raman scatter from the intense, elastically scattered light (Raleigh scattering), which has the same frequency as the incident light.

Raman spectroscopy has been recognized as a chemical analytical technique for some time, but with the development of lasers it now holds considerable promise for fluid inclusion analysis. The theory of Raman Spectroscopy concerns the vibration of bonds between atoms at different frequencies, depending on the atoms involved. In the laser-excited technique, an intense laser light source (a few μm in diameter) is focused inside a sample and the resulting scattered Raman radiation is analyzed. The equipment incorporates a standard optical microscope and allows the analysis of selected inclusions in the transparent minerals.

2.8.4 Renishaw InVia Raman spectrometer for HCFIs study

Fluid inclusions in minerals can be characterized by Laser Raman microspectrometry for which doubly polished wafers used for fluid inclusion microthermometry are suitable. Micro-Raman Spectrometry is done with a 785 nm (fixed wavelength) NIR Laser with 300 mW output. With the use of motorized neutral density filters, one can work with 16 different power levels from 0.00005% to 100% of the actual Laser power. The spectral range of the equipment is from 50 to 4000 cm^{-1} shift from the Laser line, accomplished with an edge filter. The Raman scattered light is dispersed

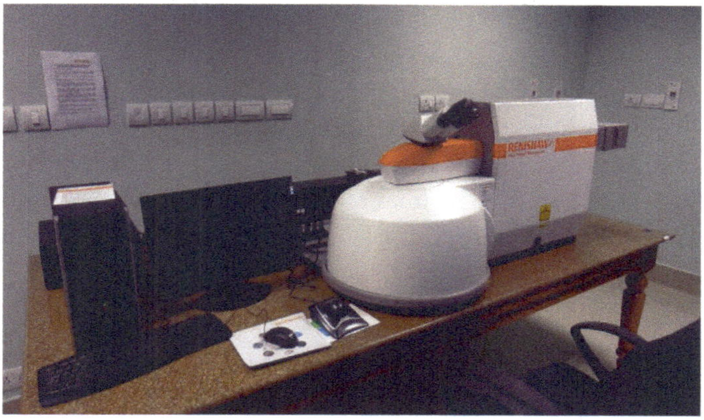

Fig. 2.7 Renishaw laser Raman spectrometer used for HCFI analysis.

with a grating and is having dual grating 1200 and 2400 L/mm. The detection is done by a Peltier cooled CCD detector with 576 × 384 pixels, with a spectral resolution of 1 cm^{-1}. The Raman system is fitted with an XYZ mapping stage as well as confocal arrangement enabling imaging studies with a spatial and depth resolution of 1 and 2 μm, respectively. The Renishaw Raman system has the flexibility to perform photoluminescence (PL) studies at two more laser wavelengths, 405 and 325 nm. The system is fully automated and self-validating with auto aligning and optimization of input laser power (Fig. 2.7). The operation of the equipment is fully software controlled.

2.9 Importance of fluid inclusion analysis

Aqueous fluids are dominating in sedimentary systems, as well as actively participating in diagenetic processes. Important data for petroleum inclusions that can be established through analysis and modeling are textures which give indirect information on the relative timing of fluid entrapment, fluid composition, fluid PVT properties, and pressure–temperature of trapping (Munz, 2001). The fluid inclusion Th data can be used as a proxy for petroleum system modeling where conventional parameters such as VRo and Tmax have been used to determine the thermal maturity of a basin.

2.9.1 Textures

The textures give important information on the relative timing of fluid entrapment while considering fluid inclusion petrography. The objective of studying textures is to establish the time relationship between diagenetic growth of minerals and trapping of inclusions, as well as between different populations of inclusions. The textural relationship between the petroleum inclusions and the host mineral is essential to further go ahead with petroleum migration and the diagenetic evolution.

2.9.2 Composition

The composition of petroleum fluids is complex, and emphasis is given to the characterization of composition. Compositional information is fundamental both in the geological interpretation of the fluid source and maturity and in the modeling of the fluid phase behavior. A procedure for compositional characterization of inclusion petroleum is possible with destructive Gas Chromatography techniques where nondestructive Raman spectroscopy along with microthermometric inputs can be used for petroleum composition identification (Jayanthi et al., 2017).

2.9.2.1 Pressure-volume-temperature (PVT) properties

This relates to the physical properties of the fluid and can be used for modeling the fluid phase behavior under any pressure and temperature conditions. The phase changes in microthermometric analysis can lead to compositional identification, and the temperature of homogenization leads to the estimation of PVT characteristics with further analytical steps. The PVT data obtained through fluid inclusion analysis can be used for the temperature and compositional calibration in petroleum system modeling and a Th data calibration plot, and the resulting thermal history plot is described in Chapter 8.

2.9.3 Pressure and temperature of trapping

An important objective of most fluid inclusion studies is to calculate fluid densities as discussed earlier in this chapter and to estimate the pressure and temperature conditions of trapping, either by combining information from solids or from coexisting fluid phases. Isochores for petroleum inclusions can be calculated when the PVT properties are known. Additional data from coeval aqueous inclusions are necessary for simultaneous determination of the pressure and temperature conditions of trapping.

Petrographic, microthermometric, and spectroscopic analytical methods are used to explore the relevant information from fluid inclusions. Petrographic, microthermometric, fluorescence, and Raman spectroscopic analysis on fluid inclusions from two offshore wells (as examples of nonproducing wells from both proven and nonproven basins—Mumbai and Kerala offshore basins, India, respectively) have been used to characterize HCFIs from sedimentary basins and are discussed in the coming chapters.

References

Aplin, A.C., Macleod, G., Larter, S.R., Pedersen, K.S., Sorensen, H., Booth, T., 1999. Combined use of confocal laser scanning microscopy and PVT simulation for estimating the composition and physical properties of petroleum in fluid inclusions. Mar. Pet. Geol. 16, 97–110.

Baba, M., Parnell, J., Muirhead, D., Bowden, S., 2019. Oil charge and biodegradation history in an exhumed fractured reservoir, Devonian, UK. Mar. Pet. Geol. 101, 281–289.

Bakker, R.J., 2003. Package FLUIDS 1. Computer programs for analysis of fluid inclusion data and for modelling bulk fluid properties. Chem. Geol. 194, 3–23.

Barbilla, J., Bougeard, D., Buntinx, G., Delhaye, M., Dhamelin Court, P., Fillaud, F., 1999. Spectrometric Raman. Techniques P2865, 1–31.

Barres, O., Burneau, A., Dubessy, J., Pagel, M., 1987. Application of micro-FT-IR spectroscopy to individual hydrocarbon fluid inclusion analysis. Appl. Spectrosc. 41, 1000–1008.

Barwise, T., Hay, S., 1996. Predicting oil properties from core fluorescence. In: Abrams, M.-A. (Ed.), Schumacher, D. Hydrocarbon Migration and its Nearsurface Expression, AAPG Memoir, pp. 363–371.

Bhullar, A.G., Karlsen, D.A., Backer-Owe, K., Seland, R.T., Le Tran, K., 1999. Dating reservoir filling - a case history from the North Sea. Mar. Petrol. Geol. 16, 581–603.

Blamey, N.J.F., Ryder, A.G., 2007. Hydrocarbon fluid inclusion fluorescence: A review. In: Geddes, C.D. (Ed.), Reviews in Fluorescence 2007. Springer, New York, USA, pp. 299–334.

Blanchet, A., Pagel, M., Walgenwitz, F., Lopez, A., 2003. Microspectrofluorimetric and microthermometric evidence for variability in hydrocarbon fluid inclusions in quartz overgrowths: implications for inclusion trapping in the Alwyn north field, North Sea. Org. Geochem. 34, 1477–1490.

Bodnar, R.J., 1990. Petroleum migration in the Miocene Monterey formation, California, USA: constraints from fluid inclusion studies. Mineral. Mag. 54, 295–304.

Bourdet, J., Eadington, P.J., 2012. Fluorescence and infrared spectroscopy of inclusion oil. CSIRO Report number EP129625, Perth, p. 60.

Bourdet, J., Pironon, J., 2008. Strain response and re-equilibration of CH4-rich synthetic aqueous fluid inclusions in calcite during pressure drops. Geochim. Cosmochim. Acta 72, 2946–2959.

Bourdet, J., Pironon, J., Levresse, G., Tritlla, J., 2008. Petroleum type determination through homogenization temperature and vapour volume fraction measurements in fluid inclusions. Geofluids 8, 46–59.

Bourdet, J., Pironon, J., Levresse, G., Tritlla, J., 2010. Petroleum accumulation and leakage in a deeply buried carbonate reservoir, Nispero field (Mexico). Mar. Pet. Geol. 27, 126–142.

Bourdet, J., Burruss, R.C., Bodnar, R.J., Eadington, P.J., 2011. Assessment of UV-Raman for analysis of petroleum inclusions. In: European Current Research on Fluid Inclusions (ECROFI-XXI), 50, Montanuniversität Leoben, Austria, 9–11 August, 2011, pp. 50–51.

Bourdet, J., Eadington, P., Volk, H., George, S.C., Pironon, J., Kempton, R., 2012. Chemical changes of fluid inclusion oil trapped during the evolution of an oil reservoir: jabiru-1A case study (Timor Sea, Australia). Mar. Pet. Geol. 36, 118–139.
Bourdet, J., Burruss, R.C., Chou, I.M., Kempton, R., Liu, K., Hung, N.V., 2014. Evidence for a palaeo-oil column and alteration of residual oil in a gas-condensate field: integrated oil inclusion and experimental results. Geochim. Cosmochim. Acta 142, 362–385.
Burruss, R.C., 1981. Hydrocarbon fluid inclusions in studies of sedimentary diagenesis. In: Hollister, L.S., Crawford, M.L. (Eds.), Fluid Inclusions: Applications to Petrology, pp. 138–156.
Burruss, R.C., 1991. Practical aspects of fluorescence microscopy of petroleum fluid inclusions. In: Barker, C.E., Kopp, O.C. (Eds.), Luminescence Microscopy and Spectroscopy, pp. 1–7. Tulsa.
Burruss, R.C., 2003. Petroleum fluid inclusions, an introduction. In: Samson, I., Anderson, A., Marshall, D. (Eds.), Fluid Inclusions: Analysis and Interpretation. Short Course Series, Vol. 32. Ottawa, Mineralogical Association of Canada, pp. 159–174.
Burruss, R.C., Cercone, K.R., Harris, P.M., 1985. Timing of hydrocarbon migration: evidence from fluid inclusions in calcite cements, tectonics, and burial history. SEPM Spec. Publ. 36, 277–289.
Chen, Y., Ge, Y., Zhou, Z., Zhou, Y., 2015. Water, is it necessary for fluid inclusions forming in calcite? J. Pet. Sci. Eng. 133, 103–107.
Chen, Y., Steele-MacInnis, M., Ge, Y., Zhou, Z., Zhou, Y., 2016a. Synthetic salineaqueous and hydrocarbon fluid inclusions trapped in calcite at temperatures and pressures relevant to hydrocarbon basins: a reconnaissance study. Mar. Pet. Geol. 76, 88–97.
Chen, Y., Wang, X., Bodnar, R.J., 2016b. UV Raman spectroscopy of hydrocarbon bearing inclusions in rock salt from the Dongying sag, eastern China. Org. Geochem. 101, 63–71.
Chen, H., Zhu, X., Chen, C., Yin, W., Zhang, Q., Shi, R., 2017. Diagenesis and hydrocarbon emplacement in the upper Triassic Yanchang formation tight sandstones in the southern Ordos Basin, China. Aust. J. Earth Sci. 64, 1–24.
Clementz, D.M., 1976. Interaction of petroleum heavy ends with montmorillonite. Clay Clay Miner. 24, 312–319.
Colthup, N.B., Daly, L.H., Wiberly, S.E., 1975. Introduction to Infrared and Raman Spectroscopy, second edition. Academic Press, New York. 523 pp.
Conliffe, J., Burden, E.T., Wilton, D.H.C., 2017. The use of integrated fluid inclusion studies for constraining petroleum charge history at parsons pond, Western Newfoundland, Canada. Fortschr. Mineral. 7, 39. https://doi.org/10.3390/min7030039.
Deaton, B.C., 1987. Quantification of rock color from Munsell chips. J. Sediment. Res. 57, 774–776.
Diamond, L., 2003. Introduction to gas bearing fluid inclusions. In: Samson, I., Anderson, A., Marshall, D. (Eds.), Fluid Inclusions: Analysis and Interpretstion. Mineral Assoc. Can. Short Course Ser, Vol. 32, pp. 101–158.
Downare, T.D., Mullins, O.C., 1995. Visible and near infrared fluorescence of crude oils. Appl. Spectrosc. 49 (6), 754–764.
Dubessy, J., Buschaert, S., Lamb, W., Pironon, J., Thiéry, R., 2001. Methane-bearing aqueous fluid inclusions: Raman analysis, thermodynamic modelling and application to petroleum basins. Chem. Geol. 173, 193–205.
Duggan, J.P., Mountjoy, E.W., Stasiuk, L.D., 2001. Fault-controlled dolomitization at swan hills Simonette oil field (Devonian), deep basin west-Central Alberta, Canada. Sedimentology 48, 301–323.
Dutkiewicz, A., Volk, H., Ridley, J., George, S.C., 2003. Biomarkers, brines and oil in the Mesoproterozoic, roper Superbasin, Australia. Geology 31, 981–984.

Dutkiewicz, A., Volk, H., Ridley, J., George, S.C., 2004. Geochemistry of oil in fluid inclusions in a middle Proterozoic igneous intrusion: implications for the source of hydrocarbons in crystalline rocks. Org. Geochem. 35, 937–957.
Eadington, P.J., Kempton, R.H., 2008. New developments to investigate models of oil accumulation and fluid interactions in reservoirs using data from fluid inclusions. In: The Australian Petroleum Production and Exploration Association Journal & Conference Proceedings 48, Extended abstracts CD.
Evans, M.A., De Lisle, A., Leo, J., Lafonte, C.J., 2014. Deformation conditions for fracturing in the middle Devonian sequence of the Central Appalachians during the late Paleozoic Alleghenian orogeny. AAPG Bull. 98, 2263–2299.
Ferrill, D.A., McGinnis, R.N., Morris, A.P., Smart, K.J., Sickmann, Z.T., Bentz, M., Lehrmann, D., Evans, M.A., 2014. Control of mechanical stratigraphy on bedrestricted jointing and normal faulting: eagle ford formation, south-Central Texas. AAPG Bull. 98, 2477–2506.
Fleger, Y., Nagli, L., Gaft, M., Rosenbluh, M., 2009. Narrow gated Raman and luminescence of explosives. Jouranl of Luminescence 129, 979–983.
Frezzotti, M.L., Tecce, F., Casagli, A., 2012. Raman spectroscopy for fluid inclusion analysis. J. Geochem. Explor. 112, 1–20.
George, S.C., Krieger, F.W., Eadington, P.J., Quezada, R.A., Greenwood, P.F., Eisenberg, L.I., Hamilton, P., 1997. Geochemical comparison of oil-bearing fluid inclusions and produced oil from the Toro sandstone, Papua New Guinea. Org. Geochem. 26, 155–173.
George, S.C., Eadington, P.J., Lisk, M., Quezada, R.A., 1998. Geochemical comparison of oil trapped in fluid inclusions and reservoired in Blackback oilfield, Gippsland Basin, Australia. PESA Journal 26, 64–81.
George, S.C., Ruble, T.E., Dutkiewicz, A., Eadington, P.J., 2001. Assessing the maturity of oil trapped in fluid inclusions using molecular geochemistry data and visually-determined fluorescence colours. Appl. Geochem. 16, 451–473.
George, S.C., Ruble, T.E., Dutkiewicz, A., Eadington, P.J., 2002. Reply to comment by Oxtoby on "assessing the maturity of oil trapped in fluid inclusions using molecular geochemistry data and visually-determined fluorescence colours. Appl. Geochem. 17, 1375–1378.
George, S.C., Ahmed, M., Liu, K., Volk, H., 2004. The analysis of oil trapped during secondary migration. Org. Geochem. 35, 1489–1511.
George, S.C., Volk, H., Ahmed, M., Pickel, W., Allan, T., 2007. Biomarker evidence for two sources for solid bitumens in the subu wells: implications for the petroleum prospectivity of the east Papuan Basin. Org. Geochem. 38, 609–642.
George, S.G., Volk, H., Dutkiewicz, A., 2012. Mass spectrometry techniques for analysis of oil and gas trapped in fluid inclusions. In: Lee, M. (Ed.), Applied Mass Spectroscopy Handbook. Wiley & Sons, pp. 647–673.
Goldstein, R.H., 1993. Fluid inclsuions as microfabrics: A petrographic method to determine diagenetic history. In: Rezak, R. (Ed.), Frontiers in Sedimantary Geologu. Springer-Verlag, New York, pp. 279–290.
Goldstein, R.H., 2001. Fluid inclusions in sedimentary and diagenetic systems. Lithos 55, 159–193.
Goldstein, R.H., 2003. Petrographic analysis of fluid inclusions. In: Samson, I., Anderson, A., Marshall, D. (Eds.), Fluid Inclusions: Analysis and Interpretation. Short Course Series, Vol. 32. Ottawa, Mineralogical Association of Canada, pp. 9–55.
Goldstein, R.H., Reynolds, T.J., 1994. Systematics of fluid inclusions in diagenetic minerals. SEPM Short Course 31, 1–199.
Greenwood, P.F., George, S.C., Hall, K., 1998. Applications of laser micropyrolysis–gas chromatography–mass spectrometry. Org. Geochem. 29, 1075–1089.

Grimmer, J.O.W., Pironon, J., Teinturier, S., Mutterer, J., 2003. Recognition and differentiation of gas condensates and other oil types using microthermometry of petroleum inclusions: oil-cracking processes evidence from synthetic petroleum inclusions. J. Geochem. Explor. 78–79, 367–371.
Guilhaumou, N., Szydlowskii, N., Pradier, B., 1990. Characterization of hydrocarbon fluid inclusions by infra-red and fluorescence micro spectrometry. Mineral. Mag. 54 (375), 311–324.
Guo, X., Liu, K., Song, Y., Zhao, M., Liu, S., Zhuo, Q., Lu, X., 2016. Influences of hydrocarbon charging and overpressure on reservoir porosity in Kela-2 gas field of the Kuqa depression, Tarim Basin. Oil Gas Geol. 37, 935–943.
Hagemann, H.W., Hollerbach, A., 1986. The fluorescence behaviour of crude oils with respect to their thermal maturation and degradation. Org. Geochem. 10, 473–480.
Hall, D.L., Shentwu, W., Sterner, S.M., Wagner, P.D., 1997. Using fluid inclusions to explore for oil and gas. Hart's Petroleum Engineer International 11, 29–34.
Heinrich, C.A., Pettke, T., Halter, W.E., Aigner-Torres, M., Audetat, A., Gunther, D., Hattendorf, B., Bleiner, D., Guillong, M., Horn, I., 2003. Quantitative multielement analysis of minerals, fluid and melt inclusions by laser-ablation inductively-coupled-plasma mass-spectrometry. Geochim. Cosmochim. Acta 67, 3473–3497.
Hode, T., Zebuhr, Y., Broman, C., 2006. Towards biomarker analysis of hydrocarbons trapped in individual fluid inclusions: first extraction by ErYAG laser. Planetary and Space Science 54, 1575–1583.
Horn, I., Von Blanckenburg, F., Schoenberg, R., Steinhoefel, G., Markl, G., 2006. In situ iron isotope ratio determination using UV-femtosecond laser ablation with application to hydrothermal ore formation processes. Geochim. Cosmochim. Acta 70, 3677–3688.
Horstad, I., Larter, S.R., Dypvik, H., Aagaard, P., Bjørnvik, A.M., Johansen, P.E., Eriksen, S., 1990. Degradation and maturity controls on oil field petroleum column heterogeneity in the Gullfaks field, Norwegian North Sea. Org. Geochem. 16, 497–510.
Jayanthi, J.L., Nandakumar, V., Anoop, S.S., 2017. Feasibility of a 785 nm diode laser in Raman spectroscopy for characterizing hydrocarbon-bearing fluid inclusions in Mumbai Offshore Basin, India. Pet. Geosci. 23 (3), 369–375.
Jensenius, J., Burruss, R.C., 1990. Hydrocarbon-water interactions during brine migration: evidence from hydrocarbon inclusions in calcite cements from Danish North Sea oil fields. Geochim. Cosmochim. Acta 54, 705–713.
Jones, D.M., Macleod, G., 2000. Molecular analysis of petroleum in fluid inclusions: a practical methodology. Org. Geochem. 31, 1163–1173.
Karlsen, D.A., Nedkvitne, T., Larter, S.R., Bjørlykke, K., 1993. Hydrocarbon composition of authigenic inclusions: application to elucidation of petroleum reservoir filling history. Geochim. Cosmochim. Acta 57, 3641–3659.
Karr, C., 1975. Infrared and Raman Spectroscopy of Lunar and Terrestrial Minerals. Academic Press, New York. 375 pp.
Kihle, J., 1995. Adaptation of fluorescence excitation-emission micro-spectroscopy for characterization of single hydrocarbon fluid inclusions. Org. Geochem. 23, 1029–1042.
Killops, S.D., Reyes, A., Funnell, R.H., 2009. Filling history of the Maui B field, New Zealand: new information from oil inclusions in authigenic minerals from the oil leg in the Maui-B1 well F Sands. J. Pet. Geol. 32, 271–286.
Leefmann, T., Heim, C., Kryvenda, A., Siljeström, S., Sjövall, P., Thiel, V., 2013. Biomarker imaging of single diatom cells in a microbial mat using time-of-flight secondary ion mass spectrometry (ToF-SIMS). Org. Geochem. 57, 23–33.
Lisk, M., Eadington, P.J., 1994. Oil migration in the Cartier Trough, Vulcan Sub-basin. In: Purcell, P.G., Purcell, R.R. (Eds.), The Sedimentary Basins of Western Australia. Proceedings of the WA Basins Symposium. Perth, pp. 301–312.

Liu, K., Eadington, P., 2005. Quantitative fluorescence techniques for detecting residual oils and reconstructing hydrocarbon charge history. Org. Geochem. 36, 1023–1036.
Liu, K., George, S.C., Eadington, P.J., 2003. Predicting abundances and n-alkane profiles of oil inclusions from bulk fluorescence spectrophotometry. In: 21st International Meeting on Organic Geochemistry, Book of Abstracts Part II, Kraków, pp. 177–178.
Liu, K., Eadington, P., Middleton, H., Fenton, S., Cable, T., 2007. Applying quantitative fluorescence techniques to investigate petroleum charge history of sedimentary basins in Australia and Papuan New Guinea. J. Pet. Sci. Eng. 57, 139–151.
Liu, N., Qiu, N., Chang, J., Shen, F., Wu, H., Lu, X., Wang, Y., Jiao, Y., Feng, Q., 2017. Hydrocarbon migration and accumulation of the Suqiao buried-hill zone in Wen'an slope, Jizhong subbasin, Bohai Bay basin, China. Mar. Pet. Geol. 86, 512–525.
Macleod, G., Petch, G.S., Larter, S.R., Aplin, A.C., 1993. Investigations of the composition of hydrocarbon fluid inclusions. In: Abstracts of the 205th ACS National Meeting, Division of Geochemistry 86.
Martyshkin, D.V., Ahuja, R.C., Kudriavvstev, Mirov, S.B., 2004. Effective suppression of fluorescence light in Raman measurements using ultrafast time gated charge coupled device camera. Review of Scientific Instruments. American Institute of Physics 75 (3), 630–635.
Matapour, Z., Karlsen, D.A., 2017. Geochemical characteristics of the Skrugard oil discovery, Barents Sea, arctic Norway: a "palaeo-biodegraded – gas reactivated" hydrocarbon accumulation. J. Pet. Geol. 40, 125–152.
McLimans, R.K., 1987. The application of fluid inclusions to migration of oil and diagenesis in petroleum reservoirs. Appl. Geochem. 2, 585–603.
Munz, I.A., 2001. Petroleum inclusions in sedimentary basins: systematics, analytical methods and applications. Lithos 55, 195–212.
Murray, R.C., 1957. Hydrocarbon fluid inclusions in quartz. AAPG Bull. 41, 950–956.
Nedkvitne, T., Karlsen, D.A., Bjørlykke, K., Larter, S.R., 1993. Relationship between reservoir diagenetic evolution and petroleum emplacement in the Ula field, North Sea. Mar. Pet. Geol. 10, 255–270.
Noah, M., Volk, H., Schubert, F., Horsfield, B., 2018. First Analysis of Polar Compounds Trapped in Fluid Inclusions Using Ultra High Resolution Mass Spectrometry – A Proof of Concept Demonstrated on a Case Study from the Pannonian Basin (Hungary). Latin American Association of Organic Geochemistry (ALAGO). Salvador, Bahia, Brazil, p. 3.
Oxtoby, N.H., 2002. Comments on: assessing the maturity of oil trapped in fluid inclusions using molecular geochemistry data and visually-determined fluorescence colours. Appl. Geochem. 17, 1371–1374.
Pan, A., Qin, J., Yao, S., Tenger, A., Shen, B., 2017. Application of single hydrocarbon inclusions in petroleum geochemistry: a case study of the Tahe oilfield. Petroleum Geology and Experiment 39, 675–681 (in Chinese).
Panczer, G., De Ligny, D., Mendoza, C., Gaft, M., Seydoux-Guillaume, A.-M., Wang, X., 2012. Raman and Fluorescence, EMU Notes in Mineralogy, Vol. 12. pp. 61–82. Chapter 2.
Pang, L.S.K., George, S.C., Quezada, R.A., 1998. A study of the gross compositions of oil-bearing fluid inclusions using high performance liquid chromatography. Org. Geochem. 29, 1149–1161.
Parnell, J., 2010. Potential of palaeofluid analysis for understanding oil charge history. Geofluids 10, 73–82.
Parnell, J., Middleton, D., Chen, H.H., Hall, D., 2001. The use of integrated fluid inclusion studies in constraining oil charge history and reservoir compartmentation: examples from the Jeanne d'Arc Basin, offshore Newfoundland. Mar. Pet. Geol. 18, 535–549.
Peters, C.A., George, S.C., 2018. Hydrocarbon biomarkers preserved in carbonate veins of potentially Paleoproterozoic age, and implications for the early biosphere. Geobiology 16, 577–596.

Peters, C.A., Hallmann, C., George, S.C., 2018. Phenolic compounds in oil-bearing fluid inclusions: implications for water-washing and oil migration. Org. Geochem. 118, 36–46.
Ping, H., Chen, H., Song, G., Régis, T., 2012. Individual oil inclusion composition prediction and its application in oil and gas accumulation studies. Earth Science - Journal of China University of Geosciences 37, 815–824.
Ping, H., Chen, H., Thiéry, R., 2013. Thermodynamic modeling of petroleum inclusions: composition modeling and prediction of the trapping pressure of crude oils. Fluid Phase Equilib. 346, 33–44.
Ping, H., Chen, H., Thiéry, R., George, S.C., 2017. Effects of oil cracking on fluorescence color, homogenization temperature and trapping pressure reconstruction of oil inclusions from deeply buried reservoirs in the northern Dongying depression, Bohai Bay basin, China. Mar. Pet. Geol. 80, 538–562.
Pironon, J., Barrés, O., 1990. Semi-quantitative FT-IR microanalysis limits: evidence from synthetic hydrocarbon fluid inclusions in sylvite. Geochim. Cosmochim. Acta 54, 509–518.
Pironon, J., Pradier, B., 1992. Ultraviolet-fluorescence alteration of hydrocarbon fluid inclusions. Org. Geochem. 18, 501–509.
Pironon, J., Sawatzki, J., Dubessy, J., 1991. NIR FT-Raman microspectroscopy of fluid inclusions: comparisons with VIS Raman and FT-IR microspectroscopies. Geochim. Cosmochim. Acta 55, 3885–3891.
Pironon, J., Canals, M., Dubessy, J., Walgenwitz, F., Laplace-Builhe, C., 1998. Volumetric reconstruction of individual oil inclusions by confocal scanning laser microscopy. Eur. J. Mineral. 10, 1143–1150.
Pironon, J., Thiery, R., Ayt Ougougdal, M., Teinturier, S., Beaudoin, G., Walgenwitz, F., 2001. FT-IR measurements of petroleum fluid inclusions: methane, n-alkanes and carbon dioxide quantitative analysis. Geofluids 1, 2–10.
Potter, J., Longstaffe, F.J., 2007. A gas-chromatograph, continuous flow-isotope ratio mass-spectrometry method for d13C and dD measurement of complex fluid inclusion volatiles: examples from the Khibina alkaline igneous complex, Northwest Russia and the South Wales coalfields. Chem. Geol. 244, 186–201.
Pottorf, R.J., Hussenoeder, K.L., Petersen, K., Tseng, H., Davis, C.L., Richardson, M., Pietraszek-Mattner, S., Moore, D.W., El Agrab, A.F., Khouri, A.A., 2008. Downdipoil potential for an onshore Abu Dhabi petroleum system. SPE Reserv. Eval. Eng. 11, 395–403.
Przyjalgowski, M.A., Ryder, A.G., Feely, M., Glynn, T.J., 2005. Analysis of hydrocarbon-bearing fluid inclusions (HCFI) using time resolved fluorescence spectroscopy. Proc. SPIE 5826, 173–184.
Ralston, C.Y., Wu, X., Mullins, O.C., 1996. Quantum yield of crude oils. Appl. Spectrosc. 50 (12), 1563–1568.
Raman, C.V., 1928. A new radiation. Indian J. Phys. 387.
Roedder, E., 1984. Fluid inclusions. Mineralogical Society of America. Revues in Mineralogy 12, 1–644.
Ryder, A.G., 2005. Analysis of crude petroleum oils using fluorescence spectroscopy. In: Geddes, C.D., Lakowicz, J.R. (Eds.), Reviews in Fluorescence. Springer, New York, pp. 169–198.
Salvi, S., Williams-Jones, A.E., 2003. Bulk analysis of volatiles in fluid inclusions. In: Samson, I., Anderson, A., Marshall, D. (Eds.), Fluid Inclusions – Analysis and Interpretation. Short Course Series, Vol. 32. Vancouver, Mineralogical Association of Canada, pp. 247–278.
Schubert, F., Diamond, L.W., Tóth, T.M., 2007. Fluid-inclusion evidence of petroleum migration through a buried metamorphic dome in the Pannonian Basin, Hungary. Chem. Geol. 244, 357–381.

Shariatinia, Z., Feiznia, S., Shafiei, A., Haghighi, M., Mousavi Dehghani, A., Memariani, M., Farhadian, N., 2013. Multiple hydrocarbon charging events in Kuh-e-Mond oil field, coastal Fars: evidence from biomarkers in oil inclusions. Geofluids 13, 594–609.
Shepherd, T.J., Rankin, A.H., Alderton, D.H.M.A., 1985. A Practical Guide to Fluid Inclusion Studies. Blackie, Chapman & Hall.
Siljeström, S., Hode, T., Lausmaa, J., Sjovall, P., Toporski, J., Thiel, V., 2009. Detection of organic biomarkers in crude oils using TOF-SIMS. Org. Geochem. 40, 135–143.
Siljeström, S., Lausmaa, J., Sjovall, P., Broman, C., Thiel, V., Hode, T., 2010. Analysis of hopanes and steranes in single oil-bearing fluid inclusions using time-of-flight secondary ion mass spectrometry (TOF-SIMS). Geobiology 8, 37–44.
Siljeström, S., Volk, H., George, S.C., Lausmaa, J., Sjövall, P., Dutkiewicz, A., Hode, T., 2013. Analysis of single oil-bearing fluid inclusions in mid-Proterozoic sandstones (roper group, Australia). Geochim. Cosmochim. Acta 122, 448–463.
Sindern, S., 2017. Analysis of rare earth elements in rock and, mineral samples by ICP-MS and LA-ICP-MS. In: Handbook of Rare Earth Elements: Analytics, pp. 334–356.
Stasiuk, L.D., Snowdon, L.R., 1997. Fluorescence micro-spectrometry of synthetic and natural hydrocarbon fluid inclusions: crude oil chemistry, density and application to petroleum migration. Appl. Geochem. 12, 229–241.
Szabo, B., Schubert, F., Tóth, T.M., Steinbach, G., 2016. Palaeofluid evolution in a fractured basalt hosted reservoir in the Üllés-Ruzsa-Bordány area, southern sector of the Pannonian Basin. Geologia Croatica 69, 281–293.
Thiel, V., Sjövall, P., 2011. Using time-of-flight secondary ion mass spectrometry to study biomarkers. Annu. Rev. Earth Planet. Sci. 39, 125–156.
Thiel, V., Heim, C., Arp, G., Hahmann, U., Sjovall, P., Lausmaa, J., 2007. Biomarkers at the microscopic range: TOF-SIMS molecular imaging of archaea-derived lipids in a microbial mat. Geobiology 5, 413–421.
Thiery, R., Pironon, J., Walgenwitz, F., Montel, F., 2000. PIT (petroleum inclusion thermodynamic): a new modeling tool for the characterization of hydrocarbon fluid inclusions from volumetric and microthermometric measurements. J. Geochem. Explor. 69, 701–704.
Toboła, T., 2018. Raman spectroscopy of organic, solid and fluid inclusions in the oldest halite of LGOM area (SW Poland). Spectrochimica Acta - Part A: Molecular and Biomolecular Spectroscopy 189, 381–392.
Van den Kerkhof, A.M., Hein, U.F., 2001. Fluid inclusion petrography. Lithos 55, 27–47.
Videtich, P.E., McLimans, R.K., Watson, H.K.S., Nagy, R.M., 1988. Depositional, diagenetic, thermal, and maturation histories of cretaceous Mishrif formation, Fateh field, Dubai. AAPG Bull. 72, 1143–1159.
Volk, H., George, S.C., 2019. Using petroleum inclusions to trace petroleum systems – a review. Org. Geochem. 129 (2019), 99–123.
Volk, H., Horsfield, B., Mann, U., Suchy, V., 2002. Variability of petroleum inclusions in vein, fossil and vug cements – a geochemical study in the Barrandian Basin (lower Palaeozoic, Czech Republic). Org. Geochem. 33, 1319–1341.
Volk, H., Dutkiewicz, A., George, S.C., Ridley, J., 2003. Oil migration in the middle Proterozoic roper Superbasin, Australia: evidence from fluid inclusions and their geochemistries. J. Geochem. Explor. 78–79, 437–441.
Volk, H., George, S.C., Middleton, H., Schofield, S., 2005. Geochemical comparison of fluid inclusion and present-day oil accumulations in the Papuan foreland – evidence for previously unrecognised petroleum source rocks. Org. Geochem. 36, 29–51.
Volk, H., Fuentes, D., Fuerbach, A., Miese, C., Koehler, W., Bärsch, N., Barcikowski, S., 2010. First on-line analysis of petroleum from single inclusion using ultrafast laser ablation. Org. Geochem. 41, 74–77.

Zhang, Z., Greenwood, P., Zhang, Q., Rao, D., Shi, W., 2012. Laser ablation GC-MS analysis of oil-bearing fluid inclusions in petroleum reservoir rocks. Org. Geochem. 43, 20–25.
Zhao, X., Zhang, L., Jin, F., Wang, Q., Bai, G., Li, Z., Wang, J., 2017. Hydrocarbon charging and accumulation history in the Niudong buried hill field in the Baxian depression, eastern China. Mar. Pet. Geol. 88, 343–358.
Zhao, S., Chen, W., Zhou, L., Zhou, P., Zhang, J., 2019. Characteristics of fluid inclusions and implications for the timing of hydrocarbon accumulation in the cretaceous reservoirs, Kelasu Thrust Belt, Tarim Basin, China. Mar. Pet. Geol. 99, 473–487.

CHAPTER 3

Petroleum system and the significance of HCFI study

V. Nandakumar[a] **and J.L. Jayanthi**[b]
[a]Scientist - G, National Facility for Geofluids Research and Raman Analysis, National Centre for Earth Science Studies (NCESS), Thiruvananthapuram, Kerala, India
[b]Project Scientist - C, National Facility for Geofluids Research and Raman Analysis, National Centre for Earth Science Studies (NCESS), Thiruvananthapuram, Kerala, India

Petroleum geology deals with the origin, migration, accumulation, and exploration of petroleum in sedimentary basins. Sedimentary basins are low areas of tectonic origin where sediments accumulate over time in a greater thickness compared to their surroundings. Sediment accumulation in basins happens due to subsidence due to loading during basin development. The sedimentary basins are individually unique in the type of materials, PT conditions, and age. The petroleum system is one of the main concepts in petroleum geology where petroleum formations along with its associated processes are involved.

3.1 Petroleum formation

The transformation of organic compounds in organic matter under anoxic conditions results in the formation of petroleum. Organic matter accumulation along with sediments is the basis of oil generation/origin of petroleum (Chapman, 1976; North, 1985; Selley, 1998).

3.1.1 Accumulation of organic matter

Oil is generated from organic matter in sediments and it consists of fundamental organic compounds such as protein (amino acids), fats, waxes, humus, etc. The composition of petroleum is governed by the type of organisms contributing to organic matter, depositional environment, and the prevailing thermal conditions. Planktonic algae are the main contributors of organic matter that yields petroleum and among them diatoms that have amorphous silica (opal A) shells are the most important. Radiolaria

Hydrocarbon Fluid Inclusions in Petroliferous Basins
https://doi.org/10.1016/B978-0-12-817416-6.00003-4

(silica shells), foraminifera (shells of calcium carbonate), and pteropods (carbonate shells) are the most important zooplankton that provide organic matter for petroleum. Cyanobacteria/blue green algae also contribute significantly to the organic material in sediments. Therefore, it is estimated that 90% of the production of organic matter is from algae. The erosion of rocks on land supplies nutrients for these algae and are transported into the ocean. These organisms sink after death and decay releasing the nutrients from it at the deeper areas of the ocean. Generally, basins with limited water circulation preserve more organic matter and produce good source rocks (Bjorlykke, 2010).

Organisms such as zooplankton, algae and planktonic algae die, decompose, and mix with sand and silt. After decomposition, the organic remains are constituted mainly of carbon and hydrogen. At the bottom of the ocean, the oxygen level is insufficient for complete decomposition; so, the partially decomposed remains will form a large, gelatinous mass, and slowly gets covered by multiple layers of sand, silt, and mud. This burying process takes millions of years and the weight of the sand and silt leads to the compression of the organic mass into a thinner layer. The geothermal gradient and the intense pressure will combine to act on the mass whenever the depth of the buried decomposing layer reaches around 10,000 ft. and leads to the formation of petroleum. The temperature experienced by the organic mass is crucial, which determines the overall properties of the petroleum formed. The favorable temperature for petroleum formation is around 60–120 °C and is termed as "oil window". In general, the upper limit oil window may extend up to 140 °C as determined by fluid inclusion studies. Lower temperatures will result in a thicker, darker petroleum deposit such as bitumen. Gas will be generated independently or sometimes mixed with petroleum if there is heat fluctuation. The original biomass will be destroyed and no gas or petroleum is formed if temperature is too high (above 230 °C). The formation of naturally occurring petroleum takes millions of years.

3.1.2 Origin of petroleum

The process by which solid organic matter present in source rocks transforms into liquid/gaseous hydrocarbons is termed petroleum generation. Under natural conditions, petroleum generation starts at temperatures around 60 °C at an apparent burial depth of 2.5–3 km that depends on the geothermal gradient. The organic matter buried with fine-grained sediment (usually clay) acts as the main source rock for petroleum generation. The diagenesis of this organic matter leads to the production of kerogen and subsequently hydrocarbon generation happens (Chapman, 1976). During compaction of sediments in a basin, water is gradually expelled as organic material becomes buried by successive accumulation of sediments. During this time, complex organic compounds such as proteins are broken down into amino acids and carbohydrates are broken down into simple sugar compounds. These are able to polymerize to make larger hydrocarbon compounds and these newly formed organic structures are called kerogen (a collective name for organic material that is insoluble in organic solvents, water, or oxidizing acids). The portion of the organic material soluble in organic solvents is called bitumen, which is essentially oil in a solid state. Kerogen consists of very large molecules and is a kind of polymer, which undergoes successive stages of diagenesis, catagenesis, and metagenesis that are explained later in this chapter. As sedimentation and subsidence proceed, kerogen is subjected to a progressive increase of temperature and pressure. When it has been exposed to enough time and temperature, these large molecules will crack into smaller molecules, mostly petroleum. An anoxic environment is critical in the formation of hydrocarbons where sediments will get enough time to accumulate before too much decay happens in the presence of oxygen (Tissot and Welte, 1978).

Under significant geological and geochemical conditions, the origin of petroleum processes can vary from place to place; therefore, the composition of petroleum will vary. Petroleum composition can give insights into its origin and geological history. Total organic carbon (TOC) in the source rock, good burial environment, enough pressure–temperature conditions that help in the maturation process, migration during favorable conditions from source rock to reservoir rock, the presence of porous and permeable reservoir rock, presence of accumulation traps (structural or stratigraphic), and the presence of impervious cap rocks are the most conducive requirements for petroleum accumulation.

Petroleum originates as hydrocarbons that have been generated by thermal alteration after burial of organic matter in sediments. Crude oils are composed of organic compounds characterized by different molecular weight, size, shape, solubility, and elemental composition. The average composition is the following: 85% carbon, 13% hydrogen and 2% sulfur, nitrogen, and oxygen. Crude oils are basically composed of hydrocarbons (alkanes, naphthenes, and asphaltenes). Other nonhydrocarbon compounds are metal-porphyrin complexes and trace elements. Distillation allows the separation of petroleum in groups characterized by different molecular weight: gas from C1 to C4, gasoline from C5 to C10, kerosene from C11 to C13, light gas oil from C14 to C18, heavy gas oil from C19 to C25, lubricating oil from C26 to C40, residuum for Cn, $n > 40$, where n is the number of carbon atoms in the molecular structure. The three most important stages of petroleum formation are diagenesis, catagenesis, and metamorphism.

Diagenesis comprises all the chemical and physical alterations of organic matter excluding effects of temperature T (T in the range from 50 °C to 60 °C), catagenesis implies the thermal alterations of organic matter (T in the range from 60 °C to 200 °C) and metagenesis implies high-temperature thermal alteration ($T > 200$ °C). The largest amount of petroleum originates from organic matter in sediments subject to diagenesis and catagenesis, with temperature in the range from 60 °C to 150 °C.

Carbon in sedimentary rocks has undergone isotopic fractionation, with the heavier isotope 13C concentrating in carbonate/inorganic carbon and the lighter isotope 12C in organic carbon. Petroleum 13C/12C values are in the range of organic carbon. Geochemical methods allow recognizing the genetic relationships of crude oils with parent organic matter and genetic relationship of oils in different stratigraphic sequences. Concentration of trace elements in oil is often high because trace elements are complexed in organic molecules. In particular, vanadium and nickel are complexed in porphyrin derived from the original chlorophyll molecule. The knowledge of kerogen characteristics enables understanding oil and gas genesis. The dispersed organic matter of rocks may be classified into insoluble and soluble in organic solvents (chloroform, benzene, alcohol, methanol, acetone, and ethanol). The portion of organic matter, which is insoluble in organic solvents, is named kerogen while the soluble portion is named bitumen. The predominant pathway for the formation of crude oils is via the kerogen intermediate.

Insoluble organic material (kerogen)

Kerogen is a very heterogeneous and complex agglomerate of macerals. Macerals are discrete particles of insoluble organic materials, which can be identified under the microscope and represent residual detritus from various sources of organic material. Major components of kerogen are carbon, hydrogen, oxygen, nitrogen, and sulfur. Kerogens can be petrographically characterized in a way similar to the classification of coals. This led to the concept of classifying kerogens as types from I to IV and different kerogen types were constrained by well-defined boundaries on the [H]/[C] vs [O/[C] plot (Tissot-Welte diagram). The Tissot-Welte diagram, derived from the coal petrographers' Van Krevelen diagram is used to assign kerogen types. It provides an indication of kerogen maturity level, and the nature of products that a particular kerogen may be expected to generate at appropriate levels of maturity. Development of the Rock Eval technique for characterization of the source rocks provided a relatively rapid alternative method for determination of two indices that could be used to replace the [H]/[C] and [O]/[C] ratios. The hydrogen index (HI) and oxygen index (OI) are directly proportional to the [H]/[C] and [O]/[C] ratios.

Another important feature related to the generation of oil or gas is the maturity level of the source rock. Organic matter has to reach a certain level of maturity before starting the thermal degradation and to convert into liquid or gaseous hydrocarbons. The threshold level for oil generation varies depending on kerogen type. The determination of maturity levels is critical to the success of every oil exploration program. Recovery of immature, but organic-rich, source rocks indicates good source potential for such rocks if buried more deeply in other parts of the basin. At the other extreme, an overmature source rock indicates a mature part of the basin not capable of generating additional liquid hydrocarbons, but possibly gas. There are several available indicators that can be used to estimate the relative maturity of a source rock. The traditional method is measuring the maturity of vitrinite. The chemical composition of the maceral vitrinite, derived from higher plant debris, changes as the level of maturity increases. With increasing maturity, the ability of vitrinite to reflect light increases and hence a vitrinite reflectance scale, which correlates the degree of reflectance with maturity, has been developed. Maturity changes of vitrinite have been studied by coal chemists for a long period of time. A similar approach was adopted by petroleum geochemists. Kerogen is a key intermediate in the formation of oil and gas. Kerogen types and maturity levels play an important role in determining the characteristics of the products that will be generated from a specific kerogen.

Soluble organic material

The soluble part of the dispersed organic matter is called bitumen, which is composed of oils, resins, and asphaltenes. Soluble compounds such as steranes and terpenes in crude oils are considered biomarkers. Biomarkers are compounds in ancient sediments characterized by carbon skeletons correlated to precursor molecules present in organisms and plants living at the time of deposition. For example, sterols are very abundant in many living organisms and plants and, upon burial diagenesis and thermal maturation, are ultimately converted into steranes; the only significant change that occurs is loss of the hydroxyl group and of the double bond present in the sterols. Therefore, a precursor to product relationship can be established between the sterol and the sterane. Similar precursors to product relationships have been recognized in a large number of other compounds as in the conversion from chlorophyll to porphyrins.

Bitumen

Bitumen and solid bitumen are secondary hydrocarbon products that are found in most source and tight reservoir rocks. They are formed during different stages of the postburial organic matter evolution. Sanei (2020) proposed five distinct stages in the evolution of solid bitumen: (a) diagenetic solid bitumen (or degraded bituminite) that is derived from degradation of bituminite in the diagenesis stage (Ro < 0.5%) and not a secondary maceral resulting from the thermal cracking of kerogen; (b) initial-oil solid bitumen, a consolidated form of early catagenetically generated bitumen at the incipient oil window (Ro ~ 0.5%–0.7%); (c) primary-oil solid bitumen derived from thermally generated bitumen and crude oil in the primary oil window (Ro ~ 0.7%–1.0%); (d) late-oil solid bitumen (solid-wax) derived from the waxy bitumen separated from the mature paraffinic heavy oil in the primary- and late-oil windows; and (e) pyrobitumen, a nongenerative solid bitumen, evolved from thermal cracking of the remaining hydrocarbon residue and other types of solid bitumen in the dry gas window and higher temperature (Ro > 1.4%). Mastalerz et al. (2018) have given a detailed review on the origin, properties, and implications of solid bitumen in source-rock reservoirs.

The evolution of organic matter immediately after sinking through the water column and deposition in sedimentary basins involves a series of postburial bacterial activities and low-temperature chemical reactions during the diagenesis stage, followed by thermal maturation and thermal cracking during the catagenesis and metagenesis phases. The low-temperature reactions involving recently deposited organic matter lead to the formation of kerogen at the end of the diagenesis stage. The subsequent thermal cracking of kerogen during catagenesis and metagenesis leads to the formation of oil and gas, respectively. The nature of generated bitumen and oil defines the migration range, pore distribution, and overall physical properties of the resulting solid bitumens in source rocks and tight reservoirs.

3.2 Petroleum system

A natural system that includes an active source rock along with all geologic elements and processes that are essential for hydrocarbon generation and accumulation is called a petroleum system (Leythaeuser, 2005). Identification of organic-rich source rocks (hydrocarbon source rocks) and its geological history is important to understand a petroleum system in a sedimentary basin (Fig. 3.1). The basic elements in a petroleum system consist of a source rock, migration pathway, a porous and permeable reservoir rock, and a tight cap rock and seal.

3.2.1 Source rock

Source rocks are mainly shales from which petroleum is generated or can be generated. Source rocks are rich in organic matter that can decompose and produce hydrocarbons under adequate temperature. Source rock can be of marine, lacustrine, or terrestrial origin. The TOC content and the vitrinite reflectance of the organic matter in the source rock can be indicators that can tell about the possibility of petroleum availability in a basin. Geochemical methods are used to quantify the nature of organic matter in source rocks, which contain the precursors to hydrocarbons, and to assess the type and quality of expelled hydrocarbons. Petroleum source beds are fine grained, clay-rich siliclastic rocks (mudstones and shales) or dark colored carbonate rocks (limestones and marlstones).

Essential conditions for a good/ideal petroleum source rock are the following: (a) should have an adequate TOC content of finely dispersed organic matter of biological origin, (b) the organic matter must be hydrogen-rich, and (c) the burial depth of source rock must be suitable for proper subsurface temperatures to initiate the process of petroleum generation. Organic matter is mainly composed of organic carbon, with minor amounts of hetero-elements (nitrogen, sulfur, and oxygen). Based on empirical evidence, in source rocks the total organic carbon (TOC) concentration varies from 1.5% and 0.5% (of siliclastic and carbonate lithologies, respectively), which is recognized as the minimum concentration levels of organic matter (Hunt, 1996). The quantity of petroleum generated and the internal storage capacity of the source rock in terms of porosity is controlled by this minimum concentration of organic carbon in source rocks. Source rocks with TOC concentrations in the range 2%–10% are known to have effectively generated and expelled commercial quantities of petroleum. Most petroleum source rocks display dark brown to black colors due to the presence of finely disseminated organic matter as well as finely dispersed pyrite crystals (FeS_2).

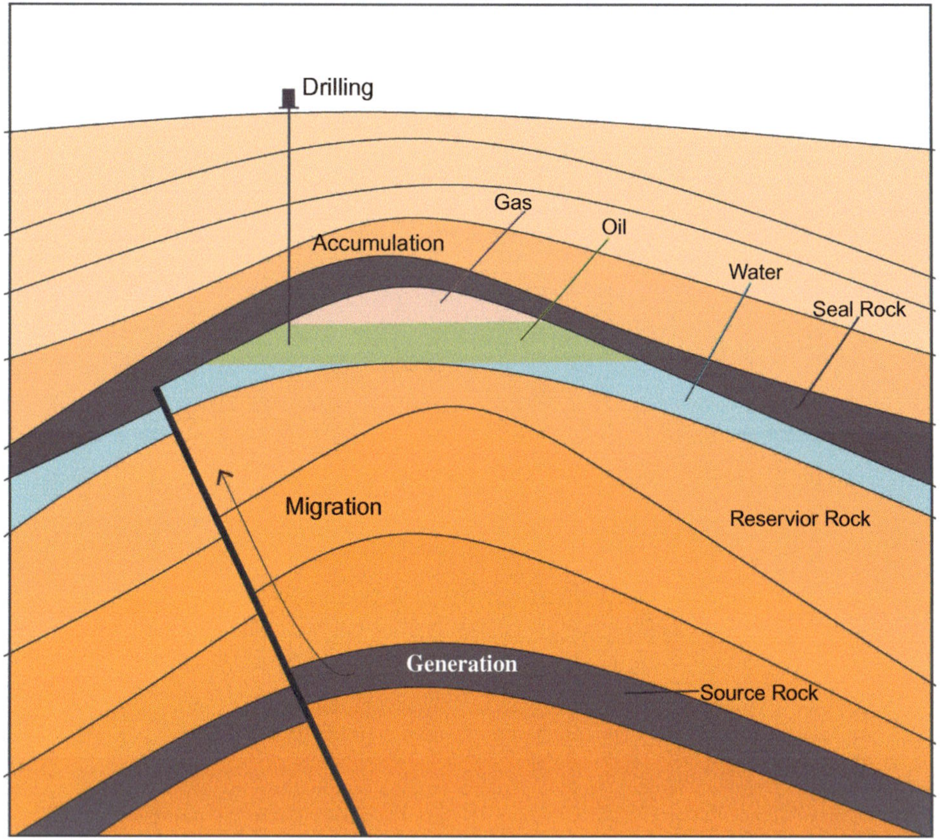

Fig. 3.1 A petroleum system and its processes (petroleum system: source rock, reservoir rock, and seal and processes: petroleum generation and migration) (after http://www.jogmec.go).

3.2.2 Migration path

Petroleum migrates from low permeability source rocks into high permeability reservoir rocks. The main driving force for petroleum migration is buoyancy since petroleum is less dense than water. There is primary migration, which is the flow of petroleum out of the source rock and secondary migration, which is the continued flow in more porous and permeable carrier beds to the reservoir rock or a fault up to the surface. The rate of migration can be a function of the rate of petroleum generation in the source rocks.

Primary migration takes place during the compaction of clay source rocks, and this compaction leads to the expulsion and displacement of large volumes of liquid/water. During primary migration, gas and oil travel together as a single liquid phase due to the high pressures in the source rock (higher than the bubble point pressure at which gases start to be liberated from liquid). Just after first expulsion, the closure of pores and fractures causes expulsion from the source rock in pulses. As time goes on, petroleum creates enough pressure to reopen fractures and pores causing a second expulsion, and the pulses continue to happen as long as the petroleum can rebuild enough pressure to reopen its path out of the source rock. After migration, the pressure decreases and the fractures and pores close. Finally, petroleum migrates out of the source rock and pressure will decrease. Usually, primary migration is vertical and secondary migration is lateral. During secondary migration, the gas and oil separate and the gas travels ahead of the oil. The largest petroleum deposits are the result of lateral migration because this provides a larger drainage volume of source rock than vertical migration. Differences in permeability between adjacent stratigraphic layers inhibit migration in most cases and this causes petroleum to flow within geologic units.

3.2.3 Reservoir rock

Reservoir rock is a rock in which the generated hydrocarbons are getting accumulated and reserved. The reservoir is a strata characterized by high porosity and permeability. Most common reservoir rocks are sedimentary rocks (e.g., sandstones, limestone, and dolomites). Reservoir rock is a vital element in a petroleum system and reservoir analysis requires an assessment of their porosity (the volume of pore space that contains fluids) and their permeability (property of the rock to transmit the fluid).

3.2.4 Seal (cap rock)

Cap rock is an impermeable rock that forms above or around a reservoir rock, and it act as a barrier to further migration of hydrocarbons so that the oil is prevented from escaping. A cap rock can be shale, anhydride, and salt. Analysis of cap rocks involves assessment of their thickness and extent, for quantifying its effectiveness.

3.2.5 Trap

After secondary migration in carrier beds, oil finally collects in a trap. Traps consist of porous reservoir rocks covered by low-permeability rocks, which do not allow oil or gas to pass. The fundamental characteristic of a trap is an upward convex form of porous and permeable reservoir rock that is sealed above by a denser, relatively impermeable cap rock (e.g., shale or evaporates). The trap may be of any shape, the critical factor being that it is a closed, inverted container. The trap is the stratigraphic or structural feature that ensures the association of reservoir and seal such that hydrocarbons remain trapped in the subsurface, rather than escaping (due to their natural buoyancy) and being lost. Structural traps are caused by structural deformation (folding, doming, or faulting) of rocks and stratigraphic traps are related to primary features in the sedimentary sequences and do not require structural deformation such as faulting or folding.

3.3 Petroleum maturation

The process by which biomolecules are converted into petroleum (kerogen transformation with increasing temperatures) is called petroleum maturation. Petroleum maturation happens over tens of millions of years where sediments initially undergo diagenesis, followed by catagenesis. During this cooking process, the original organic compounds are transformed into a far more complex mixture of hydrocarbons and asphaltics. Appropriate timing and adequate temperature are also required for the generation of hydrocarbons, its accumulation, and preservation. Analysis of maturation involves assessing the thermal history of the source rock in order to make predictions of the amount and timing of hydrocarbon generation and expulsion. Oil is formed during the diagenesis (10%–20% by volume) and more than 80% by volume during catagenesis at temperatures below 150 °C, and during metagenesis at temperatures above 150 °C most of the gaseous hydrocarbons (methane and dry gas along with nonhydrocarbon gases such as N_2, H_2S, and CO_2) are formed (Tissot and Welte, 1984).

Oil and gas are generated by the thermal degradation of kerogen in the source beds. With increasing burial, the temperature in these rocks rises and above a certain threshold temperature the kerogen begins to transform into petroleum compounds. With respect to the stage to which petroleum generation is advanced, the organic matter is labeled immature (preceding to the beginning of hydrocarbon generation), mature (if hydrocarbon generation is in progress), or overmature (when these reactions have been terminated). Heat is the main driving force in maturation and for petroleum generation the heating rate as well as the exposure time to maximum temperatures is important. The temperature interval where oil generation is in progress is referred to as the "liquid window" or "oil window" and it extends over the temperature interval of about 60–140 °C. The generation of oil and gas in source rocks is a natural consequence of the increase of subsurface temperature during geologic time.

Humic and sapropelic organic material are the two principal types of organic material that act as the organic precursors of kerogen that later on gets converted to petroleum (gas/oil). The biochemical characteristics of humic and sapropelic organic materials are very different. Humic organic material is of terrestrial origin (derived from terrestrial vegetation) and so it is associated with continental sediments. Humic organic material is largely of carbohydrate-lignin composition with some protein and is relatively high in nitrogen, but deficient in hydrogen (5% or less). The yield of liquid, volatile products is low from humic organic material and therefore it is gas prone (supplanted by CO_2 and H_2O). Thermal alteration of humic material yields humic acids that are soluble in alkalis with low hydrogen and high oxygen contents.

Sapropelic organic material is associated with aquatic sedimentary facies, both marine and lacustrine. Sapropelic organic material is structureless vegetable mud or slime (alginite) containing polymerized material derived from lipid and protein. Lipids derived from plant constituents such as spores and cuticles have high hydrogen content (more than 10%). The yield of volatiles in sapropelic organic matter is greater than that of humic organic matter. Thermal alteration of sapropelic organic material yields water-insoluble biomolecules, especially lipids. They are efficient producers of paraffin hydrocarbons; sapropelic organic material is therefore oil prone. These two organic matters yield H:C ratios as less than 0.8, predominantly humic; between 0.8 and 1.0, mixed humic and sapropelic; and above 1.0, predominantly sapropelic.

The three principal maceral groups in coal and sedimentary rocks are *liptinite (exinite), vitrinite,* and *inertinite* (Stach et al., 1982). The amount and maceral composition of kerogen determines petroleum potential and can differ vertically or laterally within a source rock. No universally accepted classification for kerogen types exists in the literature. We describe Type I, Type II, Type III, and Type IV kerogens (Tissot et al., 1974; Demaison et al., 1983; North, 1985; Peters and Cassa, 1994) based on the type of organic matter, presence of maceral groups, and the atomic hydrogen/carbon (H/C) and oxygen/carbon (O/C) values.

Higher relative hydrogen content in kerogen (atomic H/C, HI) generally corresponds to higher oil-generative potential. Gas (methane or CH_4) and oil are enriched in hydrogen compared with kerogen. During thermal maturation, generation of these products causes the kerogen to become depleted in hydrogen and relatively enriched in carbon.

Type I—Type I kerogens are oil prone and contain sapropelic organic matter, commonly of lacustrine origin. The H:C ratio is 1.6–1.8. It shows high atomic H/C (>1.5), and low O/C (<0.1). Generally, kerogens have low sulfur content. These kerogens are dominated by liptinite macerals, but vitrinites and inertinites can be present in lesser amounts.

Type II—For Type II kerogen, the organic matter is commonly marginal marine, with a mixture of continental and aquatic. Algal tissue, pollen, and spores are important contributors for it. Type II kerogens are also dominated by liptinite macerals. The H:C ratio is about 1.4 and sulfur content is generally higher in type II compared to other kerogens. Type II kerogens are oil-prone, show high atomic H/C (1.2–1.5) and low O/C compared with types III and IV.

Type III—For type III kerogen, humic organic matter of terrestrial, woody origin is prominent and is gas prone. The H:C ratio is 1.0 or less and it shows low atomic H/C (<1.0) and high O/C (<0.3). Vitrinite macerals originate from land plants and mature along the type III kerogen pathway, and therefore the vitrinite-dominated organic matter in Type III kerogen has a high potential for generating gas.

Type IV—For Type IV kerogen, the organic matter may be from any source, but it has been oxidized, recycled, or altered during some earlier thermal events. This inert carbonaceous material called "dead carbon" is now deficient in hydrogen (H:C ratio about 0.4 or less) and yields little or no hydrocarbons during maturation. These kerogens are dominated by inertinite macerals and they show very low atomic H/C (about 0.5–0.6) and low to high O/C (<0.3).

3.3.1 Diagenesis

After the burial of organic matter-rich sediments, the chemical reactions that happen (at temperatures less than 50 °C) in the first few thousand meters is defined as diagenesis and it is a process of compaction under mild conditions of temperature and pressure (Fig. 3.2). When organic aquatic sediments (proteins, lipids, carbohydrates) are deposited, they are very saturated with water and rich in minerals. Through chemical reaction, compaction, and microbial action during burial, water is forced out and proteins and carbohydrates break down to form new structures that comprise a waxy "kerogen" and a black tar such as a substance called "bitumen" and these occur within the first several hundred meters of burial. During diagenesis oxygen, nitrogen, and sulfur are removed from the organic matter leading to an increase in hydrogen content of the sedimentary organic matter. The presence of iron during diagenesis increases the probability of sulfur removal as the iron reacts with sulfur over a long burial time and this increases petroleum's economic value.

3.3.2 Catagenesis

The continuing accumulation of particles on the seafloor leads to the progressive burial of sediments that have undergone diagenesis and with progression in time these deeply buried sediments undergo abiotic chemical reactions that take place at high temperatures and pressures (Fig. 3.2). The reactions that occur between 60 °C and 140 °C (oil window) are catagenetic in nature. In marine sediments, the geothermal gradient is on the order of 20–55 °C/km. Temperatures exceeding 50 °C are attained at depths ranging from 1 to 4 km and the cooking at temperatures from 50 °C to 150 °C transforms kerogen into oil and bitumen (petroleum + asphalt). The catagenesis of sapropelic kerogens generates oil, whereas humic kerogens are converted into coal.

The organic compounds are exposed to diverse thermal degradation reactions during catagenesis that include double bond reduction by adding sulfur or hydrogen atoms, cracking reactions, and condensation reactions. As temperatures and pressures increase at greater depths/deeper burial, the process of catagenesis begins, which is the thermal degradation of kerogen to form hydrocarbon chains. The process of catagenesis is mainly catalyzed by the minerals that are deposited and persist through marine diagenesis. The conditions of catagenesis determine the product, such that higher temperature and pressure lead to more complete "cracking" of the kerogen and progressively lighter and smaller hydrocarbons. Petroleum formation thus requires a specific window of conditions; since with increased temperature/too hot, the product will favor natural gas (small hydrocarbons), but with less temperature/too cold, the plankton will remain trapped as kerogen. Sediment burial at a depth of more than 1 km is required to initiate catagenesis.

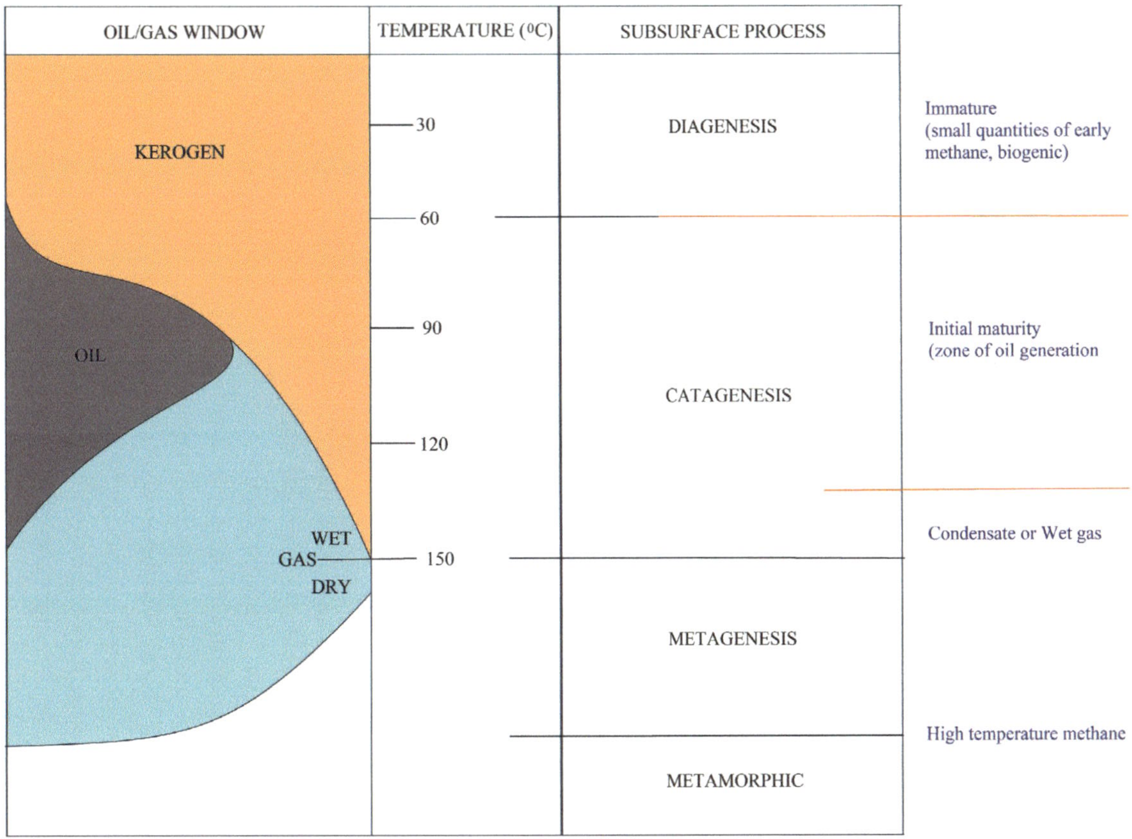

Fig. 3.2 Different stages of a thermal maturation process (after https://www.offshoreengineering.com).

3.3.3 Metagenesis

Metagenesis is the final stage of the thermal maturation process in which additional heat and chemical changes convert much of the kerogen into methane and a carbon residue (Fig. 3.2). Metagenesis occurs at temperatures of 150°–200 °C [302°–392 °F]. At the end of metagenesis, oil molecules are cracked into smaller gas molecules and methane or dry gas is evolved along with nonhydrocarbon gases such as CO_2, N_2, and H_2S.

3.4 Petroleum migration/migration of HCFIs

The presence of different populations of inclusions in different generations of cements (Burruss et al., 1983) or the compositional differences between inclusions in successive growth zones of fracture cement samples (Bodnar, 1990) implies petroleum migration. Munz et al. (1999) reported an example of a recent reservoir filling evidenced from subparallel trails of petroleum inclusions postdating most of the quartz cementation in a clastic reservoir. These subparallel trails were continuous throughout the samples and were cutting across both detrital and authigenic part of the quartz grains. The inclusions give important constraints on petroleum migration and reservoir filling and can be used as calibration points for modeling of petroleum migration and reservoir filling processes (Munz et al., 1999).

The principles for secondary migration and filling of reservoirs have been described in a series of publications (England, 1990; England & Mackenzie, 1989; England et al., 1987, England and Cubitt, 1995). The most important mechanism for migration is bulk transport driven by the gradients in petroleum fluid potential. During reservoir filling, the petroleum pore saturation increases, and the reservoir fluids may start to redistribute by processes such as diffusion and density-driven convection. The models of England et al., (1995) show that lateral variations in fluid compositions reflecting the process of filling can be preserved to the present day and the timescale of these processes is substantial.

3.5 Physical properties of petroleum

3.5.1 Petroleum composition

An understanding of the nature of petroleum, especially its physical properties, is important to realize the processes involved in the formation, migration, and accumulation of petroleum in sedimentary basins. The physical properties of crude oils are influenced by its chemical composition. There are many physical properties for petroleum such as boiling point, odor, color, optical activity, fluorescence, refractive index, specific gravity, density, etc. Among them the most significant physical properties are specific gravity, density, viscosity, the color, and refractive index of oils (North, 1985).

The hydrocarbon fluids in HCFIs may have highly complex compositions but are principally liquid with variable amounts of gas (light alkanes, CO_2, N_2, and H_2S), paraffins, naphthalene, aromatics, resins, and sometimes waxes. The apparent HCFI color depends on a combination of the wavelengths absorbed by the various chromophores present in the trapped hydrocarbon liquid and the associated mineral environment. Consequently, their color, when viewed under a microscope with plain white light illumination, may vary considerably from transparent to yellow, dark brown, or black. The mineralogical properties which can also influence the perceived color include birefringence, polarization, reflectivity, and variations in refractive index.

3.5.2 Specific gravity

The specific gravity of oil lies between 0.73 and just above 1.0. Paraffin-based oils are commonly light; asphalt-based oils are always heavy. The gravity is expressed by the Greek letter "ρ" (rho).

Density of petroleum

The density of petroleum is described using API gravity which is American Petroleum Institute's (API) scale of "lightness" or "heaviness" of crude oil and other liquid hydrocarbons. It is dependent on the density of a petroleum-based liquid versus that of water with units of degrees. The API value is a prime factor that determines the market value of crude oil.

It is defined as $API = 141.5/\rho - 131.5$, where ρ is the density/specific gravity in g/cm^3 at 60 °F (15.5 °C) and atmospheric pressure. API gravity is inversely related to the specific gravity of oil.

Based on this scale, fluids with low densities have high API gravities and fluids with high densities have low API gravities. Therefore, the scale of APIG indicates crude's relative density but inversely (i.e., the lighter the crude the higher the APIG) and vice versa. In this scheme, oil with an API value of >30° is termed light, between 22° and 30° termed medium, <22° heavy; and <10° extra heavy. The lighter the crude, the better will be the market value. The range of values for typical reservoir fluids is about

10–60° API. This scale is used to describe fluid under surface conditions that have lost all low molecular weight gas components. The liquid in fluid inclusions contains dissolved gases, making the density of the liquid slightly lower (API gravity higher) than the petroleum liquid if the fluid in the inclusion was expanded to room temperature and pressure conditions (Burrus, 2003).

3.5.3 Viscosity of oils

Viscosity is the internal friction of a fluid, causing its resistance to change of form and is an essential physical property by which a fluid differs from an elastic solid.

$$\text{Viscosity } (\eta) \text{ is defined by the ratio} = (\text{force} \times \text{distance})/(\text{area} \times \text{velocity}).$$

Viscosities vary directly with densities and therefore viscosities of oils are a function of the number of carbon atoms and of the amount of gas dissolved in the oil. Viscosities of light oils are below 30 mPa s (millipascal-seconds); the typical value is between 5.0 and 0.6 mPa and the latter is the viscosity of gasoline. Heavy asphaltic oils have viscosities of around thousands of millipascal-seconds. Viscosity is the ratio of stress to shear per unit time. Shear in liquid is not a constant but is proportional to time. "Pour point" the lowest temperature at which the crude will flow under prescribed, controlled conditions is a useful indicator of the viscosity of a crude oil. Pour points above 40 °C are relatively common among crudes having high contents of paraffin wax.

3.5.4 The colour and refractive index of oils

Paraffinic oils are usually light in color; yellow to brown in transmitted light and the familiar green of automobile engine oil in reflected light. Asphalt-based oils are commonly brown to black; many of them are known as black oils. The refractive indices of the oils vary with the relative density, between 1.42 and 1.48 and lower indices are for the light oils. The indices are also lower at low temperatures. Within any one molecular weight, the refractive indices increase from paraffin, through naphthene to aromatics (North, 1985).

3.6 Chemical properties of petroleum

Petroleum exploration is largely concerned with the search for oil and gas, two of the chemically and physically diverse group of compounds termed as the "hydrocarbons." Physically, hydrocarbons grade from gases, via liquids and semisolid substances, to solids. The hydrocarbon gases include dry gas (methane) and the wet gases (ethane, propane, butane, etc.). Condensates are hydrocarbons that are gaseous in the subsurface, but condense to liquid when they are cooled at the surface. Liquid hydrocarbons are termed oil, crude oil, or just crude, to differentiate them from refined petroleum products. The semisolid hydrocarbons include asphalt and related substances. Solid hydrocarbons include coal and kerogen. Gas hydrates are ice crystals with peculiarly structured atomic lattices, which contain molecules of methane and other gases.

Crude oil is defined as "a mixture of hydrocarbons that existed in the liquid phase in natural underground reservoirs and remains liquid at atmospheric pressure after passing through surface separating facilities" (joint definition by API, AAPG, and SPE). Crude oils vary from straw yellow, green, and brown to dark brown or black in color. Oils are naturally oily in texture and have widely varying viscosities. Oils on the surface tend to be more viscous than oils in warm subsurface reservoirs. Most oils are lighter than water. The density of oil may be measured as the difference between its specific gravity and that of water and is usually expressed in gravity units defined by the API. API degrees are inversely proportional to density. Thus light oils have API gravities of more than 40° (0.83 specific gravity), whereas heavy oils have API gravities of less than 10° (1.0 specific gravity). Heavy oils are thus defined as those oils that are denser than water. Oil viscosity and API gravity are generally inversely proportional to one another.

Crude oil consists largely of carbon and hydrogen with traces of vanadium, nickel, and other elements. Elemental composition of oils is relatively straightforward; there may be an immense number of molecular compounds. No two oils are identical either in the compounds contained or in the various proportions present. However, certain compositional trends are related to the age, depth, source, and geographical location of the oil. The compounds found in oil may be divided into two major groups: (1) the hydrocarbons, which contain three major subgroups—paraffins, naphthenes, and aromatics and (2) the heterocompounds, which contain elements other than hydrocarbons (Selley & Sonnenberg, 2015).

Paraffins—The paraffins often called alkanes are saturated hydrocarbons, with a general formula C_nH_{2n+2}. For values $n = 15$ they grade from viscous liquids to solid waxes. Two types of paraffin molecules are present within the series, both having similar atomic compositions (isomers), which increase in molecular weight along the series by the addition of CH_2 molecules. One series consists of straight-chain molecules, the other of branched-chain molecules. The paraffins occur abundantly in crude oil, the normal straight chain varieties dominating over the branched-chain structures. Individual members of the series have been recorded up to $C_{78}H_{158}$. For a given molecular weight, the normal paraffins have higher boiling points than do equivalent weight isoparaffins.

Naphthenes—The second major group of hydrocarbons found in crude oils are the naphthenes or cycloalkanes. This group has a general formula C_nH_{2n}. Like the paraffins they occur in a homologous series consisting of five- and six-membered carbon rings termed the cyclopentanes and cyclohexanes, respectively. Unlike the paraffins, all the naphthenes are liquid at normal temperatures and pressures. They make up about 40% of both light and heavy crude oil.

Aromatics—Aromatic compounds are the third major group of hydrocarbons commonly found in crude oil. Their molecular structure is based on a ring of six carbon atoms. The simplest member of the family is benzene (C_6H_6). One major series of the aromatic compounds is formed by substituting hydrogen atoms with alkane (CnH2n + 2) molecules. This alkyl benzene series includes ethyl benzene (C_6H_5, C_2H_5) and toluene ($C_6H_5CH_3$). Another series is formed by straight- or branched-chain carbon rings. This series includes naphthalene ($C_{10}H_8$) and anthracene ($C_{14}H_{10}$). The aromatic hydrocarbons include asphaltic compounds too. These compounds are divided into the resins, which are soluble in n-pentane, and the asphaltenes, which are not. The aromatic hydrocarbons are liquid at normal temperatures and pressures (the boiling point of benzene is 80.5 °C). They are present in relatively minor amounts (about 10%) in light oils, but increase in quantity with decreasing API gravity to more than 30% in heavy oils. Toluene ($C_6H_5CH_3$) is the most common aromatic component of crude oil, followed by the xylenes ($C_6H_4(CH_3)2$) and benzene.

Heterocompounds—Crude oil contains many different heterocompounds that contain elements other than hydrogen and carbon. The principal ones are oxygen, nitrogen, and sulfur, together with rare metal atoms, commonly nickel and vanadium. The oxygen compounds range between 0.06% and 0.4% by weight of most crudes. They include acids, esters, ketones, phenols, and alcohols. The acids are especially common in young, immature oils and include fatty acids, isoprenoids, and naphthenic and carboxylic acids. The presence of steranes in some crudes is an important indication of their organic origin. Nitrogen compounds range between 0.01% and 0.9% by weight of most crudes. They include amides, pyridines, indoles, and pyroles.

Sulfur compounds range from 0.1% to 7.0% by weight in crude oils. Elemental sulfur occurs in young, shallow oils (at temperatures above 100 °C sulfur combines with hydrogen to form hydrogen sulfide). Five main series of sulfur-bearing compounds in crude oil are: alkane thiols (mercaptans), thio-alkanes (sulfides), thio-cycloalkanes, dithio-alkanes, and cyclic sulfides.

Traces of numerous other elements have been found in crude oils, but determining whether they occur in genuine organic compounds or whether they are contaminants from the reservoir rock, formation waters, or production equipment is difficult to perceive. Trace elements in crude oil include common rock-forming elements, such as silicon, calcium, and magnesium, together with numerous metals, such as iron, aluminium, copper, lead, tin, gold, and silver. Nickel and vanadium occur as actual organometallic compounds, generally in porphyrin molecules. The porphyrins contain carbon, nitrogen, and oxygen, as well as a metal radical. The presence of porphyrins in crude oil is of particular genetic interest because they may be derived from the chlorophyll of plants. Metals tend to be associated with resin, sulfur, and asphaltene fractions of crude oils. Metals are rare in old, deep marine oils and are relatively abundant in shallow, young, or degraded crudes. Petroleum products have different boiling points that allow them to be separated by distillation. During distillation, crude oil is heated to very high temperatures until the petroleum is converted into a gas phase that contains a variety of hydrocarbons. The typical oil has more aromatics and asphaltics in the residuum and more paraffins in the gasoline fraction.

In the oil industry, natural gas is defined as "a mixture of hydrocarbons and varying quantities of nonhydrocarbons that exist either in the gaseous phase or in solution with crude oil in natural underground reservoirs" (definition by API, AAPG, and SPE). Natural gas is subdivided into dissolved, associated, and nonassociated gas. Dissolved gas is in solution in crude oil in the reservoir. Associated gas overlies and is in contact with crude oil in the reservoir. Nonassociated gas is in reservoirs that do not contain significant quantities of crude oil. Gases are classified as dry or wet according to the amount of liquid vapor that they contain. A dry gas may be arbitrarily defined as one with less than 0.1 gal/1000 ft^3 of condensate; chemically, dry gas is largely methane. A wet gas is one with more than 0.3 gal/1000 ft^3 of condensate; chemically, these gases contain ethane, butane, and propane. Gases are also described as sweet or sour, based on the absence or presence, respectively, of hydrogen sulfide.

The major constituents of natural gas are the hydrocarbons of the paraffin series. Methane is the most abundant; ethane, butane, and propane are quite common, and paraffins with a molecular weight greater than pentane are the least common. Methane forms in three ways: (1) derived from the mantle, (2) from the thermal maturation of buried organic matter, and (3) from the bacterial degradation of organic matter at shallow depths. Geochemical and isotope analysis can differentiate the source of methane in a reservoir. The other major hydrocarbons that occur in natural gas are ethane, propane, butane, and occasionally pentane. Unlike methane, these heavier members of the paraffin series do not form abiogenically. They are only produced by the thermal maturation of organic matter.

Helium is a common minor accessory in many natural gases—a nonhydrocarbon gas. Helium is known to be produced by the decay of various radioactive elements, principally uranium, thorium, and radium. The helium found in natural gas is widely believed to have emanated from deep-seated basement rocks, especially granite. Although the actual rate of production is slow and steady, the expulsion of gas into the overlying sediment cover may occur rapidly when the basement is subjected to thermal activity or fracturing by crustal arching. Apart from helium, traces of other inert gases such as argon and radon have been found in the subsurface. Argon and radon are of little economic significance. Nitrogen is another nonhydrocarbon gas that frequently occurs naturally in the earth's crust. It is commonly associated both with the inert gases and with hydrocarbons. Nitrogen in natural gases is predominantly of inorganic origin, although organic processes may be significant generators of atmospheric and shallow nitrogen. Some atmospheric nitrogen may have been trapped in sediments during deposition. Carbon dioxide (CO_2) is often found as a minor accessory in hydrocarbon natural gases and is commonly recorded in natural gases. Carbon dioxide is a normal product of the thermal maturation of kerogen, generally being expelled in advance of the petroleum. Carbon dioxide is also given off during the fermentation of organic matter as well as by the oxidation of mature organic matter due to either fluid invasion or bacterial degradation. Specifically, methane in the presence of oxygenated water may yield carbon dioxide and water. Hydrogen is commonly dissolved in petroleum as traces, but it is seldom recorded in conventional analyses (Hunt, 1979, 1996). Subsurface hydrogen is probably produced by the thermal maturation of organic matter. The amount of hydrogen relative to the amount of organic carbon present in a sample is termed as the hydrogen index (HI). HI is used to determine source rock quality (kerogen type) of immature rocks. Hydrogen sulfide (H_2S) occurs in the subsurface both as free gas and, because of its high solubility, in solution with oil and brine (Selley & Sonnenberg, 2015).

3.7 Biomarkers in crude oil

The concentration of individual biomarkers in a crude oil are relatively low but their information content is significant and related to source, depositional environments, maturity, age of migration, and extent of biodegradation. Biomarker fingerprints are extensively utilized for the purpose of undertaking oil and source rock correlation studies (Martinelli, 2003).

3.7.1 Source rocks and depositional environment

Source rocks are usually composed of fine-grained sediments such as shales and carbonates. Sedimentary rocks are subjected to the effect of time and temperature. Hydrocarbon generation from kerogen of sediments is closely related to temperature, kerogen type, and time. Peak oil generation occurs when the carbon content of kerogen is between 77% and 87% while peak gas generation occurs when carbon content is between 85% and 92%.

The utility of biomarkers as indicators of depositional environments arises from the fact that certain types of compounds are associated with organisms or plants that grow in specific types of depositional environments. In some situations, it has been possible to characterize depositional environments through a composite of biomarker parameters. For instance, gammacerane has long been used as an indicator of hypersalinity. Oils or extracts from evaporative environments, particularly lacustrine environments, typically have terpene chromatograms where gammacerane is the most abundant component in the chromatogram. Relatively high concentrations of pristane (Pr) are found in oxic-type environments and high concentrations of phytane (pH) are found in reducing type environments. Thus, the [Pr]/[pH] ratio evolved as an indicator of the possible toxicity of the depositional environment. This information can be used to predict the location of better source rocks in vertically drained basins, especially in deltaic-type environments with relatively young source rocks. With long-distance vertical migration, some of the biomarker parameters may become less effective.

3.7.2 Maturity indices and biodegradation processes

During crude oil maturation, the percentage of naphthenes decreases with respect to paraffins plus aromatics, and the percentage of four-, five- and six-ring naphthenes decrease, while one-ring naphthenes increases. Selected biomarkers have been used to evaluate the relative maturity of suspected source rocks and the oils they may have generated. The maturity parameters can be divided into a number of groups. The first group includes those parameters based on changes in distributions of stereoisomers or optical isomers; the second group includes the parameters resulting from the thermal degradation of certain molecules in a predictable manner (e.g., the cleavage of side chains from the tri-aromatic steroid hydrocarbon); finally, the third group contains various porphyrin ratios that depend on ring opening of the porphyrin, which occurs with increasing maturity. The most recent set of biomarker maturity parameters are those based on changes in the relative distribution of various methylated diamantoid structures. Biomarker ratios can be particularly important in samples that may not contain vitrinite, preventing the determination of a vitrinite reflectance value.

Various hypotheses concerning mechanisms of crude oil biodegradation have been introduced. These include: (a) the concept of paleopasteurization of reservoirs; (b) anaerobic, rather than exclusively aerobic degradation of crude oils in reservoirs; and (c) different mechanisms for degradation of surface seeps versus deep reservoirs. Paleopasteurization explains why it is possible to observe reservoirs of nondegraded crude oils at relatively shallow depths in the subsurface where one might expect the oils to be degraded. In general, biodegradation in crude oils has not been reported in reservoirs that have experienced temperatures above 80 °C. For paleopasteurization to occur, a reservoir should have been exposed to temperatures above 80 °C at some time during its history. Above that temperature, the reservoir is effectively sterilized and subsequent uplift, or overburden erosion, leading to the temperature falling below 80 °C, will mean that the oil which enters the reservoirs at lower temperatures will not experience any additional biodegradation. In this way, the sterilized sediments are not recolonized by hydrocarbon-degrading bacteria.

3.7.3 Age of oil and migration processes

Many attempts have been made to age-date oils using geochemical parameters. The main idea behind these attempts has been to find a compound, or class of compounds, that can be associated with a particular plant or organism, based on the fossil record. One method is based on the changes in the percentage of naphthenes, paraffins, and aromatics in gasoline-range hydrocarbons related to paraffin and aromatic hydrocarbons. The age of oil is calculated by the extent to which paraffins and aromatics have been formed from the naphthenes. A further method experimented in some test areas considers the relative proportions of steranes in crude oils derived from marine source rocks with known ages. A calibration based on the ratio of the [C27]/[C78] steranes vs geologic time has been set up. In this ratio, a strong increase has been evidenced as the samples became progressively younger as a result of the increase in the concentration of the C_{23} steranes. Comparison with information on the increase in diversity of phytoplankton with time shows a good correlation with the sterane ratio and provides an explanation for this increase. Oil compositions change with migration distance. Changes occur as a result of geochromatography due to the fact that smaller molecules move faster and further than large molecules. Oils that have migrated for the greatest distance are enriched in the lighter components and numerous attempts have been made to quantify such changes. The lighter components of a crude oil are in general isotopically heavier and, therefore, if the molecular distribution of crude oils changes one can expect isotopic changes to occur.

3.7.4 Thermal maturation indicators

In oil exploration, it is of prime importance to determine the maximum temperature to which the kerogen or a source rock has been subjected. Thus, it is necessary to have paleothermometers. There are several kerogen maturation parameters. The various techniques that can be used can be grouped into two categories: chemical paleothermometers and biological paleothermometers (Selley, 1985). The chemical methods include studies of organic and inorganic matter. The most important techniques for studying the maturity of organic matter are the carbon ratio method, the electron spin resonance (ESR) technique, and pyrolysis. Inorganic chemical paleothermometeres include clay mineral analysis and studies of fluid inclusions. Both organic and inorganic indicators are important for the reconstruction of thermal histories of sedimentary basins. The carbon ratio gives an estimate of the paleotemperature because there is a gradual decrease in the ratio of organic carbon and an increase in the amount of fixed carbon during the maturation of kerogen (Selley, 1985). The electron spin resonance method is based on the number and distribution of free electrons in the atomic fraction of kerogen. With increasing maturity, the electron characteristics change. Pyrolysis, i.e., heating of kerogen, provides immediate information on maturity and source rock potential. It is based on the fact that hydrocarbons are expelled from the kerogen with increasing temperature, and hydrocarbons expelled at different temperature intervals are determined in order to compare the amount of petroleum already generated and the total amount capable of being generated (Selley, 1985; Hunt, 1996). With increasing temperatures, argillaceous rocks are subjected to various diagenetic reactions and changes. Studies of clay mineral reactions have increased during the past decades along with intensified petroleum exploration, because clay minerals undergo various diagenetic changes during burial. These clay minerals include in particular smectites (montmorillonites), illite, and kaolinite. With increasing depth, pressure, and temperature, the clay will recrystallize and smectites will progressively change to illite. Smectites generally disappear at temperatures over 95 °C. Thus, the change in the chemical structure of clay from smectites to illites indicates the required temperature for oil generation. The kaolinite is also destroyed during diagenesis, but the disappearance of kaolinite is not indicative of a certain temperature or burial depth (Olsson-Borell, 2003). As temperatures increase to the upper limits of oil and gas generation and incipient metamorphic processes, kaolin, and illite will recrystallize into mica (muscovite) or chlorite (Selley, 1985; Olsson-Borell, 2003).

Biological paleothermometers are largely based on the color and the optical properties, for instance the light reflectance and fluorescence, of organic matter (kerogen) and fossils, such as pollen, spores, acritarchs, graptolites, scolecodonts, and conodonts. These fossils have an organic composition or have an organic component in a mineralized skeleton (conodonts). With increasing temperature and burial depth, their organic matter alters progressively, mainly because hydrogen and oxygen are lost in excess to carbon, and the consistent color changes through thermal gradients have made microfossils invaluable for the prediction of hydrocarbon windows. Aromatic and aliphatic hydrocarbons also serve as thermal maturity indicators (Naeser & McCulloh, 1989). Aromatic hydrocarbons are ubiquitous components of recent and ancient sediments. The precursor biochemicals for aromatic hydrocarbons in sediments and petroleum include terpenoids, carotenoids, quinones, and phenols. Tetra- and pentacyclic aliphatic hydrocarbon biological marker measurements are now commonly used in organic geochemical studies. Various ratios of these biomarkers have been utilized as empirical indicators of thermal maturity. The recognition of aliphatic hydrocarbon biomarker compounds as some of the most sophisticated of the current molecular thermal maturity indicators in source rocks arises from observed correlations between stereoisomeric molecular ratios and commonly accepted nonbiomarker thermal maturity indicators (e.g., vitrinite reflectance).

3.8 Trapping of petroleum inclusions

During diagenesis, many physical and chemical reactions occur in the sedimentary environment. Because of the thermal gradient, oil is generated from suitable type of kerogen. At the same time, when the diagenetic temperature is enough to break the structure of the smectite clay (the most prevalent clay in the marine sedimentary environment) into illite clay, it prompts the expulsion of huge quantities of structurally locked water from the smectites into the system. This will exert overpressure in the system leading to crack developments. Simultaneously, this petroleum oil will start migrating. Petroleum fluids, otherwise viscous and phlegmatic in nature, will start moving out of the source beds into the carrier beds (primary migration) that too along the secondary fractures (hydro-fractures) in grain scale. Within due course of time, the fractures will get healed and sealed entrapping oil as hydrocarbon fluids along with copious quantity of brine inclusions containing hypersaline water. The entrapped oil will transform to hydrocarbon fluid inclusions (HCFIs), usually with a gas bubble, the size of which depends on the difference in coefficient of thermal expansion of the fluids and the host mineral grains. Trapping of petroleum inclusions/HCFIs can occur along migration pathways or in reservoirs/pools. Mostly, trapped HCFIs are associated with reservoirs, and are trapped in diagenetic cements, overgrowths, or secondary fractures in quartz, calcite, and feldspar. Trapping of petroleum inclusions in reservoirs may occur from the beginning of reservoir filling to the present or to the time of drainage in case of wells. Trapping mechanism is highly active during reservoir filling than at a later time due to the general inhibiting effect of petroleum on cementation (Munz et al., 1998; Munz et al., 1999). The trapping of oil within quartz cement may take millions of years whereas healing of secondary fractures is much more rapid and may preserve the later stages of reservoir filling (George et al., 2007). The abundance of HCFIs in clastic sediments may be correlated with porosity and permeability (Oxtoby et al., 1995) while in chalk or limestone reservoirs the HCFIs may be more irregular favoring cemented fractures, recrystallized fossils, or coarse-grained cement (Munz, 2001; Jensenius & Burruss, 1990).

3.9 Significance of HCFI study

The study of HCFIs is important from a geologic viewpoint since it offers a unique understanding into the formation and development of petroleum basins. A combined study of the relative chronology of inclusions (both aqueous and HCFI) as well as their composition, and the temperature-pressure-volume (PVT) characteristics of individual inclusions can offer valuable information about the thermal history and migration of hydrocarbons in petroleum basins (Roedder, 1984). An understanding of the thermal history and migration mechanism can result in the discovery of economically recoverable crude oil. It provides predictive models and key indicators that may lead to further discoveries or to better reservoir management (Blamey & Ryder, 2007).

Fluid inclusions are found in many different types of natural minerals and rocks. They are small sealed vials that store host fossil fluids, which existed when minerals grew or healed after fracture; often less than 10 μm in size. Fluid inclusion composition varies considerably depending on the fluid source and temperatures, and may be host to aqueous or hydrocarbon liquids, solids, and gaseous phases. Aqueous fluid inclusions comprise mainly water with dissolved salts (brines) or daughter minerals along with minor dissolved inorganic, organic, and trace noble gases whereas hydrocarbon-bearing fluid inclusions (HCFI) generally comprise liquid hydrocarbons, low carbon-number gases, and occasionally dark solid phases.

Petroleum is the generic term of hydrocarbon-rich fluids produced during the thermal evolution of sedimentary basins. The presence of crude oils as fluid inclusions in sedimentary rocks show the importance of fluid inclusion studies in such rock types. The rapid proliferation of HCFI studies happened over the years because these inclusions are "hidden petroleum shows" (Lisk et al., 2002). The presence of petroleum inclusions in diagenetic cements or in healed fractures demonstrate that oil was present at the time the cement was formed or when the fractures were healed. This simple observation traces the time of oil generation and migration in the context of diagenetic and tectonic history of the rock. The petrographic and microthermometric data of petroleum inclusions combined with burial history reconstruction can provide information on the time of oil migration (Burruss et al., 1985) and charging of the reservoirs. When this approach is combined with current state-of-the-art methods of basin analysis, including thermo-chronology from fission track analysis, microthermometry of petroleum and aqueous inclusions, and modeling of the PVT properties of petroleum, we can get a detailed picture of the time, temperature, and pressure of petroleum migration (Ceriani et al., 2002; Tseng & Pottorf, 2002). HCFI are not limited to petroleum reservoirs and they may also be found in a diversity of geological environments. For example, there is increasing evidence for the presence of HCFI within metal deposits (Gonzalez-Partida et al., 2003; Kendrick et al., 2002).

Information that is very useful for petroleum exploration or resource evaluation could be obtained through analyzing/observing HCFIs. Stratigraphic intervals in which oil migration occurred can be identified by a comparison of the distribution of petroleum inclusions in regional outcrop samples in an unexplored basin with subsurface samples from exploration drill holes adjacent to the unexplored area (Burrus, 1999). The variations in the abundance of oil inclusions in known reservoir intervals can be used to identify former oil–water or oil-gas contacts that indicate changes in petroleum charge over time (Lisk et al., 2002). Microthermometric analysis of fluid inclusions provides homogenization temperature of two-phase, liquid–vapor HCFIs that gives minimum trapping temperatures. The pressure-volume-temperature (PVT) interpretation of HCFIs leads to the estimation of both trapping temperature and pressure. The PVT interpretation of HCFIs and estimation of the actual trapping temperatures are possible provided the composition and density data are available (Aplin et al., 1999; Burrus, 1989; Narr & Burrus, 1984, Thiery et al., 2002; Teinturier et al., 2002).

An important objective of most fluid inclusion studies is to calculate fluid densities constructing isochores, and to estimate the pressure and temperature conditions of trapping, either by combining information from solids or from coexisting fluid phases. For petroleum inclusions, estimating fluid properties at pressure and temperature conditions other than the trapping conditions is also possible. This may include saturation pressure at specified temperature or the gas-to-oil ratio (GOR) at any specified pressure and temperature. These calculations are all done using an equation of state. Geochemical analysis of petroleum inclusion yields unique information on the composition of petroleum during migration. Comparison of petroleum in inclusions to economic accumulations can tell us about variations in composition during the accumulation process (George et al., 1997, 1998, 2001; Jensenius and Burruss, 1990; Karlsen et al., 1993; Nedkvitne et al., 1993). Composition of petroleum in low-temperature ore deposits indicates extreme levels of alteration of petroleum (Mauk and Burruss, 2002; Pearcy and Burruss, 1993).

HCFIs are useful for learning about fluid compositions, temperatures, and pressure conditions in geologic systems and processes such as the migration of hydrocarbon fluids in petroleum basins. HCFIs generally occur in diagenetic cements or grains and contain complex mixtures of mainly organic compounds depending on their source/s. Accurate analysis of the chemical composition of the entrapped hydrocarbons in HCFI can yield vital information about the history, evolution, and migration of petroleum fluids, and is thus crucial data for the petroleum exploration industry. Studying HCFI is advantageous because the trapped fluids are representative of the actual hydrocarbon fluids that existed when the inclusions were sealed in the mineral. This sealing-in process preserves the petroleum fluid, thus isolating it from subsequent infiltration of petroleum fluids and events in oil reservoirs such as loss of charge, water washing, or biodegradation. It also preserves the fluid from contamination during the drilling processes used to extract samples from the ground (Blamey & Ryder, 2007). Fluid inclusions, despite their small size, are highly valuable to understanding many geological processes. When such cavities within the rock were sealed, they trapped the original fluid at the fossil pressure-volume-temperature (PVT) conditions. This PVT data can be used for modeling fluid phase behavior and give an indication of the "oil window" at which oil formation occurs (Munz, 2001). Geologists can learn much about the processes, fluid compositions, temperatures, and pressure conditions in geologic systems from fluid inclusions that form at a key time, such as during ore formation in base metal or gold deposits, or from the migration of hydrocarbon fluids in petroleum basins.

The petrographic, microthermometric, and spectroscopic analysis of HCFIs along with other biphase fluid inclusions and their futuristic applications in the petroleum exploration industry are detailed in the chapters that follow.

References

Aplin, A.C., Macleod, G., Larter, S.R., Pedersen, K.S., Sørensen, H., Booth, T., 1999. Combined use of confocal laser scanning microscopy and PVT simulation for estimating the composition and physical properties of petroleum in fluid inclusions. Mar. Pet. Geol. 16 (2), 97–110.

Bjorlykke, K., 2010. Petroleum Geoscience: From Sedimentary Environment to Rock Physics. Springer, New York. ISBN: 978-642-02331-6.

Blamey, N.J.F., Ryder, A.G., 2007. Hydrocarbon fluid inclusion fluorescence: A review. In: Geddes, C.D. (Ed.), Reviews in Fluorescence. Vol. 2007. Springer, New York, pp. 299–334.

Bodnar, R.J., 1990. Petroleum migration in the Miocene Monterey formation, California, USA. Constraints from fluid inclusion studies. Mineral. Mag. 54 (375), 295–304.

Burrus, R.C., 1989. Paleo temperatures from fluid inclusions: Advances in theory and technique. In: Naeser, N.D., McCulloh, T. (Eds.), Thermal History of Sedimentary Basins: Methods and Case Histories. Springer-Verlag, pp. 119–131.

Burrus, R. C, 1999. Evidence for petroleum occurrence and timing of migration: Petroleum fluid inclusions, dead oil, stains, and seeps. In The Oil and Gas Potential of the Arctic National Wildlife Refuge 1002 Area, Vol. US Geological Survey Open-File Report 98–34-FI (ed. Team, AA), pp. FI-1 to FI-25.
Burrus, R.C., 2003. Petroleum fluid inclusions an introduction. In: Samson, I., Anderson, A., Marshall, D. (Eds.), Fluid Inclusions: Analysis and Interpretation. Mineralogical Association of Canada, Short Course Series, Vol. 32, pp. 159–169.
Burruss, R.C., Cercone, K.R., Harris, P.M., 1983. Fluid inclusion petrography and tectonic burial history of the Al Ali no. 2 well; evidence for the timing of diagenesis and oil migration, northern Oman Foredeep. Geology 11 (10), 567–570.
Burruss, R.C., Cercone, K.R., Harris, P.M., 1985. Timing of hydrocarbon migration: Evidence from fluid inclusions in calcite cements, tectonics and burial history. In: Schneidermann, N., Harris, P.M. (Eds.), Carbonate Cements. Vol. 36. Special Publication – Society of Economic Paleontologists and Mineralogists, pp. 277–289.
Ceriani, A., Di Giulio, A., Goldstein, R.H., Rossi, C., 2002. Diagenesis associated with cooling during burial: an example from lower cretaceous reservoir sandstones (Sirt Basin, Libya). AAPG Bull. 86 (9), 1573–1591.
Chapman, R.E., 1976. Petroleum Geology a Concise Study. Elsevier, New York, ISBN: 0-444-41117-8.
Demaison, G.J., Hoick, A.J.J., Jones, R.W., Moore, G.T., 1983. Predictive source bed stratigraphy; a guide to regional petroleum occurrence. In: Proceedings of the 11th World Petroleum Congress, London. Vol. 2, pp. 17–29.
England, W.A., 1990. The organic geochemistry of petroleum reservoirs. Org. Geochem. 16 (1–3), 415–425 (1990).
England, W.A., Cubitt, J.M., 1995. Geochemistry of reservoirs, an introduction. Geological Society London Special Publications Vol. 86 (1), 1–3. https://doi.org/10.1144/GSL.SP.1995.086.01.01.
England, W.A., Mackenzie, A.S., 1989. Some aspects of the organic geochemistry of petroleum fluids, July 1989. Geologische Rundschau (International Journal of Earth Sciences) 78 (1), 291–303. https://doi.org/10.1007/BF01988365.
England, W.A., Mackenzie, A.S., Mann, D.M., Quigley, T.M., 1987. The movement and entrapment of petroleum fluids in the subsurface, April 1987. J. Geol. Soc. 144 (2), 327–347. https://doi.org/10.1144/gsjgs.144.2.0327.
England, W.A., Muggeridge, A.H., Clifford, P.J., Tang, Z., 1995. Modelling density-driven mixing rates in petroleum reservoirs on geological time-scales, with application to the detection of barriers in the Forties Field (UKCS). In: Cubitt, J.M., England, W.A. (Eds.), The Geochemistry of Reservoirs, Geological Society. vol. 86. Special Publications, London, pp. 185–201.
George, S.C., Krieger, F.W., Eadington, P.J., Quezada, R.A., Greenwood, P.F., Eisenberg, L.I., Hamilton, P.J., Wilson, M.A., 1997. Geochemical comparison of oil-bearing fluid inclusions and produced oil from the Toro sandstone, Papua New Guinea. Org. Geochem. 26 (3–4), 155–173.
George, S.C., Lisk, M., Summons, R.E., Quezada, R.A., 1998. Constraining the oil charge history of the south pepper oilfield from the analysis of oil-bearing fluid inclusions. Org. Geochem. 29 (1–3), 631–648.
George, S.C., Ruble, T.E., Dutkiewicz, A., Eadington, P.J., 2001. Assessing the maturity of oil trapped in fluid inclusions using molecular geochemistry data and visually-determined fluorescence olours. Appl. Geochem. 16 (4), 451–473.
George, S.C., Volk, H., Ahmed, M., 2007. Geochemical analysis techniques and geological applications of oil-bearing fluid inclusions, with some Australian case studies. J. Pet. Sci. Eng. 57 (1–2), 119–138.

Gonzalez-Partida, E., Carrillo-Chavez, A., Grimmer, J., O, W., Pironon, J., Mutterer, J., Levresse, G., 2003. Fluorite deposits at Encantada-Buenavista, Mexico: products of Mississippi Valley type processes. Ore Geol. Rev. 23 (3–4), 107–124.
Hunt, J.M., 1979. Petroleum Geochemistry and Geology. Freeman, San Francisco (617 pp).
Hunt, J.M., 1996. Petroleum geochemistry and geology. In: Handbook of Geochemistry Applied to Oil and Gas Formation Processes, second Edition. W.H. Freeman, New York. 743 pp.
Jensenius, J., Burruss, R.C., 1990. Hydrocarbon-water interactions during brine migration: evidence from hydrocarbon inclusions in calcite cements from Danish North Sea oil fields. Geochim. Cosmochim. Acta 54 (3), 705–713.
Karlsen, D.A., Nedkvitne, T., Larter, S.R., Bjorlykke, K., 1993. Hydrocarbon composition of authigenic inclsuions; applications to elucidation of petroleum reservoir filling history. Geochim. Cosmochim. Acta 57, 3641–3659.
Kendrick, M.A., Burgess, R., Pattrick, R.A.D., Turner, G., 2002. Hydrothermal fluid origins in a fluorite-rich Mississippi Valley-type district: combined noble gas (he, Ar, Kr) and halogen (cl, Br, I) analysis of fluid inclusions from the south Pennine ore field, United Kingdom. Econ. Geol. 97 (3), 435–451. http://petroleum.co.uk.
Leythaeuser, D., 2005. Origin, migration and accumulation of petroleum. In: Beccari, M., Romanao, U. (Eds.), Geosciences. Encyclopedia of Hydrocarbons, Volume 1, Exploration, Production and Transport, pp. 65–84.
Lisk, M., O'Brien, G.W., Eadington, P.J., 2002. Quantitative evaluation of the oil-leg potential in the Oliver gas field, Timor Sea, Australia. AAPG Bull. 86 (9), 1531–1542.
Martinelli, G., 2003. Petroleum geochemistry. In: Petroleum Engineering—Upstream. Encyclopedia of Life Support Systems (EOLSS). Reggio Emilia.
Mastalerz, M., et al., 2018. Origin, properties, and implications of solid bitumen in source-rock reservoirs: a review. Int. J. Coal Geol. 195, 14–36.
Mauk, J.L., Burruss, R., 2002. Water washing of Proterozoic oil in the midcontinent rift system. AAPG Bull. 86 (6). https://doi.org/10.1306/61EEDC38-173E-11D7-8645000102C1865D.
Munz, I.A., 2001. Petroleum inclusions in sedimentary basins: systematics, analytical methods and applications. Lithos 55 (1–4), 195–212.
Munz, I.A., Iden, K., Johansen, H., Vagle, K., 1998. The fluid regime during fracturing of the Embla field, central trough. North Sea. *Mar. Pet. Geol.* 15 (8), 751–768.
Munz, I.A., Johansen, H., Johansen, I., 1999. Characterization of Composition and PVT Properties of Petroleum Inclusions: Implications of Reservoir Filling and Compartmentalization, October 1999. Soc. Pet. Eng. J. https://doi.org/10.2118/56519-MS.
Naeser ND and McCulloh TH, Editors. Thermal History of Sedimentary Basins Methods and Case Histories, 1989 by Springer-Verlag New York.
Narr, W.M., Burrus, R.C., 1984. Origin of reservoir fractures in little knife field, North Dakota. Am. Assoc. Pet. Geol. Bull. 68, 1087–1100.
Nedkvitne, T., Karlsen, D.A., Bjorlykke, K., Larter, S.R., 1993. Relationship between reservoir diagenetic evolution and petroleum emplacement in the Ula field, North Sea. Mar. Petr. Geol. 10, 255–270.
North, F.K., 1985. Petroleum Geology. Allen & Unwin Inc., USA, ISBN: 0-04-553003-3.
Olsson-Borell, 2003. Thermal history of the Phanerozoic sedimentary succession of Skåne, southern Sweden, and implications for applied geology. Vol. 2003 Department of Geology, Lund University, Olsson, Ingela, p. 23.
Oxtoby, N.H., Mitchell, A.W., Gluyas, J.G., 1995. The filling and emptying of the Ula oilfield: fluid inclusion constraints. In: Cubitt, J.M., England, W.A. (Eds.), The Geochemistry of Reservoirs: Special Publication. Vol. 86. Geological Society, London, pp. 141–157.

Pearcy, E.C., Burruss, R.C., 1993. Hydrocarbons and gold mineralization in the hot-spring deposit at Cherry Hill, California. In: Bitumens in Ore Deposits, pp. 117–137 (Special Publication of the Society for Geology Applied to Mineral Deposits book series (MINERAL DEPOS., volume 9).

Peters, K.E., Cassa, M.R., 1994. Applied source rock geochemistry. In: Magoon, L.B., Dow, W.G. (Eds.), The Petroleum System—From Source to Trap: AAPG Memoir 60PN, pp. 93–120.

Roedder, E., 1984. Fluid inclusions. Mineralogical society of America. Rev. Mineral. 12, 1–644.

Sanei, H., 2020. Genesis of solid bitumen. Sci. Rep. 10, 15595.

Selley, R.C., 1985. Elements of Petroleum Geology. Academic Press, An imprint of Elsevier, California, USA.

Selley, R. C., 1998. Elements of Petroleum Geology, 2nd edition, Academic press. An imprint of Elsevier, California, USA. ISBN-10: 0-12-636370-6.

Selley, R.C., Sonnenberg, S.A., 2015. Elements of Petroleum Geology, 3rd edition. Elsevier, USA.

Stach, E., Mackowsky, M.-T., Teichrniiller, M., Taylor, G.H., Chandra, D., Teichmuller, R., 1982. Coal Petrology. Gebriider Borntraeger, Berlin, p. 535.

Teinturier, S., Pironon, J., Walgenwitz, F., 2002. Fluid inclusions and PVTX modelling: examples from the garn formation in well 6507/2-2, haltenbanken. Mid-Norway. Mar. Petr. Geol. 19, 755–765.

Thiery, R., Pironon, J., Walgenwitz, F., Montel, F., 2002. Individual characterization of petroleum fluid inclusions (composition and P-T trapping conditions): by microthermometry and confocal laser scanning microscopy: inferences from applied thermodynamics of oils. Mar. Petr. Geol. 19, 847–859.

Tissot, B.P., Welte, D.H., 1978. Petroleum Formation and Occurrence: A New Approach to Oil and Gas Exploration. Springer-Verlag, Berlin, Heidelberg, New York. xviii + 538 pp.

Tissot, B.P., Welte, D.H., 1984. Petroleum Formation and Occurrence-a New Approach to Oil and Gas Exploration. Springer-Verlag, New York, ISBN: 3-540-08698-6.

Tissot, B., Durand, B., Espitalié, J., Combaz, A., 1974. Influence of nature and diagenesis of organic matter in formation of petroleum. Am. Assoc. Pet. Geol. Bull. 58, 499–506.

Tseng, H.Y., Pottorf, R.J., 2002. Fluid inclusion constraints on petroleum PVT and compositional history of the greater Alwyn-South Brent petroleum system, northern North Sea. Mar. Pet. Geol. 19 (7), 797–809.

CHAPTER 4

HCFIs—Examples from Mumbai and Kerala offshore basins in India

V. Nandakumar[a] and J.L. Jayanthi[b]
[a]Scientist - G, National Facility for Geofluids Research and Raman Analysis, National Centre for Earth Science Studies (NCESS), Thiruvananthapuram, Kerala, India
[b]Project Scientist - C, National Facility for Geofluids Research and Raman Analysis, National Centre for Earth Science Studies (NCESS), Thiruvananthapuram, Kerala, India

4.1 Introduction

India was part of Rodina and Gondwanaland, and later most of the landmasses in the Gondwana assembly drifted away from each other and India rapidly moved towards northeast and collided with the Asian subcontinent leading to the formation of the Himalayas. The ensuing tectonic activities lead to the formation of Arabian Sea and the Bay of Bengal. The west coast sedimentary province was formed along the periphery of the Indian Peninsula and is broadly divided into Kerala-Konkan Basin, Mumbai Offshore Basin, Kutch Basin, and Saurashtra basin. East northeast (ENE)-trending ridges—the Saurashtra Arch, Vengurla Arch, and Tellechery Arch (Fig. 4.1)—separate these basins.

A total of 38 sedimentary basins were documented in onshore and offshore areas of the Indian plate including western and eastern Deep Ocean basins; among them, eight basins are petroleum-producing basins. Basins with proven hydrocarbon deposit and basins with good indications of hydrocarbon prospect are termed as petroliferous basins. Based on the exploration status petroliferous basins are categorized into four groups (Rai et al., 1998): Basins with proven commercial production (category 1), established basins expecting commercial production (category 2), basins with considerable geological prospect along with hydrocarbon shows (category 3), and basins that are not having established commercial production now but maybe prospective in the future (category 4). In India, Mumbai offshore basin is the most productive basin with a prognostic reserve estimation of 9190 million metric tons (MMT) as reported in April 2014, which accounts for nearly two-thirds of the annual petroleum production in India (Goswami et al., 2007, ndrdgh.gov.in). Kerala-Konkan basin is a genetically continuous

Hydrocarbon Fluid Inclusions in Petroliferous Basins
https://doi.org/10.1016/B978-0-12-817416-6.00001-0

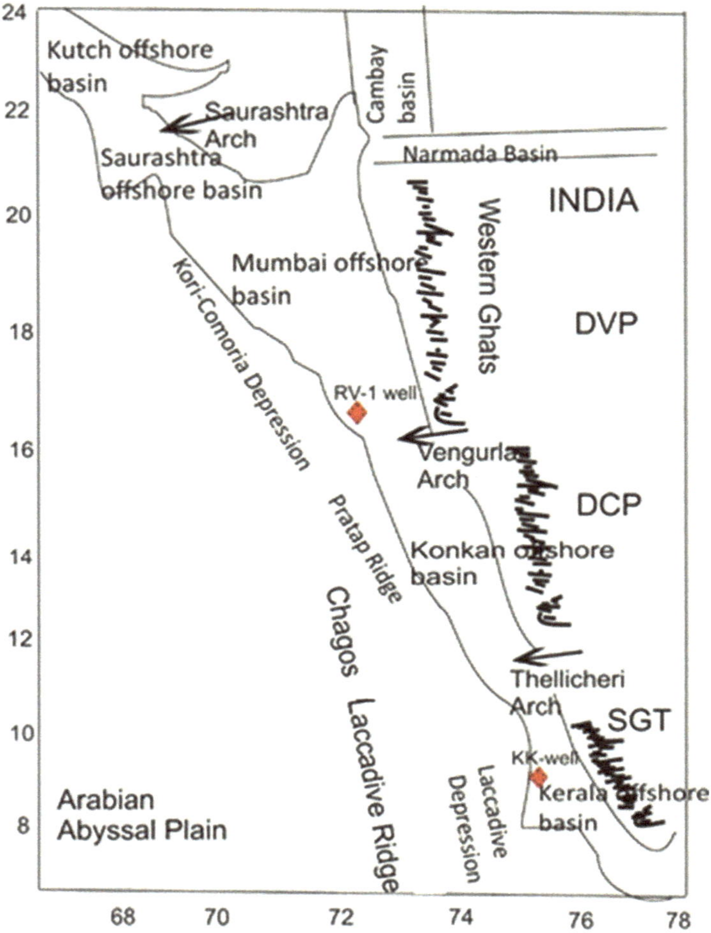

Fig. 4.1 Indian west coast margin and offshore basins; study areas (dry wells) are shown with red squares. DVP—Deccan Volcanic Province, DCP—Dharwar Craton Proince, and SGT—southern Granulite Terrain. *Modified after Goswami B.G., Singh H., Ashok Kumar Bhatnagar A.K., Sinha A.K., Ram Raj Singh, R.R., 2007. Petroleum Systems of the Mumbai Offshore Basin, India. Search and Discovery Article #10154 (2007). Adapted from extended abstract prepared for AAPG Annual Convention, Long Beach California, April 1–4, 2007.*

basin of Mumbai basin in the western offshore and both the basins are genetically comparable, but with provenance difference as far as the sediment composition are considered. In these basins, sediment thickness is around 5–7 km. In this book, we are detailing the characteristics of hydrocarbon fluid inclusions from two dry wells in these two selected basins: (1) Mumbai offshore basin (Hydrocarbon Proven Basin) and (2) Kerala-Konkan basin (hydrocarbon nonproven basin). Mumbai offshore is a category 1 basin and the Kerala-Konkan basin is a category 3 basin (Biswas, 2012). The fluid inclusions from these two dry wells (locations marked with red square symbols in Fig. 4.1) were used for petrographic, microthermometric, and spectroscopic analyses and further discussion.

Most of the Indian sedimentary basins are pericratonic, bordering the Indian plate radially. The northern basins are bordered by orogenic belts of Himalaya and Assam-Arakan and are the peripheral foreland Basins subducting under the orogenic belt. The coastal pericratonic basins, especially in the south are extending from onshore to offshore and are rifted Basins, separated by horsts. Petroleum occurs mainly in the marginal and pericratonic rift Basins. At present India's estimated hydrocarbon resource is of the order of 28 billion tons of oil equivalent (BTOE) and an initial in-place volume of about 11.18 billion tons of oil equivalent (BTOE) (Dwivedi, 2016).

4.2 Geologic evolution of western Indian continental margin and its impact on hydrocarbon occurrences

The largest hydrocarbon accumulations of India are localized in the western Indian continental margin by a unique convergence of tectonic events. The passive margin basins of the Indian coast is formed due to Mesozoic rifting along with the structural trends of Proterozoic mobile belts formed. During the Campanian rifting of Madagascar from India, north–south striking normal faults propagated into the area of the Cambay Graben. Late Maastrichtian doming over the incipient Deccan/Reunion hotspot imparted extensional stresses to the northwestern Indian coast and formed the fault block that became the Mumbai High. This extension continued in the weakened crust after the eruption of the Deccan flood basalts in the early Paleocene and this resulted in enhanced subsidence of the Cambay Graben and Surat Depression, rifting of the Seychelles microcontinent from India, and reactivation of normal faults on the emergent Bombay High. The Cambay Graben and the Surat Depression filled with organic-rich source shales as they underwent thermal subsidence. Shallow-water Eocene to Miocene carbonates developed on the slowly subsiding Bombay Platform, and sea level oscillations fostered secondary porosity development. Late Tertiary maturation of the Surat Depression shales generated hydrocarbons that migrated into the carbonate reservoirs on the Mumbai High. Konkan-Kerala, and the older basins of the east coast of India, were unaffected by the post-Campanian tectonic events, and lack the favorable play elements that characterize the giant fields.

The Indian Shield is an ancient fragment of Gondwana and it is composed of Precambrian gneiss and granite which has a long and varied history with Precambrian rocks as old as 3800 Ma are reported (Goodwin, 1991; Biswas, 1982; Kumar, 1985; Naqvi and Rogers, 1987; Gombos et al., 1995). About two-thirds of the shield consists of Precambrian rocks exposed at the surface (Kumar, 1985) and a large part of the remainder of the shield is covered by Late Cretaceous to earliest Paleocene Deccan volcanics. Late Paleozoic to Mesozoic 'Gondwana' sediments are preserved mainly in rift valleys such as Mahanadi; Krishna-Godavari; Rewa-Son and Narmada grabens. Patches of unmetamorphosed Proterozoic sediment are also encountered, like the Cuddapah Depression in east-central India.

The Mumbai Platform is a rifted continental margin platform that occupies the widest reach of the continental shelf around India. It extends for about 300 km west of Mumbai and 350–400 km south of the Saurashtra Horst. It is characterized by normal fault blocks which are oriented N–S, parallel with the Precambrian Dharwar trend (Biswas, 1982; Biswas and Singh, 1988). These faults establish the typical high-side fault-dependent traps in the carbonate reservoir section. This is the most prolific oil province in India (Rao and Talukdar, 1980). The Konkan-Kerala coast is a long, straight rifted passive margin. It formed in response to the separation of Madagascar from India about 85–90 Ma. The rifting event set up a system of north–south (N–S)-trending horsts and grabens parallel with the coastline. Rifting was focused along the ancient Dharwar structural trend, i.e., north-northwest to south-southeast (NNW–SSE). This established a shelfal horst-graben complex, which characterizes the area today. Unlike the areas to the north, such as the Mumbai Platform and Cambay Graben, the Konkan-Kerala coast did not undergo later structuring (Mitra et al., 1983; Duncan, 1990; Agrawal et al., 1992).

The breakup of Gondwana began about 150 million years ago. The split resulted in the creation of two smaller continents: West Gondwana consisted of Africa, and East Gondwana consisted of Madagascar, the Seychelles, India, Antarctica, Australia. About 128–130 Ma (Holmes and Watkins, 1992), early in the Cretaceous, conjoined Antarctica-Australia rifted from the smaller Madagascar-Seychelles-India fragment of East Gondwana. This separation occurred along the old eastern Ghats trend of eastern India. The rifting defined the present coastline configuration of eastern India. Uplift, associated with the rifting of conjoined Antarctica-Australia from India, oriented the drainage system of peninsular India so that major river drainage was from east to west (Cox, 1989; Kent, 1991). This drainage pattern persisted until the rifting of Madagascar from western India. Madagascar began rifting from western India about 90 million years ago, in the mid-Cretaceous. Madagascar rifting proceeded along the NNW–SSE-trending Dharwar basement. Madagascar drifted to the southwest leaving the western margin of India with the configuration of a long, relatively straight, rifted passive margin. As with the rifting of Antarctica-Australia from eastern India, the Madagascar rifting event was preceded by uplift that reoriented the fluvial systems on the subcontinent (Subramanian, 1987; Cox, 1989). The rivers, which once drained from east to west into the Arabian Sea, drain from west to east into the Bay of Bengal nowadays and comprise a dome-flank drainage system (Cox, 1989). The major, east-flowing rivers of the subcontinent, including the Mahanadi, Krishna, Godavari, Palar, and Cauvery, originated at this time. The drainage divide of peninsular India lies close to the west coast (Cox, 1989) and the drainage divide is between 40 and 160 km from the coastline along most of the length of western India. The proximity of the continental divide to the west coast of India greatly reduces the drainage area available to the small river systems draining into the Arabian Sea. Therefore, the clastic sediment influx to the offshore basins of western India had been small compared to that of the basins off the east coast, which received contributions from the large rivers draining the greater expanse of peninsular India.

The major rivers feeding into the western offshore areas of India are the Narmada and Tapti rivers, which flow west along the Narmada Graben, and the Sabarmati and Mahi rivers, which flow south out of the Aravalli Highlands and whose courses are controlled to a large degree by northwest-southeast (NW-SE)-trending faults associated with the Cambay Graben. These rivers drain limited areas of northern India and debouche at the head of the Gulf of Cambay. Thus, the main sediment supply for offshore western India is mainly focused at this one location. Sediment discharged there is channeled into the submarine Surat Depression. The Panna Depression lies between the Mumbai Platform and the coastline of India. The Panna Depression may receive some sediment from the rivers to the north, but the Vengurla Arch (Biswas and Singh, 1988), an east–west (E-W)-trending basement high, eventually blocks the southward transport of this sediment. The Bombay Platform touches the offshore extension of Vengurla Arch and forms a passage between itself and mainland India. Therefore, the Konkan-Kerala coast of India is effectively cut off from any sediment supply from the big rivers to the north. Similarly, the Mumbai Platform is isolated from clastic sediment discharged into its vicinity with its high relief relative to the surrounding troughs.

Following the rifting of Madagascar from India, fluvial sandstones of the Himatnagar formation covered most of northwestern India, during Late Cretaceous, including the area now occupied by the Cambay Graben (Akhtar and Aquil, 1984). These sands were derived from the long-lived Aravalli Highlands, which are still elevated today. The presently submerged, offshore Mumbai Platform was emergent, much as the huge Saurashtra Horst as it is today. The evidence for this comes from fragments of weathered gneiss basement recovered in wells drilled on the Bombay Platform. Such fragments are frequently recovered from the floors of grabens (Rao and Talukdar, 1980; Roychoudhury and Deshpande, 1982; Mitra et al., 1983). Also, fragments of subaerially deposited Deccan basalt have been encountered in wells drilled on the Bombay Platform. Furthermore, Paleogene sediments are seen to onlap the flanks of the Mumbai High (Rao and Talukdar, 1980; Roychoudhury and Deshpande, 1982).

The most significant tectonic event that affected western India was the initiation and eruption of the Deccan/Reunion hotspot about 66 million years ago (Courtillot et al., 1986; Duncan, 1990; Vandamme et al., 1991; Baksi and Farrar, 1991). Doming associated with the mantle plume changed stress in the Mumbai Platform area, which enhanced the already established horst-graben structure. The Mumbai High basement structure (a large normal fault block) was uplifted during this time, possibly by reactivation of older faults and new rifts were initiated in response to the mantle plume. The Cambay Graben began to subside at this time, along the old NE–SW Aravalli trend. Between marine magnetic anomalies 28 and 27, or about 63 million years ago, the Central Indian Ridge axis shifted to the east, most likely in response to weakened lithosphere in the vicinity of the mantle plume and this caused the separation of Seychelles from mainland India. The Deccan/Reunion mantle plume reached a depth at which decompression melting took place and one of the largest subaerial eruptions of flood basalt known quickly followed this. Within a span of a few million years (Duncan and Pyle, 1988; Jaeger et al., 1989; Hooper, 1990; Baksi and Farrar, 1991; Richards et al., 1991; Gallagher and Hawkesworth, 1992; Hill et al., 1992), an area of about $2 \times 10^6\, km^2$ was covered with numerous, nearly horizontal, basalt flows and these basalt flows are known as the Deccan Traps (Gombos et al., 1995).

The Bombay Platform experienced a period of subaerial erosion before further subsidence caused the platform to drown. This maybe spanned throughout the Paleocene and possibly part of the early Eocene (Rao and Talukdar, 1980). Deccan volcanics, as well as Precambrian granitic and gneissic crust, underwent further weathering during this time. Coarse clastics accumulated in topographic lows on the platform and draped against highs (horsts). These deposits occasionally form basal, clastic reservoirs. In a study on documenting the Cenozoic sedimentation history along the Western Continental Margin of India (WCMI) and its role in evolutionary geodynamics using new multichannel seismic (MCS) reflection profiles in the Mumbai offshore basin, Nair and Pandey (2018) reported that in Mumbai offshore basin, large volumes of sediment deposition are seen during Cenozoic period and it revealed the tectonic processes that led to the development of sedimentary basins during Cenozoic evolution.

4.3 Characteristics of petroleum inclusions

4.3.1 Types of inclusions

Petroleum fluids can range from heavy oils with API gravity 10 through heavy, medium, and light oil (API gravity varies from below 10–60) and to "dry" gases that have no associated liquid hydrocarbons. Usually, oil-rich inclusions, gas-rich inclusions, and oil-gas inclusions are observed where the distinction is based on the volume ratio of liquid to gas in the inclusions. Petroleum can also occur as multiphase inclusions. In addition to hydrocarbon liquid and vapor, these inclusions may contain hydrocarbon solid phase too. Black or brown solids maybe present within the hydrocarbon liquid phase (see Chapter 1). Hydrocarbon solids in oils can form in several ways: precipitation on cooling, precipitation during the addition of gas, and as a cracking by-product, or during the transformation of oils to gas at high levels of thermal maturity.

4.3.2 HCFI fluorescence

The petroleum fluids entrapped as HCFI fluoresce because of the presence of polycyclic aromatic hydrocarbons. However, the complex chemical composition of the entrapped oils ultimately determines the precise absorption and fluorescence characteristics of HCFI. The presence of aromatic molecules with more than 10 carbon atoms in the petroleum only will be given petroleum fluoresce visible at wavelengths to the human eye under a typical fluorescence microscope (Burruss, 1991; Stasiuk and Snowdon,

1997). These molecules are not present in gas inclusions ($CH_4 > 95$ mol%). Natural gas inclusions that are one phase at room temperature commonly do not fluoresce and as the fraction of liquid phase hydrocarbons increase, the intensity of fluorescence increases. Typical, two-phase, liquid-rich petroleum inclusions can fluoresce in the full range of colors in the visible spectrum, although colors can be altered somewhat by the internal design of the microscope (Burruss, 1991; George et al., 2002; Oxtoby, 2002). The basic color of the fluorescence is due to the concentration and composition of aromatic hydrocarbons in the trapped oil (Oxtoby, 2002; Stasiuk and Snowdon, 1997). There is a general correlation of fluorescence emission color with the density of the liquid petroleum phase. Low-density (high API gravity) fluid fluoresce at short wavelengths (blue) and as density increases the fluorescence emission shifts to longer wavelengths in the orange and red (see the section fluorescence of crude oils/HCFIs in Chapter 2). As mentioned, gas inclusions do not fluoresce and in some gas-rich inclusions, the fluorescence is difficult to detect because the color (commonly dark blue) and the low intensity of the emission make these inclusions hard to distinguish from the background. Such inclusions typically occur in relatively high-temperature environments (Evans and Battle, 1999; Levine et al., 1991; O'Reilly and Parnell, 1999).

HCFIs may easily be mistaken for aqueous-bearing fluid inclusions during optical petrography in the transmitted light mode because many HCFIs are colorless or slightly brown with a vapor bubble and resemble aqueous-dominated fluid inclusions. Excitation by UV illumination is the routine method for identifying HCFI in thin section and discriminating it from aqueous inclusions. Aqueous fluid inclusions generally do not fluoresce.

The first work of its kind to describe HCFI fluorescence was done by Murray in 1957, but later during the 1980s, the increased availability of fluorescence microscopes empowered the studying of HCFI as an active research using fluorescence methods (Burruss et al., 1980; Burruss, 1981). The linking of crude oil fluorescent properties with that of fluid inclusion was not done in a widespread manner due to the following reasons (George et al., 2001, 2002; Ryder, 2005). The prime reason is that the use of 'dead' oils where the light alkane and gaseous fractions lost from the oils before analysis. When one considers HCFI, because they are sealed, should retain a significant light gaseous fraction and therefore the oil is termed "live." Therefore, a comparison of fluid inclusions with that of bulk oils is not attempted previously. Another consideration for HCFI studies is that the reference oils should be sourced from the same locality, to avoid the wide fluorescence variances between oils from different sources.

4.4 Mumbai basin

Mumbai offshore basin (Fig. 4.1) is on the west coast of India (between Saurashtra basin in NNW and Kerala-Konkan in the south) which contributes around 55% of domestic crude oil and 70% of the gas production for India. The basin falls under the category I, which implies that the basin has proven commercial productivity. The Mumbai High field is the largest in the basin. It covers an area of about 116,000 km^2 from coast to 200 m isobath. The Mumbai offshore is demarcated from south of Saurashtra arch up to Vengurla arch. The basin accumulated Cenozoic sedimentary succession. The age of the basin ranges from late Cretaceous to Holocene with thick sedimentary fill ranging from 1100 to 5000 m. While vugs and molds represent the reservoir facies, the major hydrocarbon accumulation is controlled by a combination of stratigraphic and structural plays, channel and fracture porosity also being present intermittently. The Mumbai Offshore Basin is the most prolific of the entire hydrocarbon-bearing provinces in India. Continued and sustained exploratory and development efforts in Mumbai Offshore Basin since the last 4 decades have shown remarkable results and led to the conversion of nearly 50% of the prognosticated resource of 9190 MMT in to in-place volume of hydrocarbons (ndrdgh.gov.in, Biswas, 1982).

4.4.1 Stratigraphy of Mumbai offshore basin

The Mumbai offshore shelf consists of Surat depression, Mumbai platform, Heera-Bassein block, and Ratnagiri block. The Cenozoic sedimentary succession in the basin has been divided into seven chronostratigraphic sequences: Paleocene-Early Eocene, Middle and Upper Eocene, Lower Oligocene, Upper Oligocene, Lower Miocene, Middle Miocene, and Middle Miocene-Holocene. These were phases of clastic deposition in the basin. The oldest was in the Late Paleocene-Early Eocene. A major part of these clastic sediments was deposited in the restricted shelf. The second phase of clastic supply was in Late Oligocene- Middle Miocene in the Shelf margin Basin, which is in the Kori-Comorin Depression. The last phase of clastic supply in the middle Late Miocene Holocene times covered the whole basin. The shelf carbonate deposition was divided into two phases; the first phase covering Middle Eocene to Early Oligocene covers the entire platform area except for Mumbai High. The second phase covering Late Oligocene to Middle Miocene covers Mumbai High. The lithostratigraphy of Mumbai Offshore Basin along with Kerala-Konkan basin is shown in Fig. 4.2.

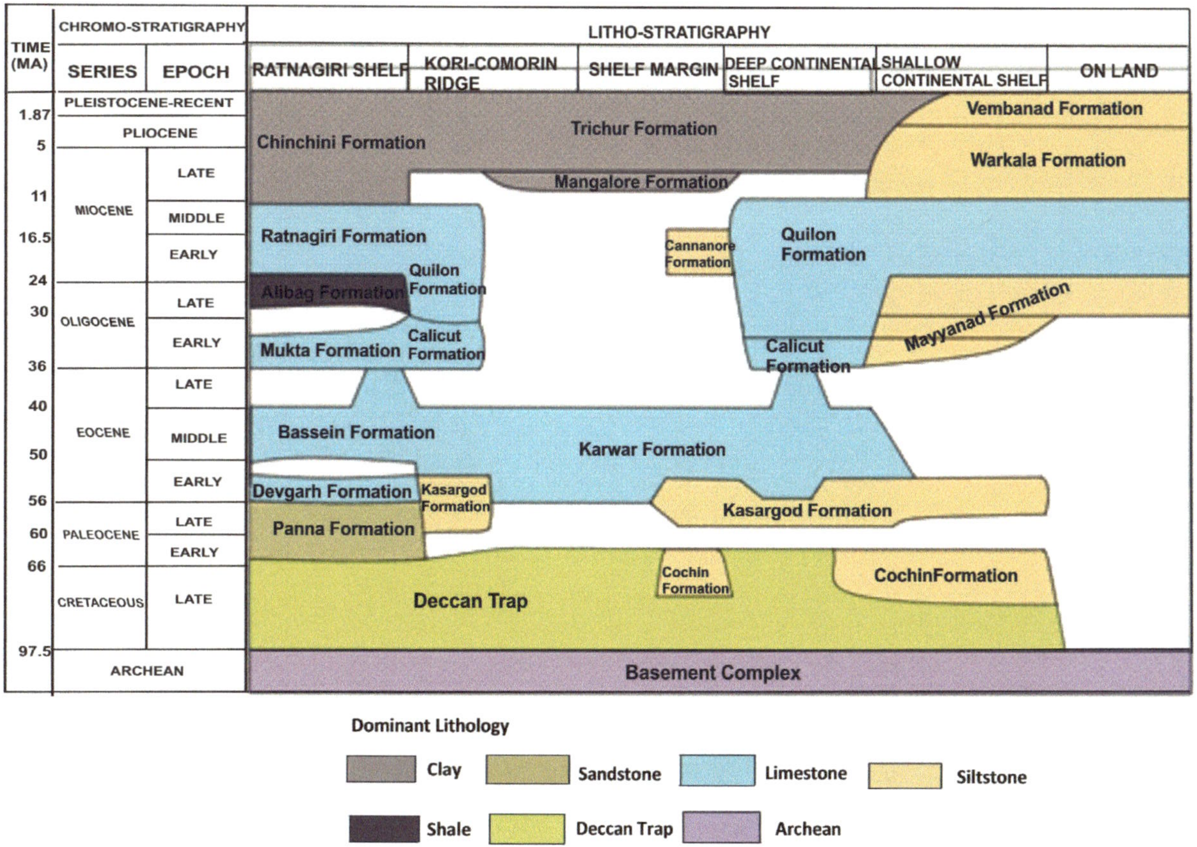

Fig. 4.2 Lithostratigraphy of Kerala-Konkan and Mumbai offshore basins (after RB Mathur et al., 1993).

Stratigraphic description of Mumbai offshore basin is as follows:

Panna Formation: The formation is spread over the entire Mumbai Offshore Basin excepting the paleo-highs and is composed of sandstone, siltstone, and clay stone at the bottom, overlain by a section of coal-shale alternation in the middle and succeeded by shale unit at the top. It is a source cum reservoir horizon due to intercalations of sandstone-siltstone-shale combination.

Bassein Formation: This formation consists of wackestone (mud-supported limestone with more than 10% carbonate grains).

Mukta Formation: This Formation mostly consists of argillaceous limestone/muddy limestone underlain by a 15–20-m thick zone characterized by radioactive shales.

Panvel Formation: It consists dominantly of limestone interspersed with a few thin shale layers. The formation has been assigned Upper Oligocene to Basal Miocene age.

Ratnagiri Formation: This Formation consists dominantly of biomicritic limestone occasionally interspersed with thin layers of gray, fissile shales. Lower-Middle Miocene age has been assigned to this.

Chinchini Formation: The Formation consists of sticky clays with frequent intercalations of shell debris. Upper Miocene-Recent age has been assigned to this formation.

4.4.2 Petroleum geology of Mumbai offshore basin

Mumbai offshore basin, a divergent passive continental margin basin, is located on the continental shelf off the west coast of India. The basin was formed due to extensional tectonics at the time of rifting of the Indian plate from Madagascar during the Late Jurassic-Early Cretaceous period. Large-scale volcanic eruptions followed this episode that covered most of the basin. As the rifting continued, the immature sediments deposited at the toe of faults as alluvial fans, filled the initial morphotectonic depressions during Paleocene. This was followed by the first marine incursion towards the close of the Paleocene and the beginning of the early Eocene. Thus, the early Eocene marks a widespread transgression. Sediments were deposited in deltaic to restricted marine to shallow marine environments. Sedimentation during this period caused some adjustments in the basin. The early Oligocene transgression covered most parts of the basinal area and inundated parts of Mumbai high. A major unconformity is noted at the top of the lower Oligocene. Sea level rise during the early Miocene submerged large areas of the basin and terminated the Oligocene delta progradation. The middle Miocene transgression marks the last phase of the widespread carbonate sedimentation in the Mumbai high–Deep Continental Shelf (DCS) area (Basu et al., 1982; Zutshi et al., 1993). The basin has NW-SE-trending horst-graben geometry. The grabens are bounded by normal faults, and the horsts/ridges are dissected by NE–SW-trending cross faults. Based on its structural configuration and its nature, as well as the type of sediment fill, the basin is divided into six tectonic blocks: Tapti-Daman, Diu, Heera-Panna-Bassein, Mumbai high-DCS, Ratnagiri, and Shelf Margin blocks. The main Mumbai high block is surrounded by three depressions: Surat depression (Daman, Purna, and Navsari lows) and its southward extension through Mahim depression in the east, Saurashtra low in the northwest, and Southern paleosink and Murud depression in the southwest. The Shelf Margin block bounded on its west by Kori-Comorin ridge and east by the Paleogene hinge and its northern part includes Saurashtra offshore (Fig. 4.1). Surat depression and its southward extension through Mahim depression to Vijaydurg depression in Ratnagiri block are the prime depocenter of the clastic sediments of early Eocene to Paleocene age. Murud depression and Saurashtra low had relatively more open marine environment due to minor shielding provided by the west basement arch compared to Surat-Mahim depression (Fig. 4.3). The main reservoir rocks in the basin are the limestones ranging in age from Eocene to middle Miocene. The clastic sequence of Paleogene also hosts the hydrocarbons. The extensive post-Miocene shale acts as a regional cap rock in the basin. The local shale interbeds within limestones act as a local cap rock for different pay zones. However, in Ratnagiri block, compact and tight limestones may also act as cap rocks for hydrocarbon accumulations in fractured limestone reservoirs.

Mumbai Offshore Basin, a pericratonic rift basin in the western continental shelf of India, covers about 148,000 km^2 from coast to 200 m isobath. Several large oil and gas fields have been discovered in this basin, and the presence of hydrocarbons has been established in the multiple pay zones belonging to L-III limestone reservoir of Miocene age (only in Mumbai high), Mukta (early Oligocene), Bassein (middle Eocene), Panna (Paleocene to early Eocene) reservoirs, and Daman (early Miocene-late Oligocene) and Mahuva (early Oligocene) formations in Tapti Daman block. Studies indicate that Paleocene to early Eocene Panna Formation has good to excellent source-rock characteristics in the basinal part of every block in the entire basin. The middle of the Panna Formation attained the maturity of 0.6% VRo equivalent at about 12–20 Ma. The Panna source rocks have

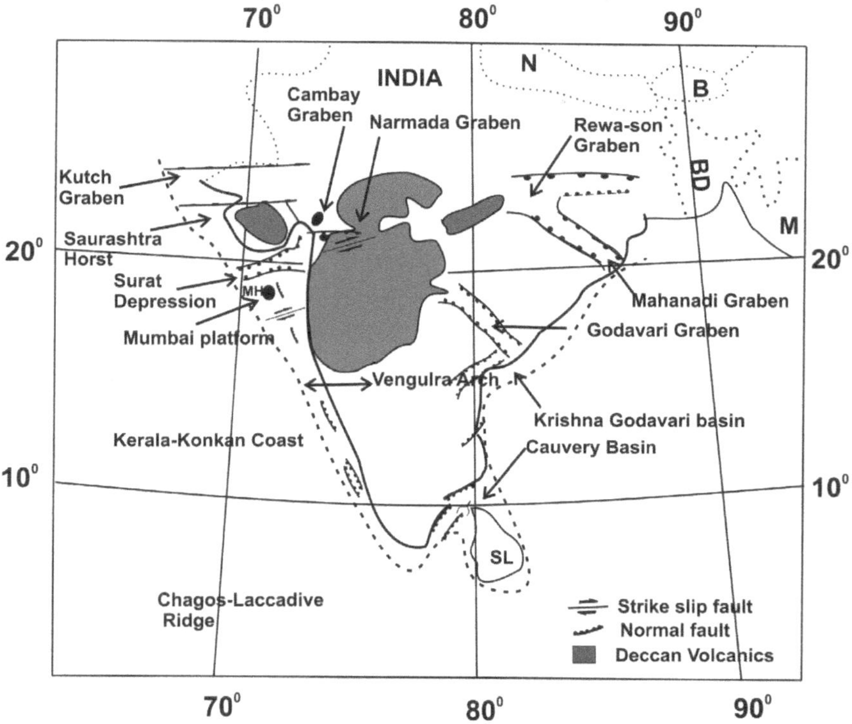

Fig. 4.3 Generalized tectonic map of India. Locations of major hydrocarbon occurrences are indicated by oval-shaped symbols. AV = Ankleswar field; BH = Bombay High field; C = Cambay field. A = Afghanistan; B = Bhutan; BD = Bangladesh; M = Myanmar (Burma); N = Nepal; PRC = People's Republic of China; SL = Sri Lanka. *Modified after Gombos, A.M., Powell, W.G., Norton, I.O., 1995. The tectonic evolution of Western India and its impact on hydrocarbon occurrences—an overview. Sediment. Geol., 96, 119–129.*

predominantly Type III kerogen in lower layers and Type II and III mixed kerogen in upper layers. However, few source-rock layers in the Mahim graben and Deep Continental Shelf (DCS) area contain Type I kerogen. Based on the variations in organofacies, the oils have been placed into two groups. The first group oils, having only terrestrial source input, correlate with the source-rock extracts of lower layers, whereas the second group oils, with terrestrial and marine mixed input, correlate with the source-rock extracts of upper layers of the Panna Formation. Based on the reservoir and source combinations, major petroleum systems in the basin are Panna-L-III, Panna-Mukta, Panna-Bassein, and Panna-Panna. Mumbai offshore basin accounts for nearly two-thirds of the annual petroleum production of India.

The mature source rocks are present in the lower Eocene-Paleocene Panna Formation. Further, marginally mature potential source rocks within Oligocene in Tapti-Daman area and within Neogene in DCS and deeper part of the basin also exist. Hydrocarbons have been discovered in multiple reservoirs in this basin, ranging from the fractured basement to the middle Miocene. The Mumbai offshore basin has three major depressions: Surat and its southward extension to Ratnagiri in the east, Saurashtra low in the northwest, and Murud and Rajpur lows in the southwest. Due to the multiplicity of depressions, source rocks, and reservoirs, the oil-source genetic relationship is a challenge. Understanding of genetic correlation among oils and source rocks is a prerequisite to model hydrocarbon generation, expulsion, and entrapment (Goswami et al., 2007).

The Mumbai offshore basin is the most prolific hydrocarbon-bearing basin among all Indian sedimentary basins. The first well on the Mumbai highway structure was drilled in 1974 and the oil was struck in limestone facies (L-II and L-III reservoirs of Miocene age). Surat depression and Murud depression are the two major depocenters of source rocks. Previously, it was believed that most of the oil generated in the two major source rock centers migrated along the unconformities to various structures and traps. Though, recent research shows that migration was short and localized (from adjacent local depressions or graben along the bounding faults of the structures). Middle Eocene and Middle Miocene limestones form major reservoirs in most fields in Mumbai. Secondary porosity development below unconformities plays a major role in these reservoirs. Mostly, transgressive shale or a tight limestone forms the effective cap rocks. Extensional faulting leading to differentiation of the shelf in narrow horsts and grabens together with thermal events created ideal habitat for petroleum generation, migration, and entrapment. Thermal regimes during episodic rift cycles helped cooking of sediments to maturity in this basin (Biswas, 1987). Thermo-tectonic adjustment following mantle updoming aided structure formation while subsequent tectonic pulses triggered migration. Mitra et al. (1983) considered that generation of Mumbai High oil took place in Paleogene.

The source rock potential of the Mumbai offshore basin

The clastic sediments in the lower Eocene to Paleocene sedimentary sequences (Panna Formation) are the principal source rocks across the basin. The thickness of the source rock varies from 30 to 1000 m depending on location. The excellent source rocks of restricted marine to lagoonal deposits within the Panna Formation in the Central graben and adjoining area are the prime source of hydrocarbon accumulation in the basin. In the Mahim graben, a 400-m-thick sequence in the Panna Formation contains very good/excellent oil-prone effective source-rock facies, which account for the commercial petroleum reservoirs within Bassein, Mukta and Heera formations in the east of Panna and Bassein fields (average TOC = 2.3–15.4%; average S2 = 3.5–50.1 mg HC/g rock; average HI = 112–277 mg HC/g TOC). Organic-rich mature source-rock sequences in the Panna Formation occur in depressions across the DCS area and west-southwest of Mumbai high (average TOC = 1.5%–5.6%; average S2 = 2.6–11.6 mg HC/g rock; average HI = 94–270 mg HC/g TOC). Source-rock data from the deepest exploratory well in the Vijaydurg graben of Ratna depression show good, mature source-rock section in the lowermost unit of the Panna Formation and thin coal and coaly shale layers with very good source-rock quality at the top of Panna Formation. The sedimentary column in Shelf Margin areas is dominated by clastics, except in the middle Eocene, which has carbonates. The source-rock potential of the Paleogene sediments is moderate, but some good organic carbon-rich source-rock layers are present in Neogene sediments (Goswami et al., 2007).

Paleocene-Early Eocene sequence is characterized by the richest concentration of organic carbon (TOC) within the Tertiary succession of the Mumbai basins. The Hydrogen Index (HI) order of 150–250 in the Heera-Bassein of the northern Ratnagiri areas indicates high TOC. Good concentration of TOC has been observed in the sediments of the *Middle to late Eocene* sequence in the Surat depression and in a narrow linear area in the western part of the Mumbai High platform. This sequence is matured adequately and the Time–Temperature Index (TTI) values are varying between 10 and 100. The sediments of *Oligocene* sequence in the Murud depression shows good to moderate TOC concentration. Studies indicate that the organic matter in this sequence is of Type III and the sediments are in the early stage of oil generation. The sediments of the *Early to Middle Miocene* sequence contain a good concentration of TOC, which is more than 1.5% in the Kori-Comorin depression. There is a moderate concentration of TOC in the central part of the Surat depression. The sediments in these areas are in the early stage of oil generation.

Reservoir rocks in the Mumbai offshore basin

In the Bombay-High area, oil and gas are produced from the fractured basement through middle Miocene reservoirs, with the most prolific being the platform carbonates such as the lower Miocene L-III limestone. Other significant reservoirs are the middle Miocene L-II limestone, L-I limestone, S-sandstone, middle Eocene Bassein limestone, and lower Eocene Panna sandstone. Approximately 70% of the reservoirs in the Bombay-High area (Fig. 21) are carbonates. The primarily intergranular porosities range from less than 10% to greater than 30%, with an average between 15% and 20%. Permeabilities range from 0.01 millidarcies (mD) to 1000 mD with averages in the range of 100–250 mD. The Surat Depression reservoirs are primarily Eocene, Oligocene, and Miocene sandstones and siltstones. Porosities range from less than 10% to more than 30% and permeabilities range from 10 to 8000 mD. The Cambay Graben reservoirs include fractured Cretaceous basement rocks; Paleocene basalts, conglomerates, and sandstones; Eocene sandstones, siltstones, shales, and coals; Oligocene siltstone and sandstone; and Miocene sandstones and siltstones. In the Cambay Graben, approximately 70% of the reservoirs are Eocene sandstones and siltstones. Porosities are generally good, ranging from less than 10% to greater than 40%, with an average between 20% and 30%. Permeabilities range from 0.1 to 20,000 mD, with averages of 100–200 mD. The Kutch Basin potential reservoir rocks include the massive sandstones of the Upper Jurassic Jhuran Formation. The Upper Cretaceous Bhuj and Kori limestones may also have good reservoir qualities. Middle Eocene and Oligocene platform limestones analogous to those of the Bombay High and Pliocene Kanakawati sandstones are the most prospective reservoirs. Porosity and permeability averages are as high as 25% and 30 mD, respectively, were reported.

Reservoir facies and entrapment

During Cenozoic sedimentation, the Mumbai offshore basin experienced four major unconformities in addition to a few minor and localized ones. These unconformities have helped in the development of reservoir characteristics and the entrapment condition is mostly of combination (structural, stratigraphic/hydrodynamic) type.

The main reservoir facies in the *Paleocene-Early Eocene* sequence are thin sandstone and siltstone bands within basal clastics.

The extensive carbonate deposits of the *Middle to Late Eocene* sequence form the most important reservoir rocks. The Bassein "B" limestone is the most important reservoir rock, characterized by moderate to good porosity development, mostly of secondary origin. The gross thickness of this limestone on an average is about 300 m.

Limestone sequence of *Early to Middle Miocene* age forms an important reservoir over Mumbai High structure. The well-known L-III and L-II reservoir of Mumbai High belong to this sequence. The commercial hydrocarbon pools in this sequence are also present in the Mumbai platform, the Ratnagiri shelf and in the North Tapti structure of the Surat Basin. The commercial hydrocarbon pool in this sequence is also present in the Mumbai platform, the Ratnagiri shelf and in the North Tapti structure of the Surat Basin. The limestone is described as algal and foraminiferal packstone and wackestone deposited in a lagoonal environment. Porosity in the limestone is mostly intergranular, moldic, and vugular. Rare channel and fracture porosities are also encountered. Mumbai High field is a typical example of an ideal hydrocarbon trap in the western offshore. The Reservoir is a multilayered porous limestone separated by thin shale layers. The overlying shales are the cap rocks. The Late Miocene and younger sediments are immature and do not have any hydrocarbon generation potential and effective cap rock (Directorate General of Hydrocarbons (DGH), Ministry of Petroleum and Natural Gas, Govt. of India, n.d.).

Traps and seals

The most prolific traps are those located on persistent paleo-highs of the Bombay-High area and horst blocks within the Cambay Graben. These structures developed in Late Cretaceous or early Paleocene time in association with rifting events (Sankaran et al., 1995) were reactivated numerous times. Anticlinal structures are also productive in the area of the Bombay High, Surat Depression, and Cambay Graben (Fig. 24). Anticlines in the Kutch Basin are breached except offshore to the west where they may still hold potential. There are also untested anticlinal features on the outer shelf. Stratigraphic traps and pinch-outs are productive in the Cambay Graben and have recently received more attention. These traps may also have potential in the Kutch area west of the Median High.

In the Bombay offshore and Kutch areas, the most likely seals are an extensive series of thick middle to upper Miocene shales. In the Cambay Graben, interbedded Paleocene through middle Miocene shales provide seals for the various reservoirs. Generation and expulsion adjacent to the Bombay-High area may have occurred as early as the middle Miocene and continued, or began again, in the Pliocene (Basu et al., 1980). Burial history data indicate peak generation most likely occurred during the late Miocene and early Pliocene (Samanta, 2000; Biswas et al., 1994; Banerjee et al., 2002). Eocene to middle Miocene sedimentary sequences in the shelf-margin area west of the Bombay-High area show thermal-alteration-index (TAI) values >2.5 (Rawat et al., 1987), and the onset of hydrocarbon generation occurred at approximately 2.5 Ma. Generation in the Kutch area may have begun as early as Paleocene for Jurassic source rocks, and in the late Miocene and Pliocene for Tertiary source rocks (Wandrey, 2004).

Hydrocarbon occurrences, quality, and source rock characteristics

Oils mainly occur in the limestone horizons of lower Miocene age (L-III) in major hydrocarbon fields; viz., Mumbai High, Panna, S. Bassein, Heera and Ratna, Mahuva and Daman pay (Oligocene) in Tapti area, Bassein pay (middle Eocene to upper Eocene) in Panna-Bassein-Heera and Ratna areas and Ratna pay (middle Eocene to lower Eocene) in Ratnagiri area. With the recent oil occurrences in the sands of Panna Formation in Vasai east, the Panna pay is emerging as a commercial pay zone. Few oils also occur in clastic/fractured basement reservoirs and middle Miocene L-II and S1 pays. The majority of these oils show moderate API gravity (25–40°), high pour point (27–33 °C), significant wax (7%–20%), and low sulfur (0.1%–0.3%) contents. These oils are predominantly aliphatic, having high saturate/aromatic ratio (>1.5) and saturate content (>40%). Only a few oils and mostly condensates were found in the Tapti Daman block (Goswami et al., 2007).

Based on the reservoirs and the source combinations, the various petroleum systems in the basin are the Panna-Daman, Panna-Mahuva (only in Tapti Daman block) and Panna-L-III, Panna-Mukta, Panna-Bassein, and Panna-Panna. Oils of the Mumbai Offshore Basin can be broadly categorized into two groups based on their biomarkers distribution and stable isotopic composition. Group I oils containing only terrestrial organic matter from fluvial/fluvio-deltaic depositional environment generally correlate with the lower Eocene-Paleocene source-rock extracts from the wells located in the peripheral area of the lows in each block. Group II oils, which contain mixed type organic matter deposited in marginal marine conditions, correlate with the upper layers of the lower Eocene Paleocene source sequences in the wells located in the central parts of the lows in each block. Lower Eocene-Paleocene sediments have very good source-rock characteristics and good hydrocarbon generation potential throughout the basin, but the quality and quantity of the organic matter deteriorated towards the peripheral area from the basinal lows. The organic matter is dominated by input from terrigenous higher land plants. However, two potential source sequences are identified in the lower Eocene-Paleocene sediments. Specific biomarkers, particularly the relative contribution from resinous plants, flowering plants, carbon isotopic composition and marine organic matter diagnostic biomarkers indicate that these sediments exhibit intervals with different organofacies. All oils in the field are at approximately the same level of thermal maturity and were expelled from sources that still retain significant generative potential. These oils are generated locally and have not undergone long-distance migration. Maturity modeling of vitrinite indicates that in general the sedimentary sequences of the Panna Formation started oil expulsion 18–30 Ma and peak oil expulsion 12 Ma and continues to the present day (Goswami et al., 2007).

4.4.3 HCFIs from Mumbai offshore basin

From the RV-1 well (an exploratory well sampled for fluid inclusion studies) of Mumbai offshore basin fluid inclusions (Figs. 4.4–4.6) have been identified from five different Formations.

Ratnagiri, Panvel, Mukta, Bassein, and Panna are the major Formations that have been studied for fluid inclusions from Mumbai offshore basin, India. Details of the age, depth and lithology of these formations are given in Table 4.1. Ratnagiri Formation at a depth of 750–1250 m with limestone, clay, and shale lithologies and age of the Formation is early to middle Miocene. At Panvel Formation, HCFIs are observed at a depth of 1345–1350 m with limestone, shale lithology with late Oligocene age. Again, HCFIs were observed at the Mukta Formation with Early Oligocene age at a depth of 1355–1360 m with limestone lithology. Most of the HCFIs in RV-1 well

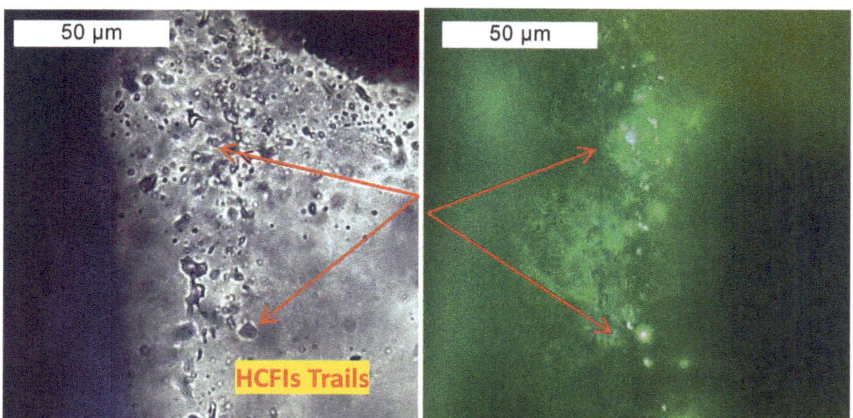

Fig. 4.4 Fluorescence of HCFI trails observed in Panna formation, RV-1 well, Mumbai offshore basin at a depth 3495–3500 m depth with sandy siltstone and shale lithology. Bright field (A) fluorescent (B) images.

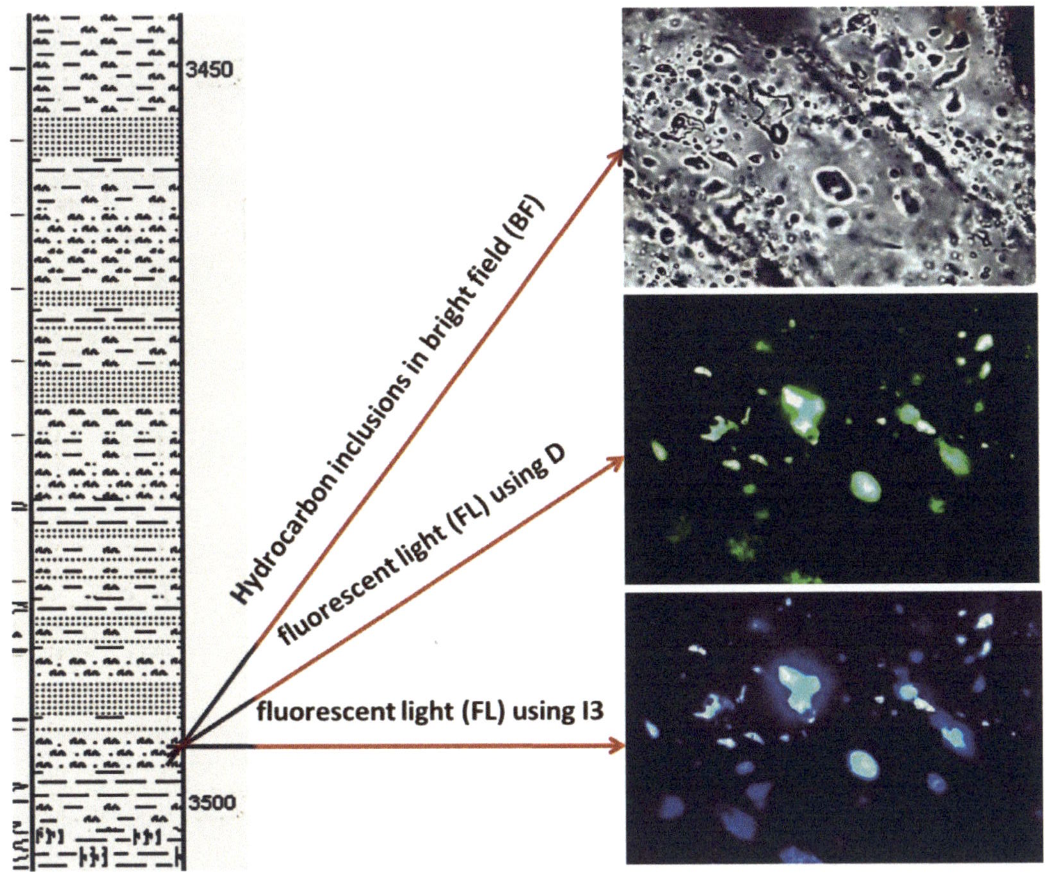

Fig. 4.5 Fluorescing hydrocarbon fluid Inclusions (HCFIs), Litholog (Panna formation, offshore to Ratnagiri, Mumbai Basin). D and I3 are two different filters used to observe fluorescence in Leica Microscope.

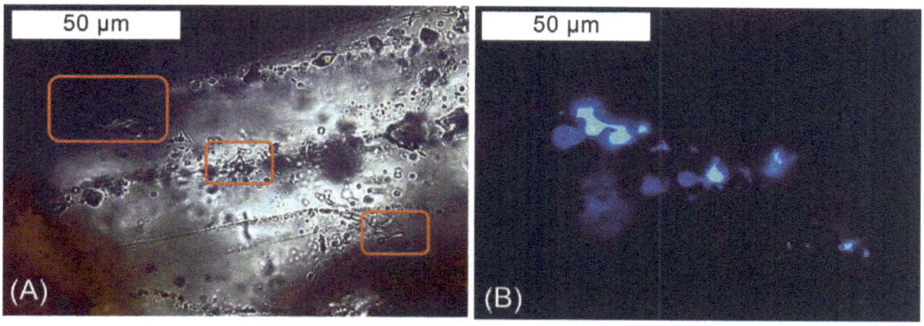

Fig. 4.6 Fluorescence of HCFIs in Panna Formation, Mumbai offshore basin at a depth 3495–3500 m depth with sandy siltstone and shale lithology. Bright field (A) fluorescent (B) images.

of Mumbai offshore were observed at the Panna Formation with different depths ranging from 3055 to 3690 m with different lithologies such as sandstone, sandy siltstone and silty shale. The different depths at which HCFIs observed at Panna Formation are 3055–3070, 3270–3275, 3345–3350, 3395–3400, 3445–3470, 3495–3500, and 3675–3690 m. Further analysis on the characteristics of HCFIs from Mumbai offshore basin is given in Chapters 5 and 6. HCFIs were not detected at Chinchini, Ratnagiri, and Bassein Formations.

Detailed analysis on the HCFIs (fluorescence and Raman) showed that the oils in HCFIs in Panna formation are different from those found in Mukta and Panvel both in terms of API gravity, chemical constituents, etc. The characteristics of HCFIs in Mumbai offshore basin, India were detailed in the coming chapters of this book.

The size, shape, and phases in the inclusions depend on the cooling rate of the entrapped liquid. In the case of immiscible compounds (alkanes, alkenes, aromatics, organic solvents, etc.), the inclusion formation efficiency increases in accordance with the interface size, which is controlled by the emulsion rate of the solution. The presence of oil droplets in the aqueous phase makes the formation of hydrocarbon inclusions easier (Pironon, 1990).

The Panna formation has its maximum thickness in the Ratnagiri area, is highly fractured, and cracks could be seen even at the granular levels. In the quartz grains of the Panna formation, the predominance of secondary and pseudo-secondary fluid inclusions are seen in the sandstone and siltstone lithology, largely seen in the healed postcrystallization fractures. Secondary inclusions are recognized by their occurrence in trails or cluster that often

Table 4.1 Details of age, lithology and depth of formations in Mumbai offshore basin where fluid inclusions are identified and studied.

Sl. no.	Formation	Age	Depth (m)	Lithology	Major fluids present
1	Ratnagiri	Early to middle Miocene	750–1250	Limestone, clay, shale	Non-HCFIs
2	Panvel	Late Oligocene	1250–1350	Limestone, shale	HCFIs (1345–1350 m)
3	Mukta	Early Oligocene	1350–1500	Limestone, shale	HCFIs (1355–1360 m)
4	Bassein	Middle to late Eocene	1500–2200	Limestone, shale, siltstone	Non-HCFIs
5	Panna	Paleocene to early Eocene	2200–3700	Shale, siltstone, sandstone, sandy siltstone	HCFIs (at different depths between 3055 and 3690 m)

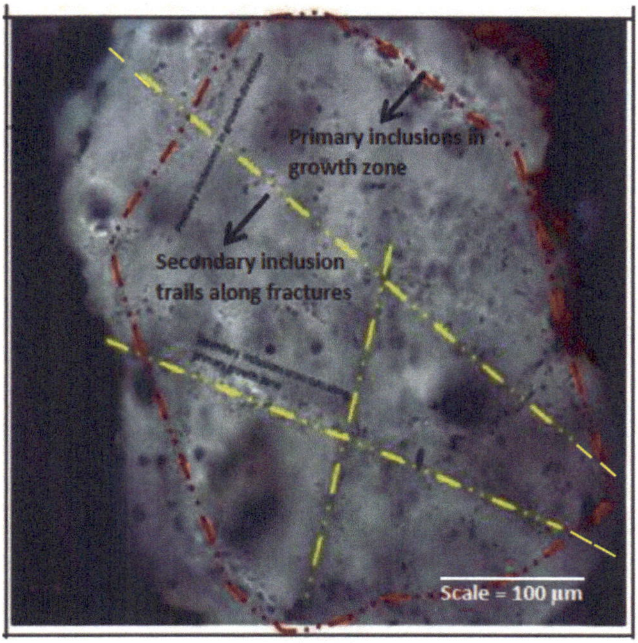

Fig. 4.7 Characteristics appearance of primary fluid inclusions along the authigenic growth zone and secondary fluid trails along the fractures in quartz (sandstone and siltstone lithology) at depth 3345–3500 m in Mumbai offshore basin, India.

cut across grain boundaries (Fig. 4.7). FIs shapes vary from rounded, subrounded to anhedral in nature and the size of FIs range from 2 to 20 μm. Gas bubbles show Pseudo-Brownian movement in most of the two-phase FIs and they appear as dark globules. Two-phase fluid inclusions are abundant than monophase fluid inclusions in quartz and feldspars. The secondary fluid inclusions were showing cross-cutting relationship and are seen confined in fractures within the host crystals revealing the conduits of petroleum migration in the Panna formation.

Microthermometric observation in the fluid inclusions in Fig. 4.8. shows first melting range from −76.2 to −76.9 °C, last melts varies between −27.3 and −21.4 °C and the Th of inclusions in quartz is around 114 °C and the state of transformation is from liquid + vapor to liquid. The salinity value of the aqueous inclusions varies from 23.3 to 27.05 eq.wt% NaCl.

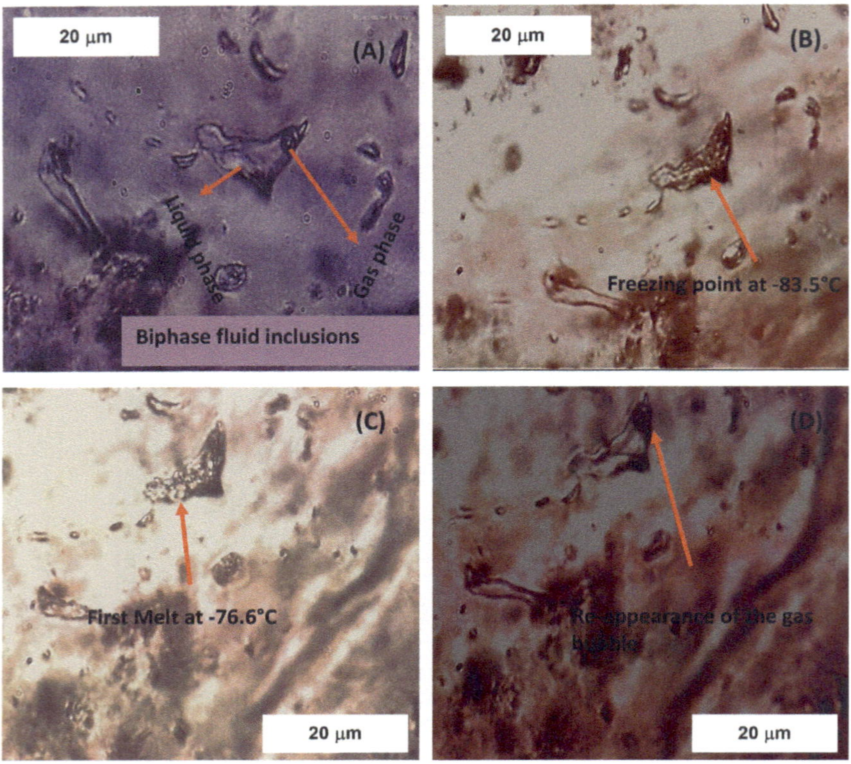

Fig. 4.8 Microthermometric observation shows two-phase (liquid + gas) fluid inclusions in the siltstone-shale. (A) Biphase fluid inclusions appear at room temperature (~23–27 °C), (B) freezing point at −83.5 °C, (C) T_{FM} observed at −76.6 °C, and (D) reappearance of bubble and FIs at −21.9 °C (T_{LM}).

4.5 Kerala-Konkan basin

Kerala basin to the south of Tellichery arch (an offshore basement high) is a coast parallel longitudinal basin associated with a passive continental margin which received sediments from the erosion and deposition of the Western Ghats (Fig. 4.1). NE–SW-trending Vengurla Arch separates the Mumbai offshore Basin from the Kerala-Konkan Basin to its south. Basin fill is characteristically terrigenous (those derived from the erosion of rocks from the terrain/land) marine and shallow water marine carbonates. The total thickness of sediments varies from 200 to 3200 m reaching a maximum of 5500 m at some places (Campanilen et al., 2008). There is considerable potential for hydrocarbon reserve in the basin. The Kerala-Konkan basin has a sediment record of 65 Ma. The basin represents sag basin development on the stretched pre-Deccan continental basement during the rifting of India from Madagascar (Singh and Lal, 1993).

The basin evolved through an early rift and postrift phases and contains more than 5 km thickness of Cretaceous to Recent sediments (ndrdgh.gov.in). Postrift mature sediments with sufficient organic carbon content are present in the basin. Drilling results and adsorbed gas anomalies confirm the generation of hydrocarbons in the basin. This basin covers an area of around 580,000 km^2. This basin comes under category III. About 15 exploratory wells have been drilled so far but none of them is producing. No commercial accumulation could be located to date in the Kerala-Konkan Basin even though it is well explored. The presence of noncommercial hydrocarbons is known from these Basins but is yet to open up. In Kerala-Konkan Basin, gas shows were reported from the Tertiary sequence mostly from Miocene and Eocene formations. Though various data sets indicate the presence of sub-trappean Mesozoic Basin, its hydrocarbon potential is yet to be confirmed (Dwivedi, 2016). Gombos et al. (1995) reported that a lack of deep burial and immaturity of source rocks maybe the reasons that the Konkan-Kerala coast is not productive.

Considering the depocenter thickness of almost 5 km and assuming the geothermal gradient of 2.5–3 °C/100 m any of these rocks (sandstone, clay, and shale facies along with carbonate facies) could be potential source rocks (Clift et al., 2001). Paleocene sandstone in boreholes contains a considerable amount of primary intergranular porosity. Eocene limestone displays secondary porosity in the form of solution vugs and channels. These sandstones and limestone facies could be potential reservoirs in Kerala-Konkan basin. Abundant mud-rock/shale could act as potential seals/cap rocks and the terrain is structurally as well as stratigraphically favorable for the entrapment of hydrocarbons.

4.5.1 Lithostratigraphy of Kerala-Konkan basin

Neogene and Quaternary sand, clay, lignite, and limestone overlying the Archaean basement form the outcrops in the coastal belt stretching from Cochin in the north to Trivandrum in the south (Fig. 4.1). The rock units that are identified in Kerala-Konkan Basin are:

Basement complex: Archaean crystalline rocks of the peninsular shield composed of granites, gneisses, and charnockites are exposed along the eastern margin in the Konkan and Malabar coastal areas. The outcrops are quite close to the coast near Calicut but are up to 30–40 km from the coast elsewhere. The sedimentary rocks on land rest over the granitic basement as seen in a few boreholes.

Deccan trap: Deccan trap lies over the Archean basement complex that consists of Archean crystalline rocks of granites, gneisses, and charnockites.

Cochin formation: Cochin formation consists of sandstones, siltstones, shales, and claystones with almost 919 m thick sediments, *Kasargod formation* consists of argillaceous limestones in the lower part and sandstones and shales in the upper part with fresh water sediments on the shallow shelf and by marginal marine and shallow inner neritic deposits on the deeper continental shelf, *Karwar formation* consists of carbonate facies comprising gray, light gray, and brownish gray limestone with occasional thin shale bands, *Calicut formation* consist of limestones sandwiched between the Eocene and the Early Oligocene unconformities. Sediments were deposited under open marine inner neritic to the shallow shelf and marginal marine environments, *Mayyanad formation* comprises coarse to medium-grained sandstones with interbedded white and gray clays and lignites in the coastal area of Kerala. The sediments were deposited under shallow fluctuating marginal marine and swampy environments, *Quilon formation* comprises of brownish white to dirty white, buff, and gray foramniferous algal wackestones and packstones, which are dolomites in several intervals. The formation was deposited in very shallow marine to inner neritic environments, *Cannanore formation* comprises a sequence of shales and sandstones with a few thin limestones. The depositional environment of this formation is a fluctuating environment between middle neritic and upper bathyal regimes, *Warkali formation* consists of a sequence of gray pebbly and coarse sands variegated and mottled clays, white plastic clays, and peat/lignite. The formation is deposited under an extremely shallow marginal marine environment, *Mangalore formation* comprises a sequence of claystones and shales, *Trichur formation* comprises a sequence of light to dark gray and brownish gray, soft, sticky plastic clays, and claystones with abundant shell fragments and occasional thin limestone, dolomite, or siltstone bands. The formation was deposited under middle to outer neritic environment; *Vembanad formation* comprises sands, peat, black, sticky, and sandy calcareous clays, gravel beds with lateritic gravels, lime shell deposits, and alluvium in Kerala coastal region.

4.5.2 Geological evolution of Kerala-Konkan basin

The Kerala-Konkan basin situated south of east–west-trending Vengurla Arch and extends up to Cape Comorin in the south of the Indian subcontinent in the western offshore basin (Fig. 4.1). The sedimentary sequence is comprised of Mesozoic and Tertiary sediments. The western continental margin basin of India evolved through the breakup of eastern Gondwanaland from western Gondwanaland in the Late Triassic/Early Jurassic and the subsequent spreading history of the eastern Indian Ocean. The western margin evolved through the early rift and postrift phases of divergent margin development. A series of regional and local horsts and grabens resulted in response to rifting along the dominant basement tectonic trends. The northernmost part of the western continental margin was the first to be subjected to continental rifting and crustal subsidence in the Late Triassic. The process of rifting gradually advanced towards the south and by Cretaceous time almost all, the rift-related horsts and grabens came into existence. The deposition started with the continental environment, changed gradually to paralic and finally to pulsating marine conditions, punctuated by basic lava flows (Deccan Trap) in the terminal stages towards the end of Cretaceous and Early Paleocene. Towards the end of the early rift phase, most of the rift-related grabens and horsts, located in the deeper parts were covered up with sediments and the continental margin became less intricately differentiated.

The Deccan traps (Cretaceous-Early Paleocene) form the basement of the Tertiary Basin. For most part of the early Tertiary period, Alleppey Plateau and Laccadive ridge were subjected to relative differential rise during the entire Paleogene, thereby creating depressions around them. There were progressive transgressions, accompanied by *syn*-sedimentary positive radial movements of low-to-medium amplitude. Lakshadweep Depression was characterized by rejuvenation of faulting and subsidence of basement, where continental, deltaic, lagoonal, and shallow marine deposition was taking place. The lithological sequence shows upward gradation from continental sand-clay section to sand-shale-coal section and finally terminating in sand-shale-coal limestone (thin) section. The depocenters kept shifting from west to east during Paleocene to Mid-Eocene period. During the late Eocene-early Oligocene period finer differentiation into second-order horst and graben features became more pronounced, and in some cases, the direction of radial movement could have been reversed and resulted in minor epochs of alternating transgression and regression; and consequent lateral shifting of the depocenters. Thus, the late Eocene-Early Oligocene period witnessed a relative intensification of the oscillatory/pulsating movements of medium magnitude and dominance of paralic-to-shallow marine depositional environment in Lakshadweep Depression. The shelfal horst-graben complex formed the site for the development of shelf depositional systems, whereas the Lakshadweep Depression and Laccadive Ridge lay in slope and basinal region. The shelf depositional systems comprised deltas, clastic and carbonate tidal flats, strand plains and carbonate ramps, and platforms and banks. Lows within the shelf provided a trap for the bulk of the terrigenous clastic sediments, whereas the distal and relatively positive shoal areas favored the development of thick limestone sequences. Extensive carbonate sequences thus developed over the northern part near Vengurla arch, Alleppy platform, distal parts of paleo-shelves and around Laccadive, and Maldive group of islands (ndrgdh.gov.in).

The Konkan and Kerala Basins constitute a major depocenter for sediments from the onshore hinterland of Western India and as such provide a valuable record of the timing and magnitude of Cenozoic denudation along the continental margin. The southwestern continental margin of India is characterized as a complex Volcanic Passive Margin. Tectonic reconstructions (Rogers and Santosh, 2001; D'Silva et al., 2008) show that the western continental margin of India (WCMI) owes its origin to the Mesoproterozoic extensions of the Columbia Supercontinent (~1.5 Ga) that were responsible for the formation of Morondova, Mascarene, Mahanadi, and Pranhita-Godavari rifts. Related tectonic processes throughout geologic history, resulting in correlatable stratigraphy, had rejuvenated all these mobile belts and some were taken advantage of during the ultimate breakup of Gondwanaland. Mishra et al. (2011) reported that the subbasalt Mesozoics of the KK Offshore Basin may likely to have good hydrocarbon potential with the help of basin modeling studies.

4.5.3 Hydrocarbon potential in Kerala-Konkan basin

In Kerala-Konkan basin, the source rock is mainly Paleocene to Eocene shales. The Karwar formation (Eocene) has rich organic matter (TOC up to 29%) and Kasaragod formation (Late Paleocene) has reasonably good source potential. Paleocene to Middle Eocene shales have high TOC content mainly of type II and III organic matter and are thermally mature. Shelfal Depressions and Lakshadweep Depression are considered as the source rocks for hydrocarbon generation. Eocene to Middle-Miocene Carbonates and Paleocene to Middle Miocene Sandstones are the possible reservoirs of Kerala-Konkan. Eocene to Middle Miocene Limestone Bands serve as the cap rock in this basin.

The depositional models of the initial postrift and late postrift phases of basin evolution suggest that Kerala-Konkan Basin holds promise for hydrocarbons. The prognosticated hydrocarbon resources of Kerala-Konkan offshore basin are estimated to be 660 MMt (ndrdgh.gov.in).

Challenges and possibilities in KK-basin: Hydrocarbon has been discovered all through coastal areas of India except Kerala-Konkan basin. In Kerala-Konkan basin, a thick basaltic layer separates the Tertiary and Mesozoic sediments and the imaging below basalts is a problem in oil exploration since the basalt is opaque and it masks deeper seismic events below the basalt. Only few wells have penetrated and drilled below the basalt, till date the petroleum system has not been well established in this basin.

Even though the thermal maturity levels, depocenter, and TOC are congenial for the generation of oil in KK-basin, no oil has been reported so far and all the exploratory wells have been declared as dry. In such a situation, the fluid inclusion technique could be adopted to ascertain whether any minute quantities of oil in the form of HCFIs are available in this basin. This will help to delineate the carrier beds at various levels and further research based on migration studies coupled with burial history curve and fluid inclusion data can possibly find a reserve in the nearby locality.

4.5.4 HCFIs from Kerala-Konkan basin

In KK-4C-A-1 well of Kerala-Konkan basin, the presence of hydrocarbons in the form of bitumen (a sluggish black to brown color hydrocarbon with a very low range of API gravity) was observed at the horizon depths 3890–3895, and 3950–3955 m (Figs. 4.9.1 and 4.9.2). Secondary hydrocarbon fluid inclusion trails and some occluded bitumen fluorescence are observed in the 3065–3070 m (Fig. 4.10), 3980–3985 m (Fig. 4.11) depth

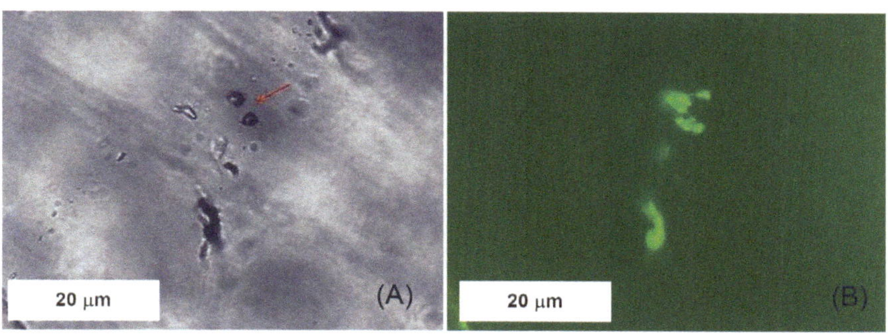

Fig. 4.9.1 Bright field (A) and fluorescent (B) images from the oil inclusion in KK-4C-A-1 well (depth 3950–3955 m) with sandstone, clay stone lithology.

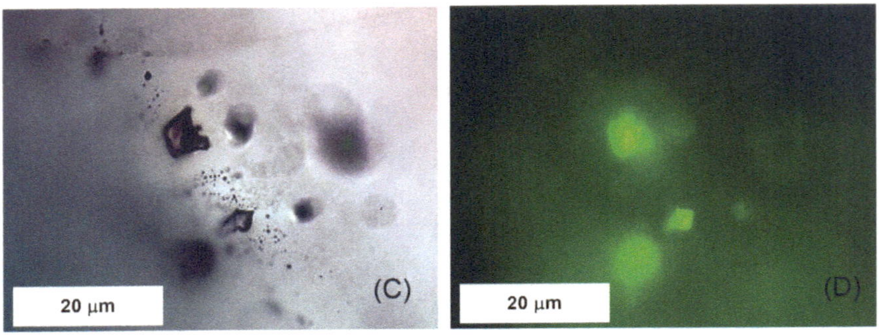

Fig. 4.9.2 Bright field (C) and fluorescent (D) images from the oil inclusion in KK-4C-A-1 well (depth 3890–3895 m) with clay stone, lime stone lithology.

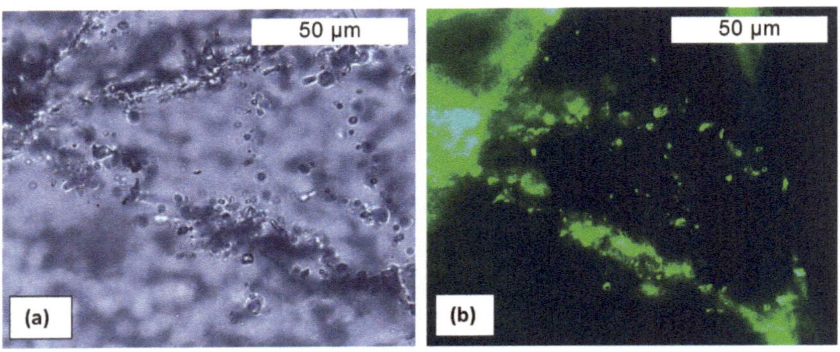

Fig. 4.10 (a) Bright filed and (b) fluorescent images of hydrocarbon fluid inclusion trails in KK-4C-A-1 well in Kerala-Konkan basin at a depth of 3065–3070 m.

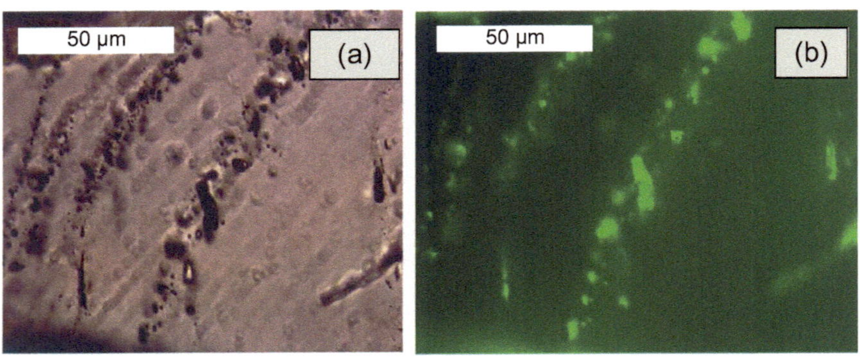

Fig. 4.11 Secondary trails of hydrocarbon fluid inclusions at a depth of 3980–3985 m with sandstone lithology (a) Bright filed image (b) Fluorescent image.

in KK well with sandstone, clay stone lithology. The secondary fluid inclusion trail observed is having migrating trend.

One method of verifying temperature up to which the Kerala Basin had been heated is by the careful cataloguing of the nature of fluid inclusions (composition, temperature of entrapment, etc.) in the carbonate cement and the authigenic enlargements around the ubiquitous quartz grains (clastic phase). Extensive petrographic and microthermometric studies of fluid inclusions from the lithologies within the Kerala basin spread over in space and time can provide essential data on the fluids, which have moved through various stratigraphic successions. Therefore, fluid inclusion studies provide information of fundamental importance to the hydrocarbon exploration and production industry in the Kerala Basin, which so far is designated as "dry" in terms of oil findings. Secondly, microthermometric analysis of fluid inclusions provides an estimate of fluid trapping temperature, composition, and fluid density. Microthermometric analysis was carried out for total of 40 fluid inclusion samples at depth of 3205–4570 m (a minimum of eight measurements for each sample) and observed a first melting temperature (T_{FM}) is ranging from −80.2 °C to −4.9 °C and temperature of last melt (T_{LM}) ranges from −34.5 °C to 0.4 °C. The temperature of homogenization (Th) shows a minimum of 40.2 °C and maximum of 213.6 °C. Among these, almost 12 coeval biphase aqueous fluid inclusion samples/assemblages given a Th between 64.8 and 139 °C that falls within the oil window, the temperature at which transformation of kerogen to oil happens. Salinity was calculated for brine inclusions based on the T_{LM} value and it ranges from 4 to 31.67 eq.wt% NaCl. Based on the mircothermometric studies most of the fluid inclusion systems identified in the KK-basin are H_2O-CO_2 (Eutectic Temperature Te-56.6 °C) H_2O-NaCl (Te-21.2 °C), H_2O-KCl (Te-10.6 °C), and H_2O-NaCl-KCl (Te-23.5 °C) based on fluid inclusion Th data and most of the primary fluid inclusions are aqueous carbonic (H_2O–CO_2) and secondary fluid inclusions mainly come in the H_2O–NaCl and H_2O–KCl systems. Fig. 4.12 shows the changes observed for two biphase fluid inclusions in a secondary trail from the KK-basin, when homogenization happens during microthermometry and the temperature of homogenization indicates oil window.

4.6 Comparison of fluid inclusions from Mumbai and Kerala offshore basins

Petroliferous basins are basins with proven hydrocarbon deposits and basins with good indications of hydrocarbon prospects. About 38 sedimentary

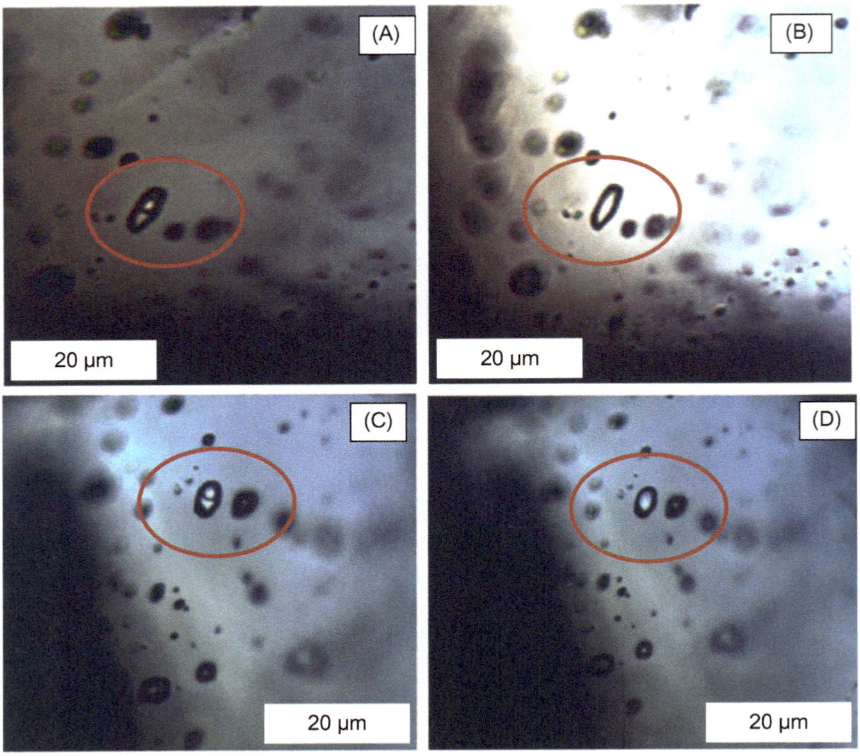

Fig. 4.12 Phase changes observed in two biphase fluid inclusions in a secondary trail from KK-basin at 4565–4570 m depth during microthermometric heating runs (A and C) fluid inclusion appearance in room temperature at 22 °C, where the gas-phase outer rim is clearly visible and (B and D) fluid inclusion appearance when homogenization temperature (Th) reached 83 °C, which falls within the oil window.

basins in a plate tectonics framework were recognized in onshore and offshore areas of the Indian plate including western and eastern Deep Ocean basins, among which eight basins are petroleum-producing basins. Basins were categorized into four groups based on the exploration status (Rai et al., 1998; Biswas, 2012). Basins with proven commercial production (category 1), established basins expecting commercial production (category 2), basins with favorable geological prospects along with hydrocarbon shows (category 3) and basins that do not have established commercial production now but maybe prospective in the future (category 4). Mumbai offshore is a category 1 basin and Kerala-Konkan basin is a category 3 basin.

In order to detect oil as entrapped fluid inclusions we selected two petroliferous basins/locations (Mumbai offshore basin and Kerala-Konkan offshore basin; category 1 and category 3 basins, respectively) based on a detailed discussion with Oil and Natural Gas Corporation (ONGC), Government of India for exploring the utility of fluid inclusion study in dry wells in proven and nonproven basins. A detailed analysis on the hydrocarbon quality/quality in terms of API gravity using fluid inclusions from a well (RV-1 in the offshore basin, Mumbai) was presented in our publication (Nandakumar and Jayanthi, 2016). The hydrocarbon quality of oils in the HCFIs from KK-Basin was also determined, based on data from a previous study. The homogenization temperature of coeval biphase aqueous fluid inclusions from Mumbai offshore basin and Kerala-Konkan offshore basin were studied using microthermometric procedure and the oil window was determined. This data provides an estimate of the paleotemperature that the rock attained in a basin. Oil window and API gravity of fluid inclusions from different depths of both the basins were described in the coming chapters.

A depthwise comparison of fluid inclusion data from both the basins shown in Fig. 4.13 is providing a qualitative depiction of oil in these basins (Jayanthi and Nandakumar, 2021). RV-1 well, even though it is a dry well in a proven basin, the oil window and the API gravity of oils from HCFIs in different Formations (especially the Panna Formation) open a demand for detailed exploration in nearby areas to find a high-value prospect, as indicated by our study (Nandakumar and Jayanthi, 2016). In the KK-Basin, even though it is a category 3 basin, almost 15 wells previously drilled did not indicate the presence of oil. Our study of the API gravity shows medium-to-heavy oil in this basin. We also observed hydrocarbon trails at some depths. The oil window in the KK well also shows that there is a conducive environment for oil generation, but there may be a lack of proper kerogen type. The western offshore sedimentary basins are post-Cretaceous after the effusion of Deccan basalts and coast parallel faulting; hence, a depthwise comparison is carried out.

Biswas (2012) reported that prospects are good in Kerala-Konkan, Saurarastra, and Bengal basins and exploration should be focused in subbasins which may have relatively high subsidence rate and local euxinic environment. The Mumbai offshore (Surat-Ratnagiri) basin is limited to the north by the Saurashtra Arch and to the south by Vengulara Arch. Surat Depression and Ratnagiri Basin are separated by the Bombay High Arch. The Paleocene-Early Eocene Panna Formation is the main source rock in the

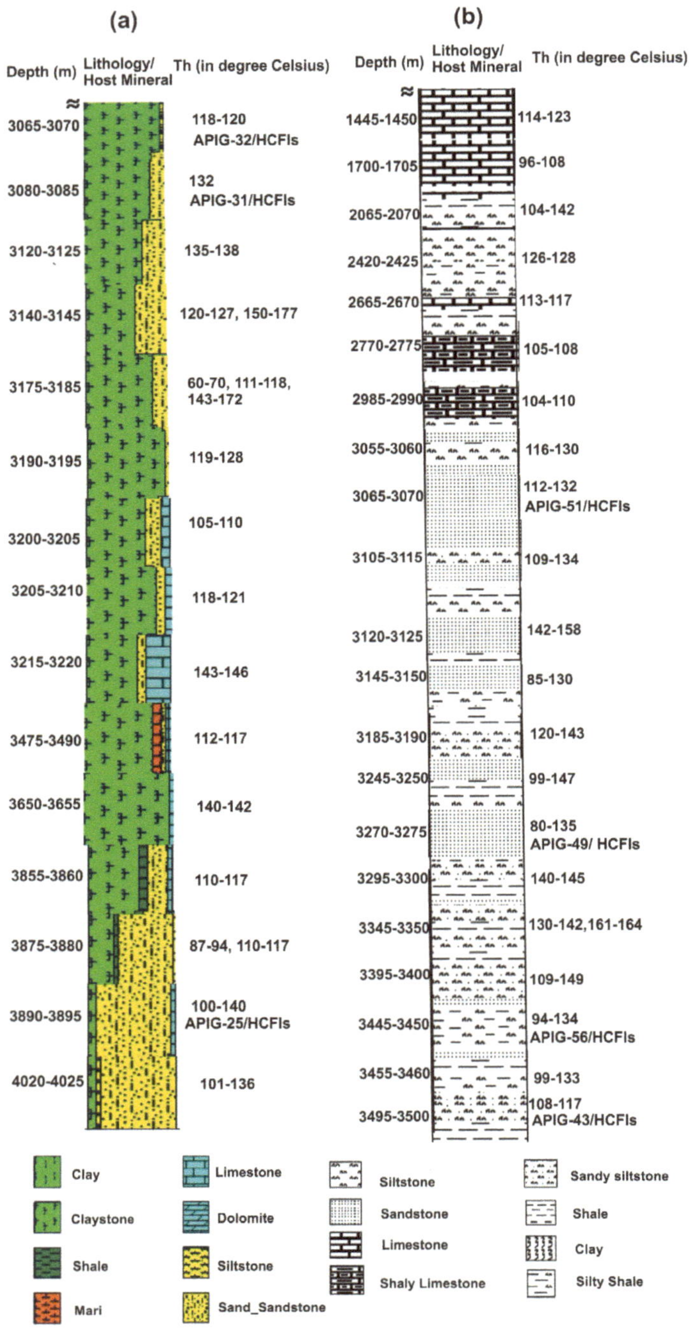

Fig. 4.13 Depthwise comparison of fluid inclusions from Kerala-Konkan (a) and Mumbai (b) offshore basins. Lithological details, oil window, and API gravity where HCFIs are identified are shown (apparent depths are given) API gravities of oils in HCFIs are determined wherever HCFIs observed.

Ratnagiri basin and the Oligocene sequence is the source rock in the Surat depression. The high geothermal gradient due to crustal stretching and shallowing of the mantle in the region of Surat depression (considered as a plume site) helped in the thermogenic processing of the organic matter for hydrocarbon generation. The middle Miocene–III limestone and Mid-Eocene Bassein Lime stone are the major hydrocarbon play zones in this basin. The basal clastics of the Panna Formation are also an important play zone. In south Mumbai High where sediments lie directly on the Precambrian basement, fractured granite/gneiss are also found to be a good reservoir. The cap rock in the area is Post-Mid Miocene shales. Besides the structural traps, the Paleocene-Eocene wedge out along the periphery of the Mumbai High offer good prospects (Biswas, 2012).

Geological Age and Formation wise data given below are available for RV-1 well from Mumbai offshore basin and KK-4C-A1 well from Kerala-Konkan offshore basin are given in Tables 4.2 and 4.3. Age wise, depthwise, Formation wise and lithology wise comparison of the geofluids present in these two wells of the basins are possible for further comparative studies.

The following chapters are detailing the fluid inclusion wafer preparation for sedimentary basins; petrographic methods to detect HCFIs; the determination of oil window using microthermometric studies; fluorescence spectroscopic methods to assess the hydrocarbon quality in terms of API gravity from micron-sized HCFIs; and identification of constituents in fluid inclusions using Raman spectroscopic analysis.

Table 4.2 Geological age and formation wise data for RV-1 well from Mumbai offshore basin, India.

Formation	Age	Top (TVDSS) (m)
Chinchini	Late Miocene to Recent	47.24
Ratnagiri	Early to Mid. Miocene	715.15
Panvel	Late Oligocene	1068.15
Mukta	Early Oligocene	1317.15
Panna	Middle to Late Eocene	1368.15
Bassein	Paleocene to Early Eocene	1736.15
Deccan trap	Late Cretaceous to Paleocene	3695.15

Note: The well RV-1 was drilled as a vertical well. All depths are in meter.

Table 4.3 Geological age and formation wise data for KK-4C-A1 well from Kerala-Konkan offshore basin, India.

Formation	Age	Top (TVDSS) (m)
Trichur	Pliocene to Recent	2135
Cannanore	Early Miocene	2725
Calicut	Equivalent Early Oligocene	3185
Karwar	Middle to Late Eocene	3350
Kasargod	Paleocene to Early Eocene	3525
Deccan trap	Late Cretaceous to Paleocene	6080

Note: The well KK-4C-A1 was drilled as a vertical well. All depths are in meter.

References

Agrawal, P., Pandey, O., Negi, J., 1992. Madagascar: a continental fragment of the paleo-super Dharwar craton of India. Geology 20, 543–546.

Akhtar, K., Aquil, M., 1984. Depositional environments of the Himatnagar Sandstone (Lower Cretaceous). Gujar. Pet. Asia J. 92–96.

Baksi, A.K., Farrar, E., 1991. Ar40/Ar39 dating of the Siberian Traps, USSR: evaluation of the ages of the two major extinction events relative to episodes of flood-basalt volcanism in the USSR and Deccan Traps, India. Geology 19, 461–464.

Banerjee, A., Pahari, S., Jha, M., Sinha, A.K., Jain, A.K., Kumar, N., Thomas, N.J., Misra, N.K., Chandra, K., 2002. The effective source rocks in the Cambay Basin. India. Am. Assoc. Pet. Geol. Bull. 86 (3), 433–456.

Basu, D.N., Banerjee, A., Tamhane, D.M., 1980. Source areas and migration trends of oil and gas in Bombay offshore basin, India. Bull. Am. Assoc. Pet. Geol. 64 (2), 209–220.

Basu, D.N., Banerjee, A., Tamhane, D.M., 1982. Facies distribution and petroleum geology of the Bombay Offshore Basin, India. J. Petroleum Geol. 5 (1), 51–75.

Biswas, S.K., 1982. Rift basins in western margin of India and their hydrocarbon prospects with special reference to Kutch Basin. Bull. Am. Assoc. Pet. Geol. 66 (10), 1497–1513.

Biswas, S.K., 1987. Regional tectonic framework, structure and evolution of western marginal basins of India. Tectonophysics 135, 307–327.

Biswas, S.K., 2012. Status of petroleum exploration in India. Proc. Indian Natl. Sci. Acad. 789 (3), 475–494.

Biswas, S.K., Singh, N.K., 1988. Western continental margin of India and hydrocarbon potential of deep-sea basins. In: 7th Offshore South East Asia Conf., Singapore, 2–5 February 1988, OSEA. 88119, pp. 170–181.

Biswas, S.K., Rangaraju, M.K., Thomas, J., Bhattacharya, S.K., 1994. Cambay-Hazad(!) petroleum system in the South Cambay Basin, India. In: Magoon, L.B., Dow, W.G. (Eds.), The Petroleum System—From Source to Trap. Am. Assoc. Pet. Geol. Mem. 60, pp. 615–624.

Burruss, R.C., 1981. Hydrocarbon fluid inclusions in studies of sedimentary diagenesis. Mineral. Assoc. Canada, Short Course Handb. 6, 138–156.

Burruss, R.C., 1991. Practical aspects of fluorescence microscopy of petroleum fluid inclusions. Luminescence microscopy and spectroscopy: qualitative and quantitative applications. In: Barker, C.E., Kopp, O.C. (Eds.), SEPM Short Course. vol. 25, pp. 1–7.

Burruss, R.C., Toth, D.J., Goldstein, R.H., 1980. Fluorescence microscopy of hydrocarbon fluid inclusions: relative timing of hydrocarbon migration events in the Arkoma basin, NW Arkansas (abst.). EOS Trans. Am. Geophys. Union 61, 400.

Campanilen, D., Nambiar, C.G., Bishop, P., Widdowsonz, M., Brownn, R., 2008. Sedimentation record in the Konkan-Kerala Basin: implications for the evolution of the Western Ghats and the Western Indian passive margin. Basin Res. 20, 3–22.
Clift, P.D., Shimizu, N., Layne, G.D., Blusztajn, J.S., Gaedicke, C., Schluter, H.U., Clark, M.K., Amjad, S., 2001. Development of the Indus fan and its signi¢cance for the erosional history of the Western Himalaya and Karakoram. Geol. Soc. Am. Bull. 113, 1039–1051.
Courtillot, V., Besse, J., Vandamme, D., Montigny, R., Jaeger, J., Cappetta, H., 1986. Deccan flood basalts at the cretaceous/tertiary boundary? Earth Planet. Sei. Lett. 80 (3–4), 361–374.
Cox, K.G., 1989. Mantle plumes and continental flood basalts: supporting evidence from drainage patterns. Nature 342, 873–877.
D'Silva, K., Nathaniel, D.M., Bastia, R., Verma, R., Tenepalli, S., Sreedhar, S.V., 2008. Southwestern continental margin of India: A Poly-Rift Basin. Society of Petroleum Geophysicists, Hyderabad, India.
Directorate General of Hydrocarbons (DGH), Ministry of Petroleum and Natural Gas, Govt. of India n.d. (dghindia.gov.in).
Duncan, R.A., 1990. The volcanic record of the Reunion hotspot. Proc. ODP Sci. Res. 115, 3–10.
Duncan, R.A., Pyle, D.G., 1988. Rapid eruption of the Deccan flood basalts at the cretaceous/tertiary boundary. Nature 333, 841–843.
Dwivedi, A.K., 2016. Petroleum exploration in India—a perspective and endeavours. Status report. Proc. Indian Natl. Sci. Acad. 82 (3), 881–903. July Special Issue.
Evans, M.A., Battle, D.A., 1999. Fluid inclusions and stable isotope analyses of veins from central Appalachian Valley and Ridge Province; implications for regional synorogenic hydrologic structure and fluid migration. Geol. Soc. Am. Bull. 111 (12), 1841–1860.
Gallagher, K., Hawkesworth, C., 1992. Dehydration melting and the generation of continental flood basalts. Nature 358, 57–59.
George, S.C., Ruble, T.E., Dutkiewicz, A., Eadington, P.J., 2001. Assessing the maturity of oil trapped in fluid inclusions using molecular geochemistry data and visually-determined fluorescence colours. Appl. Geochem. 16 (4), 451–473.
George, S.C., Ruble, T.E., Dutkiewicz, A., Eadington, 2002. Reply to comment by Oxtoby On "Assessing the maturity of oil trapped in fluid inclusions using molecular geochemistry data and visually determined fluorescence colors". Appl. Geochem. 17 (10), 1375–1378.
Gombos, A.M., Powell, W.G., Norton, I.O., 1995. The tectonic evolution of Western India and its impact on hydrocarbon occurrences—an overview. Sediment. Geol. 96, 119–129.
Goodwin, A.M., 1991. Precambrian Geology: The Dynamic Evolution of the Continental Crust. Academic Press, London. 666 pp.
Goswami B.G., Singh H., Ashok Kumar Bhatnagar A. K., Sinha A. K., Ram Raj Singh R. R., (2007). Petroleum Systems of the Mumbai Offshore Basin, India. Search and Discovery Article #10154 (2007). Adapted from extended abstract prepared for AAPG Annual Convention, Long Beach California, April 1–4, 2007.
Hill, R.I., Campbell, I.H., Davies, G.F., Griffiths, R.W., 1992. Mantle plumes and continental tectonics. Science 256, 186–193.
Holmes, M.A., Watkins, D.K., 1992. Middle and late cretaceous history of the Indian Ocean. In: Duncan, R.A., et al. (Eds.), Synthesis of Results from Scientific Drilling in the Indian Ocean. Geophys. Monogr. AGU. 70, pp. 225–244.
Hooper, P.R., 1990. The timing of crustal extension and the eruption of continental flood basalts. Nature 345, 246–249.
Jaeger, J.-J., Courtillot, V., Tapponnier, P., 1989. Paleontological view of the ages of the Deccan Traps, the Cretaceous/Tertiary boundary, and the India-Asia collision. Geology 17, 316–319.

Jayanthi, J.L., Nandakumar, V., 2021. Fluid inclusion studies to determine the paleotemperature and hydrocarbon quality of oils in petroliferous basins. J. Petrol. Sci. Eng. 197, 108082.
Kent, R., 1991. Lithospheric uplift in eastern Gondwana: evidence for a long-lived mantle plume system? Geology 19, 19–23.
Kumar, R., 1985. Fundamentals of Historical Geology and Stratigraphy of India. John Wiley and Sons, Inc., New York, NY. 254 pp.
Levine, J.R., Samason, I.M., Hesse, R., 1991. Occurrence of fracture hosted impositiet and petroleum fluid inclusions. Qubec city region, Canada. Am. Assoc. Pet. Geol. Bull. 75, 139–155.
Mathur, R.B., et al., 1993. Lithostratigraphy of Indian Petroliferous Basins. Kerala Konkan, Document VI. Unpub. ONGC Report.
Mishra S., Ravi Verma R., D'Silva K., Banerjee S., Bastia R., and Nathaniel D. M., (2011). Sub-basalt hydrocarbon prospectivity in Kerala-Konkan Offshore Basin, India: a basin modeling approach. Search and Discovery Article #30157 (2011), Adapted from extended abstract presented at GEO-India, Greater Noida, New Delhi, India, January 12–14.
Mitra, P., Zutshi, P.L., Chourasia, R.A., Chugh, M.L., Ananthanaraanan, S., Shukla, B., 1983. Exploration in western offshore basins. Pet. Asia J. VI (4), 15–24.
Nair, N., Pandey, D.K., 2018. Cenozoic sedimentation in the Mumbai Offshore Basin: implications for tectonic evolution of the western continental margin of India. J. Asian Earth Sci. 152, 132–144.
Nandakumar, V., Jayanthi, J.L., 2016. Hydrocarbon fluid inclusions, API gravity of oil, signature fluorescence emissions and emission ratios: an example from Mumbai offshore, India. Energy Fuel 30, 3776–3782.
Naqvi, S.M., Rogers, J.J., 1987. Precambrian geology of India. In: Allen, P., Charnock, H., Oxburgh, E.R., Skinner, B.J. (Eds.), Oxford Monographs on Geology and Geophysics, 6. Clarendon Press, New York, NY. 233 pp.
O'Reilly, C., Parnell, J., 1999. Fluid flow and thermal histories for Cambrian-Ordovician platform deposits, New York: evidence from fluid inclusions studies. Geol. Soc. Am. Bull. 111, 1884–1896.
Oxtoby, N.H., 2002. Comments on: assessing the maturity of oil trapped in fluid inclusions using molecular geochemistry data and visually determined fluorescence colours. Appl. Geochem. 17, 1371–1374.
Pironon, J., 1990. Synthesis of hydrocarbon fluid inclusions at low temperature. Am. Min. 75 (1–2), 226–229.
Rai, A., Chandrasekharam, P., Misra, V.N., 1998. Indian sedimentary basin and their hydrocarbon potential. In: Tiwari, R.N. (Ed.), Recent Researches in Sedimentary Basins—Implications in the Exploration of Natural Resources. Indian Petroleum Pub, Dehradun, pp. 91–103.
Rao, R.P., Talukdar, S.N., 1980. Petroleum geology of Bombay High Field, India. In: Halbouty, M.T. (Ed.), Giant Oil Fields of the Decade, 1968–1978. Am. Assoc. Pet. Geol. Mem. 30, pp. 487–506.
Rawat, M.S., Berry, C.M., Singh, R.N., 1987. Source rock palynology of shelf margin area of Bombay Offshore Basin. In: Kumar, R.K., Dwivedi, P., Banerjee, V., Gupta, V. (Eds.), Petroleum Geochemistry and Exploration in the Afro-Asian Region. Balkema, Rotterdam, pp. 387–391.
Richards, M.A., Jones, D.L., Duncan, A., DePaolo, J., 1991. A mantle plume initiation model for the Wrangellia flood basalt and other oceanic plateaus. Science 254, 263–267.
Rogers, J.W., Santosh, M., 2001. Configuration of Columbia, a mesoproterozoic supercontinent. Gondw. Res. 5 (1), 5–22.
Roychoudhury, S.C., Deshpande, S.V., 1982. Regional distribution of carbonate facies, Bombay offshore region, India. Bull. Am. Assoc. Pet. Geol. 66 (10), 1483–1496.

Ryder, A.G., 2005. Analysis of crude petroleum oils using fluorescence spectroscopy. In: Geddes, C.D., Lakowicz, J.R. (Eds.), Reviews in Fluorescence. Springer, New York, pp. 169–198.
Samanta, M.K., 2000. Cambay-Kalol(!) petroleum system in north Cambay Basin, India [abs.]. India. Am. Assoc. Pet. Geol. Bull. 84 (9), 1395.
Sankaran, V., Ram, J., Venkataraman, K., Mohapatra, P., Gupta, R., 1995. Late rift plays in the area north of Mukta field, Bombay Offshore Basin, India—a case study. In: Proceedings of the First International Petroleum Conference and Exhibition, New Delhi, PETROTECH-95, vol. 2, pp. 251–256.
Singh, N.K., Lal, N.K., 1993. Geology and petroleum prospects of Konkan–Kerala basin. In: Biswas, S.K. (Ed.), Proceedings of the Second Seminar on Petroliferous Basins of India. Indian Petroleum Publishers, Dehra Dun, India, pp. 461–469.
Stasiuk, L.D., Snowdon, L.R., 1997. Fluorescence micro-spectrometry of synthetic and natural hydrocarbon fluid inclusions: crude oil chemistry, density and application to petroleum migration. Appl. Geochem. 12 (3), 229–241.
Subramanian, K.S., 1987. Possible cause of easterly tilt of the southern part of India. J. Geol. Soc. India 29, 362–363.
Vandamme, D., Courtillot, V., Besse, J., 1991. Paleomagnetism and age determinations of the Deccan Traps (India) of a Nagpur-Bombay traverse and review of earlier work. Rev. Geophys. 29 (2), 159–190.
Wandrey, C.J., 2004. Bombay geologic province Eocene to Miocene composite total petroleum system, India. U.S. Geol. Surv. Bull. 2208-F. 26 p.
Zutshi, P.L., Sood, A., Mohapatra, P., Ramani, K.K.V., Dwivedi, A.K., Srivastava, H.C., 1993. Lithostratigraphy of Indian petroliferous basins. In: Document V, Bombay Offshore Basin. ONGC, India, 383 p.

Further reading

Bastia, R., 2007. Geologic Setting and Petroleum Systems of India's East Coast Offshore Basins—Concepts & Applications. Technology Publications, Dehradun, pp. 1–124.
Bastia, R., Radhakrishna, M. (Eds.), 2012. Basin Evolution and Petroleum Prospectivity of the Continental Margins of India, Developments in Petroleum Science. vol. 59. Academic Press, Elsevier, ISBN: 978-0-444-53604-4.
Biswas, S.K., 1989. Hydrocarbon exploration in western offshore basins of India. Geol. Survey India, Spl. Pub. 24, 184–194.
Blamey, N.J.F., Ryder, A.G., 2007. Hydrocarbon fluid inclusion fluorescence: a review. In: Geddes, C.D. (Ed.), Reviews in Fluorescence 2007.
Burruss, R.C., 2003. Petroleum fluid inclusions, an introduction. In: Fluid Inclusions: Analysis and Interpretation. Mineralogical Association of Canada, Vancouver, pp. 159–169.
Ghosh, B.N., Zutshi, P.L., 1989. Indian west coast shelf break tectonic features. Geol. Surv. India Spec. Publ. 24, 309–318.
Jayanthi, J.L., Nandakumar, V., Anoop, S.S., 2017. Feasibility of a 785 nm diode laser in Raman spectroscopy for characterizing hydrocarbon-bearing fluid inclusions in Mumbai Offshore Basin, India. Petroleum Geosci. 23 (3), 369–375.
Krieger, F.W., Eadington, P.J., Lisk, M., 1996. Fluid inclusion data for R w in reserves estimation. In: SPE-Asia Pacific Oil & Gas Conference, pp. 137–141.
Krishan, R., Joppen, T., Schmidt, R., Paul, R., Sinha, A.K., Prabhakar, V., Mayor, S., 2012. P-032 Tectonics, Structural Style and Petroleum System Modeling to Understand Generation and Migration of Hydrocarbon in Kerala Konkan Area, Western Offshore, India. In: 9th Biennial International Conference and Exposition on Petroleum Geophysics. Hyderabad 2012.

Molnar, P., 2004. Late cenozoic increase in accumulation rates of terrestrial sediment: how might climate change have affected erosion rates? Annu. Rev. Earth Planet. Sci. 32, 67–89.

Munz, I.A., 2001. Petroleum inclusions in sedimentary basins: systematics, analytical methods and applications. Lithos 55 (1–4), 195–212.

Nandakumar, V., Jayanthi, J.L., 2019. Method of detecting API gravity of oil present in hydrocarbon bearing fluid inclusions. National Centre for Earth Science Studies (NCESS), Trivandrum, India. Patent No. 315456 dated 03/07/2019.

Nisha, N., Pandey, D.K., 2018. Cenozoic sedimentation in the Mumbai Offshore Basin: implications for tectonic evolution of the western continental margin of India. J. Asian Earth Sci. 152, 132–144.

Singh, R.P., Rawat, S., Chandra, K., 1999. Hydrocarbon potential in Indian deep waters. Explorat. Geophys. 30, 83–95.

Tectonic Map of India, 1963. Geological Survey of India, Govt. of India Publication, map at the scale of 1:2,000,000.

CHAPTER 5

Petrographic and microthermometric studies on HCFIs

V. Nandakumar[a] and J.L. Jayanthi[b]
[a]Scientist - G, National Facility for Geofluids Research and Raman Analysis, National Centre for Earth Science Studies (NCESS), Thiruvananthapuram, Kerala, India
[b]Project Scientist - C, National Facility for Geofluids Research and Raman Analysis, National Centre for Earth Science Studies (NCESS), Thiruvananthapuram, Kerala, India

5.1 Introduction

Petrography is an important tool in the fluid inclusion study. The basic utility of petrography is to infer the relative timing of entrapment of fluid inclusions as to whether it is primary, secondary, or pseudosecondary and to categorize the fluid phases such as monophase, biphase, or multiphase. The analysis of the phase behavior and chemical composition of individual inclusions through microthermometry can be used to determine the pressure-volume-temperature-composition (PVTX) properties of the trapped fluids. Fluid inclusions may be entrapped during or after mineral growth. Inclusions that are trapped during mineral growth can provide the conditions of mineral growth, and those trapped subsequently can provide later conditions (Goldstein, 2003). Petrographic analysis of fluid inclusion wafers from sedimentary basins is the first step in visualizing HCFIs, and with an ultraviolet (UV) illumination (of light around 365 nm), we are able to distinguish HCFIs from other aqueous inclusions.

5.2 Petrography: Identification of petroleum inclusions

Petrographic identification of HCFIs is possible because of its property to fluoresce (Fig. 5.1). Therefore, fluorescence microscopy is used to distinguish HCFIs from others in a sample. Under normal light, petroleum inclusions and aqueous inclusions cannot be distinguished. Hydrocarbon fluids emit light in the visible range when exposed to UV light, while aqueous fluids are nonfluorescing. During petrographic observation, fluorescing "artifacts" such as green/blue boundaries caused due to the imperfections/impurities

Hydrocarbon Fluid Inclusions in Petroliferous Basins
https://doi.org/10.1016/B978-0-12-817416-6.00005-8

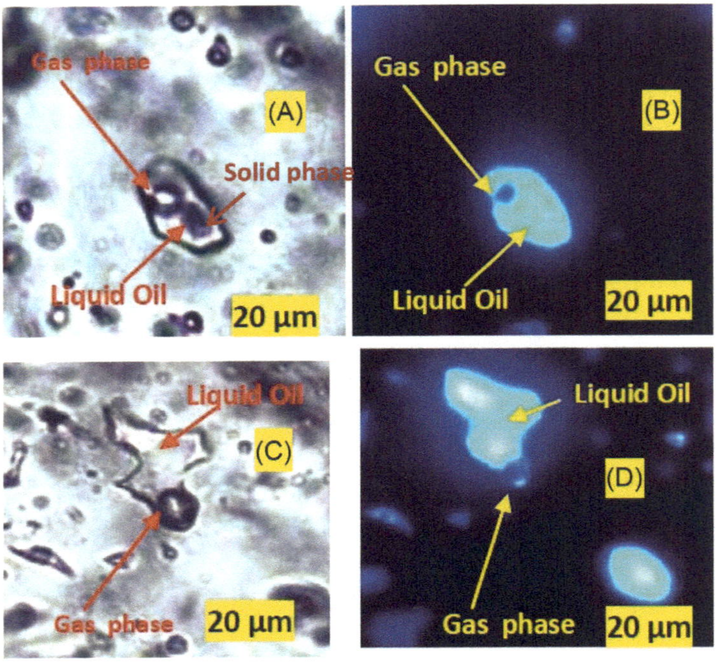

Fig. 5.1 Photomicrographs of multiphase HCFIs under normal and UV excitation from secondary fractures in feldspar grains (A and B), depth 3495–3500 m and HCFIs in secondary annealed fractures in quartz (C and D), depth 3445–3450 m from the Mumbai basin, India.

during wafer preparation can be confused with hydrocarbons. Experience in petrography and further procedures carried out with such fluid inclusions can decipher fluorescing artifacts with HCFIs.

The fluorescence color variations in inclusions can arbitrarily indicate whether the sample contains petroleum inclusions with different maturity levels. Table 5.1 shows the difference in color of fluorescence of petroleum inclusions according to the change in their API gravity. HCFIs are often colorless under normal light, but sometimes have a light bluish, red, or brown

Table 5.1 Shows the difference in color of fluorescence of petroleum oils according to the change in their API gravity.

<15	Brown
15–25	Orange
25–35	Yellow
35–45	White
>45	Blue-white to violet

color. The fluorescence color of HCFIs can hint at the maturity and API gravities of the oil (Burruss, 1981). In this study, the HCFIs from two Formations of the Mumbai offshore basins, Mukta (Early Oligocene) and Panna (Paleocene-early Eocene), showed a difference in color in their fluorescence emission, and further analysis shows that these are HCFIs of a different generation, different in composition and with different API gravities as described in Chapter 6. Petroleum inclusions are generally biphase, liquid, and gas (vapor); at room temperature, exceptions are there for dry gas compositions. The proportion of liquid-to-gas ratio at room temperature is variable and is a function of both compositions and density. Inclusions with gas condensate compositions may have large gas bubbles at room temperature and can fill most of the inclusion area, whereas the gas bubble in other oil compositions may be small and the liquid phase will be dominating in these inclusions. Petroleum inclusions sometimes contain black solid precipitated bitumen. In most of the samples from the Mumbai offshore area, we have observed bitumen as solid phase (Fig. 5.1A) and are therefore multiphase inclusions. An aqueous phase is visible in inclusions with heterogeneously trapping fluids. Hydrocarbon solids can form in several ways (Burruss, 2003) like precipitation on cooling, precipitation during addition of gas, and as a cracking by-product (pyrobitumen, that which is formed during the transformation of oils to gas at high levels of thermal maturity).

5.2.1 Fluid inclusion appearance

The appearance of the outer margin of fluid inclusions is controlled by inclusion shape and the difference in refractive index between the mineral and the trapped fluid. When observed using a transmitted light microscope at room temperature, most fluid inclusions have a rather sharp outer boundary, which marks the edge of the inclusion cavity because of a significant difference in refractive index between inclusion fluids and the host mineral. Most aqueous fluids have refractive indices between 1.33 and 1.45, whereas the host minerals have refractive indices from 1.43 to as high as 3.22. However, hydrocarbon liquids have refractive indices that may be similar to their host mineral (Burruss, 1981) and therefore are not all easily visible. Generally, HCFIs contain a large amount of bright, clear liquid and sometimes may contain a small dark bubble of vapor or gas (any liquid-to-gas ratio is possible) that is dark because of internal reflection (Fig. 5.1C). Bubbles in flat inclusions could be less dark. Some hydrocarbon fluid inclusions and fluid inclusion trails observed in the Mumbai and Kerala offshore basins are given (Figs. 5.2 and 5.3).

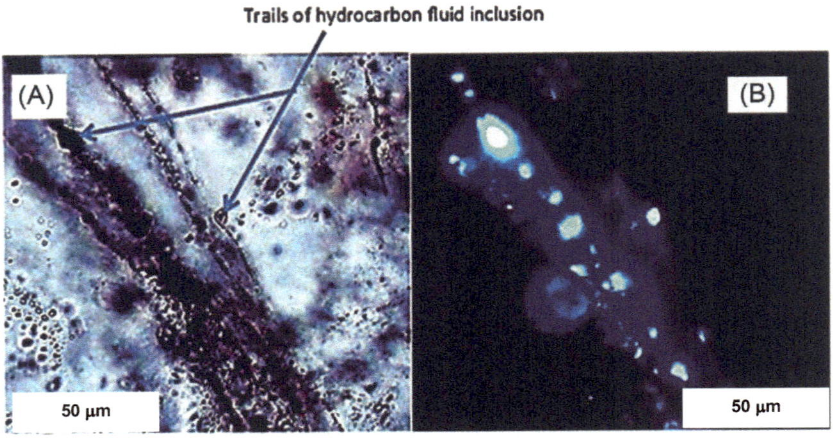

Fig. 5.2 Fluorescence of biphase hydrocarbon inclusions in a secondary trail at a depth of 3065–3070 m with sandstone lithology in the Panna Formation, Mumbai offshore basin with host mineral quartz. (A) Bright filed image and (B) fluorescing image.

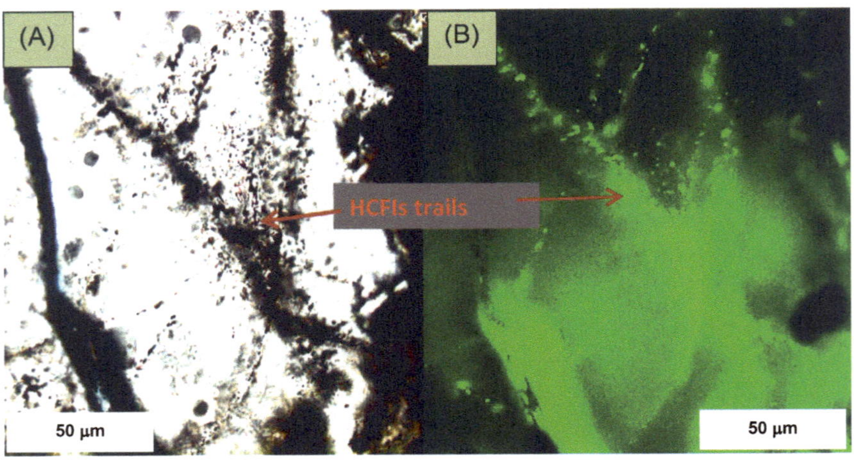

Fig. 5.3 Hydrocarbon fluid inclusion trail in the KK-4C-A-1 well in the Kerala-Konkan basin at a depth of 3615–3620 m with sandstone-clay stone lithology. (A) Bright filed image and (B) fluorescing image.

Various combinations of phases can be observed within fluid inclusions. Many inclusions contain a single observable liquid or have a gaseous phase. Generally, two fluid phases are present; the lowest density fluid floats to the center or top of the inclusion and the higher density phase appears in the exterior part of the inclusion. Aqueous inclusions consist of a dominant

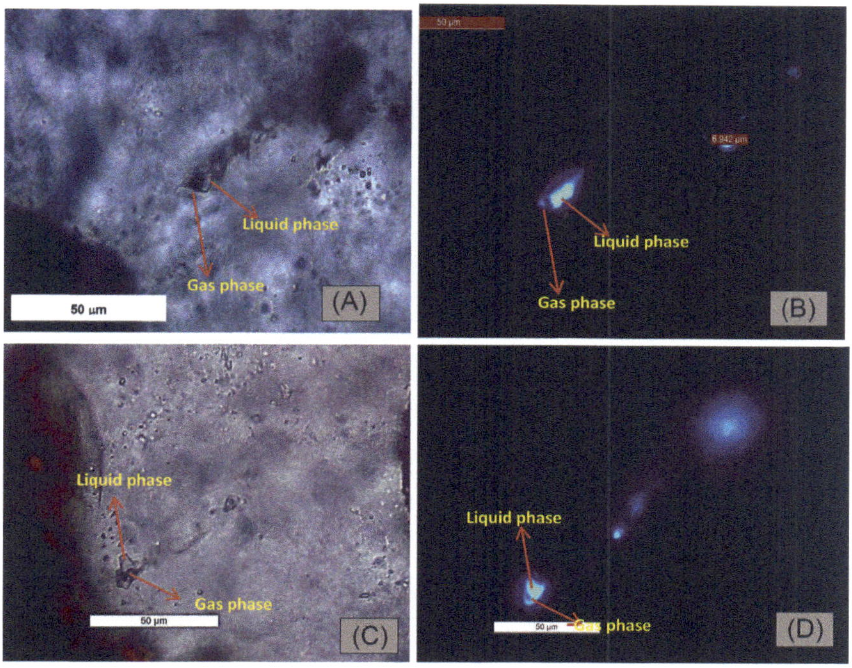

Fig. 5.4 The fluorescence of biphase hydrocarbon fluid inclusions is shown. The fluorescence of the liquid phase and the nonfluorescent gas phase is clearly visible. Figures (A and B), (C and D) are different HCFIs at a depth of 3055–3060 m from the RV-1 well of the Panna Formation, Mumbai offshore basin.

liquid phase and a small vapor or gas bubble (Figs. 5.4–5.8). The bubble normally appears quite dark compared to the surrounding liquid not only because of the refractive index contrast, but because of its common spherical shape. The bubble may be lodged in a corner of the inclusion, may have floated to the top of the inclusion, or may be dancing around the inclusion in pseudo-Brownian motion. Fluid inclusions are viewed as a two-dimensional representation of a three-dimensional object. Therefore, the area of the inclusion occupied by the gas bubble appears to be greater than the actual volume occupied by the bubble. Fluid inclusions also may contain solid phases together with the fluid. These include aqueous inclusions containing crystals that have precipitated during cooling, solid hydrocarbons precipitated from oil, and solids accidentally entrapped in inclusions as they were formed. The optical properties of inclusion fluids, enclosing minerals, and common solid phases that are found within fluid inclusions were summarized by Van den Kerkhof and Hein (2001). The appearance of an

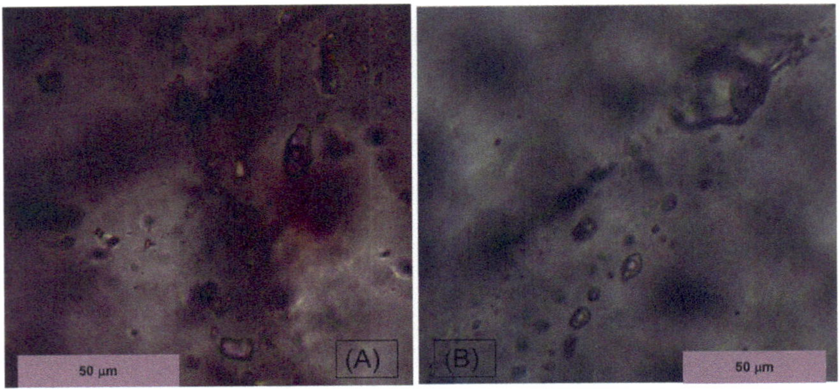

Fig. 5.5 Biphase (A) H_2O inclusions, where the gas bubble is having a dark outer appearance and monophase CO_2 inclusion with a small gas bubble at the corner (B) biphase CO_2 inclusions, where the CO_2 gas phase is clearly visible. Composition ascertained with microthermometry and Raman spectroscopy.

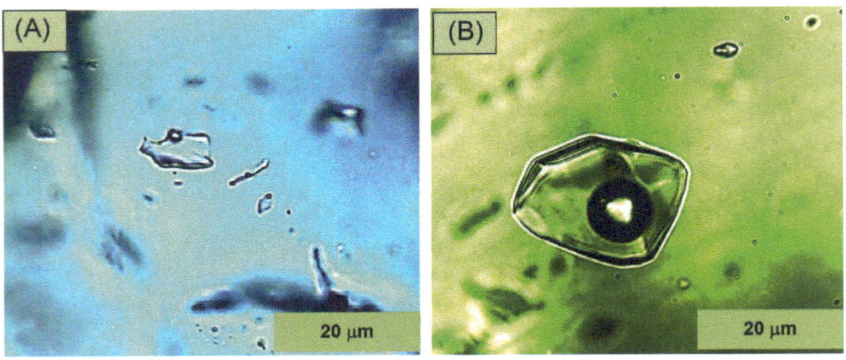

Fig. 5.6 (A) CO_2 inclusion in a sedimentary environment with a small gas bubble at the corner (B) biphase H_2O-CO_2 inclusions.

inclusion in transmitted light (bright or dark) is controlled by a combination of inclusion shape and difference in refractive index between the host mineral and the inclusion fluid. Inclusions with flat upper and lower surfaces transmit light well and appear bright, whereas irregular and globular inclusions have a dark appearance.

Most fluid inclusions have a clear-cut outer boundary, which marks the outline of the inclusion when observed at room temperature using a transmitted light microscope. This is due to the difference in refractive index between the host mineral and fluid inclusion. Usually, the range of refractive

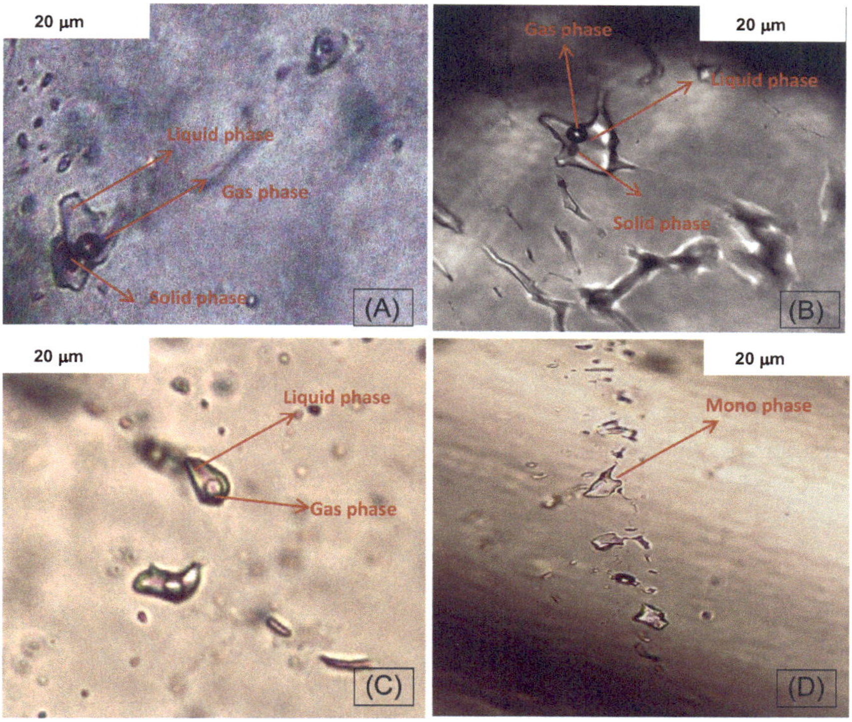

Fig. 5.7 Multiphase hydrocarbon fluid inclusion where the solid, liquid, and gas phases of hydrocarbons are visible (A and B), biphase inclusions (C), and monophase (D) are shown.

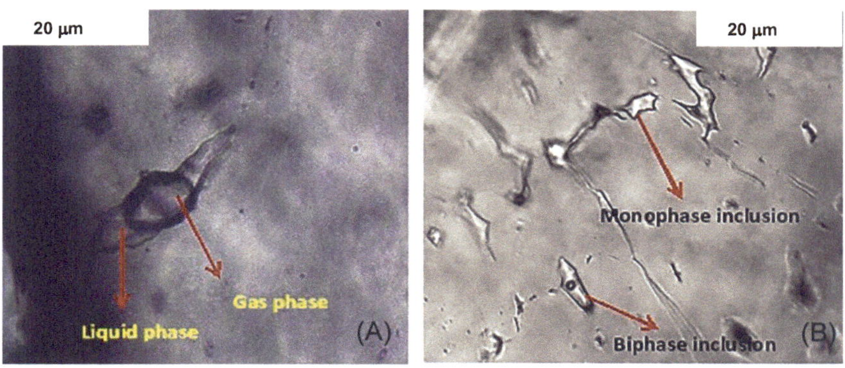

Fig. 5.8 Biphase and monophase aqueous inclusions. In figure (A) the gaseous phase is very large compared to the liquid phase (B) both monophase and biphase inclusions are seen; biphase inclusion with a small gas-phase region.

index in aqueous fluids is between 1.33 and 1.45 whereas the host minerals show a refractive index in the range of 1.43–3.22. Liquid hydrocarbons usually have a refractive index similar to the host mineral and are not clearly visible in normal conditions (Burruss, 1981, 2003). A fluid inclusion comprises a clear and bright liquid that may contain a small dark bubble of vapor or gas that is dark because of internal reflection. Bubbles in flat inclusions may not appear dark. Even though most liquids are colorless, some hydrocarbon liquids may have colors ranging from pale reddish-brown to yellow.

5.2.2 Solid phases

Fluid inclusions commonly contain solid phases along with fluids. Some solid phases were trapped accidentally, where a solid contaminant may have settled or precipitated on the surface of a growing crystal and then was incorporated as part of a fluid inclusion when the host crystal grew around it. Daughter solids could be solid phases precipitated from the inclusion fluid after entrapment as it is cooled, or as its composition changed through some process of alteration of the fluid chemistry. Fluid inclusions, bearing an aqueous phase, may have a daughter solid that has precipitated in optical continuity with the wall of the inclusion, and which has the same mineralogy as the host crystal.

Petrographically, accidental solids can be distinguished from daughter solids by examining phase ratios among inclusions along the fluid inclusion assemblages (FIA). If solid phases have precipitated as daughter objects in those fluid inclusions, all trapped at the same conditions, there would be a consistent ratio of solid to fluid among the inclusions in the FIA. If solids were trapped as accidents, then the ratios would not be consistent. The most common daughter solid in aqueous fluid inclusions can be a single crystal of halite or other salt (Goldstein, 2003), which has precipitated in the aqueous fluid inclusion as it cooled. Highly saline fluids are well known in sedimentary systems and represent residual evaporite brines or brines formed from dissolution of evaporite minerals.

5.2.3 Aqueous inclusions

Typical aqueous inclusions consisting of biphase inclusions with a large gas bubble (Fig. 5.8A) and a single liquid phase (Fig. 5.8B) which is trapped at low temperature and those trapped or re-equilibrated at higher temperature, normally consisting of a liquid phase containing a small gas bubble (Fig. 5.8B). The liquid phase typically has a bright appearance in transmitted light. The gas phase normally appears dark. Daughter solids are common in aqueous inclusions. Oil, oil-aqueous, condensates, and bitumen inclusions containing solid and liquid hydrocarbons are common in diagenetic systems. Fluid inclusions containing liquid oil also may contain a small amount of brownish bitumen-like solid clinging to the inclusion walls (Fig. 5.7A and B). The solids either were entrapped as accidental solids or formed as daughter solids. Most commonly, oil inclusions consist of two fluid phases, a dominant liquid phase and a small gas bubble (Figs. 5.1A and C). As in aqueous inclusions, the gas bubble typically appears dark in HCFIs also.

The methodology for classifying the fluid inclusions petrographically was summarized by Roedder (1979, 1984). Inclusion origin relates closely to modes of fluid inclusion entrapment (Bodnar, 2003), which is a field that is rapidly developing. Goldstein (2003) summarized the criteria for the recognition of primary, secondary, and pseudosecondary fluid inclusions and this could be very useful to those who are interested in fluid inclusion petrography.

5.3 Petrography: Relationship to the host mineral

The textural relationship between the petroleum inclusions and the host mineral can shed light on the relative timing of oil entrapment and ascertain the migration trends. Some textural occurrences of petroleum inclusions in sedimentary rocks are shown in Figs. 5.9–5.12. Some examples of the

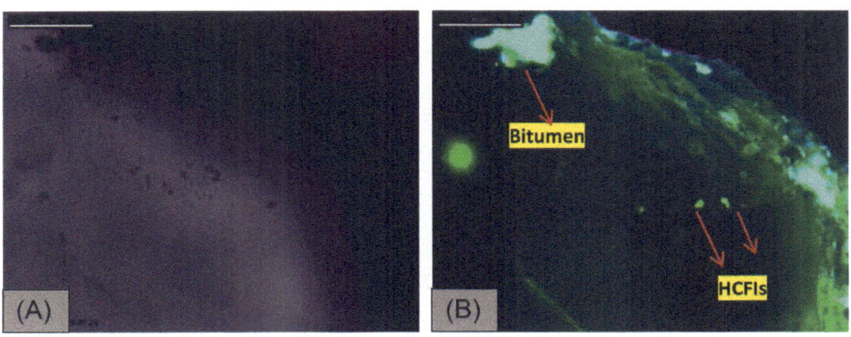

Fig. 5.9 Petroleum fluid inclusions from the Kerala-Konkan basin with clay stone limestone lithology showing a textural relationship with the host mineral at a depth of 3950–3955 m from the KK-4C-A-1 well. The blebs are bitumen sticking outside the calcite. (A) Bright field image and (B) fluorescing image.

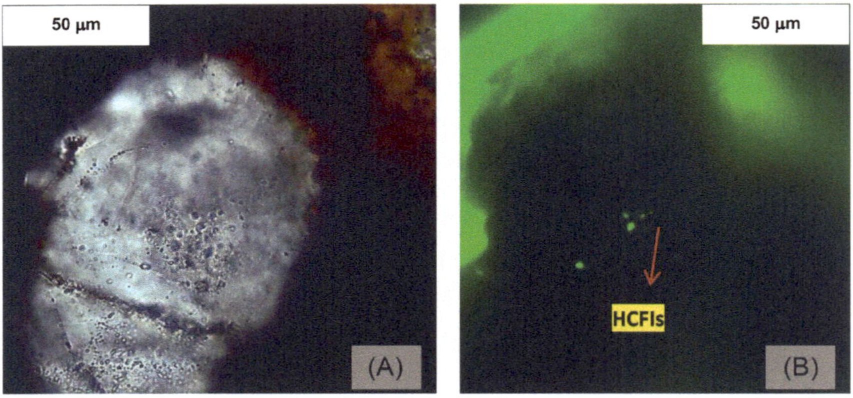

Fig. 5.10 Image showing a textural relationship with the host mineral with petroleum fluid inclusions from the Mumbai offshore basin at a depth of 3055–3060 m from the RV-1 well, Panna Formation with sandstone, siltstone lithology. (A) Bright field image and (B) fluorescing image.

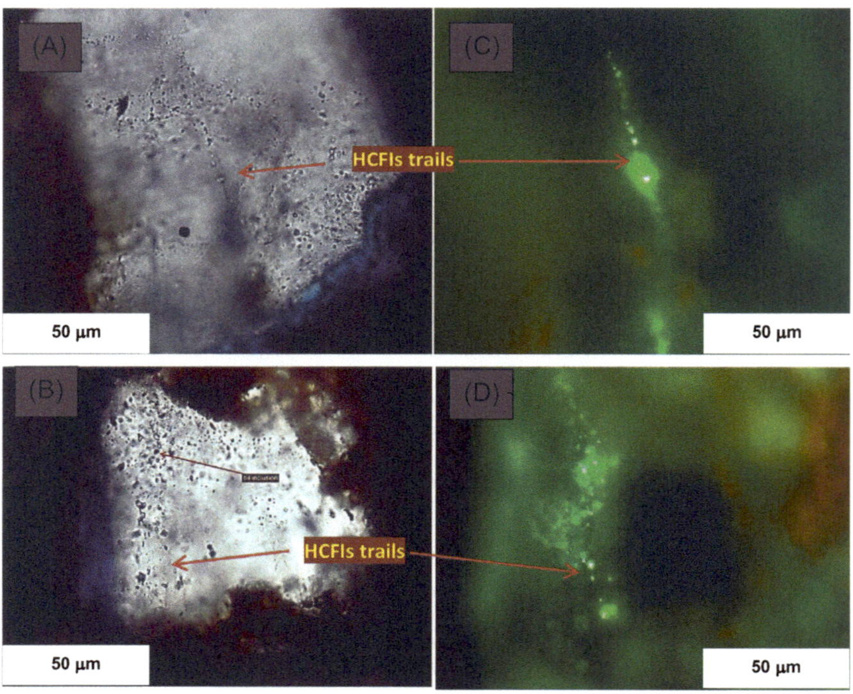

Fig. 5.11 Hydrocarbon fluid inclusion secondary trails (A and B) and HCFI fluorescence (C and D) at a depth of 3495–3500 m from the RV-1 well, Panna Formation, Mumbai offshore basin, India. The host lithology of this area is sandysiltstone, siltyshale, and shale of the Paleocene Earlyeocene age.

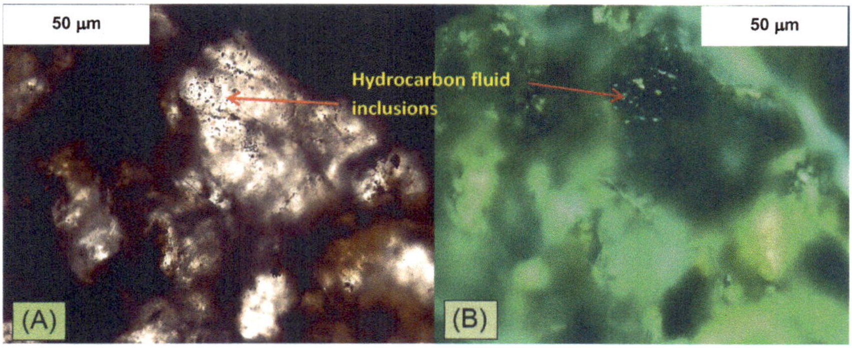

Fig. 5.12 Images showing a textural relationship with the host mineral with petroleum fluid inclusions from the Mumbai offshore basin at 3495–3500 m depth with shale and siltyshale lithology. HCFIs visible with a 10 × lens objective are marked. (A) Bright field and (B) fluorescing image.

relationship between quartz cementation and trapping of petroleum inclusions within the rim of quartz overgrowths were shown by Nedkvitne et al. (1993) and Oxtoby et al. (1995).

In Fig. 5.9, the bitumen seen around the growth zone could be of a different generation as compared to HCFIs inside the crystal in healed fractures and showing a difference in fluorescence color. The most common occurrence of petroleum inclusions in clastic reservoirs is in secondary trails crosscutting detrital quartz and feldspar. Petroleum inclusions observed in detrital feldspars cannot be used for microthermometry due to poor sealing. Petroleum inclusions in authigenic quartz are of particular interest for the relative timing between petroleum reservoir filling and quartz diagenesis (Munz et al., 1998; Burruss, 2003). Petroleum inclusions may also occur in mineralized fractures both in chalk and in clastic reservoirs. Petroleum inclusions in primary growth zones in fractured minerals can provide evidence for the presence of petroleum at the time of fracture cementation, suggesting that the fractures were conduits for petroleum migration. Primary and pseudosecondary petroleum inclusions in fracture cements may give evidence for the presence of petroleum during fracturing and subsequent cementation (Munz et al., 1998).

Most of the HCFIs from the Mumbai offshore basin were observed along secondary trails (Fig. 5.11) and provide evidence for the migration of petroleum during the time of fracture healing. The prevalence of secondary fractures in the Panna formation filled with hydrocarbon fluid inclusions points to the possibility of oil pooling in adjacent areas. The studied well from Mumbai offshore (RV-1) was a dry well, which is in a proven basin. The textural relationships of oil migration through the fractures are evidence of it.

5.4 Microthermometric analysis of petroleum inclusions

Microthermometric analysis of petroleum inclusions follows the general procedure outlined by Shepherd et al. (1985). Selection of inclusions for analysis should be based on a critical evaluation of textures. Inclusions that show textural indications of poor sealing or postentrapment changes like necking should be avoided. The most common use of fluid inclusions is to obtain a measure of the temperature of past events. Sorby (1858) showed that the bubble present in most inclusions represents the differential shrinkage of the host crystal and the fluid that was trapped in it from the temperature of trapping. By reversing this shrinkage on a heating stage until homogenization occurs, the minimum trapping temperature can be determined (Fig. 5.13).

The microthermometry of petroleum inclusions in quartz, carbonates, and barite usually gives reproducible results. Petroleum inclusions trapped in partly dissolved K-feldspars often show extreme variations in size of the vapor bubble, very irregular shapes, and variable fluorescence, which indicate poor sealing and are generally not recommended for microthermometry. Leakage of inclusions may also occur during analysis. The best sequence in the laboratory is to measure homogenization before the low-temperature phase transitions. It is important to start with the populations of inclusions

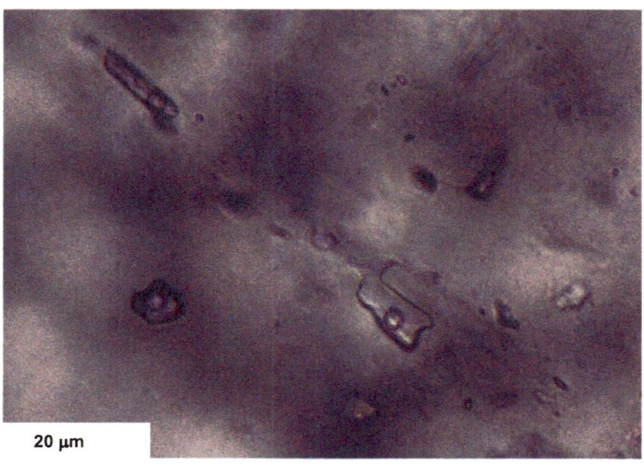

Fig. 5.13 Biphase inclusions used for microthermometric analysis.

which homogenize at lower temperatures before proceeding to homogenization measurements at higher temperatures. The measurements of homogenization temperatures for petroleum inclusions must be done before the measurements of aqueous inclusions. Freezing of ice in the aqueous inclusions very often creates microcracks in diagenetic minerals such as carbonates and barite, and this may lead to severe leakage of inclusions. Ice melting temperatures should therefore be the last analysis in the sequence (Fig. 5.14).

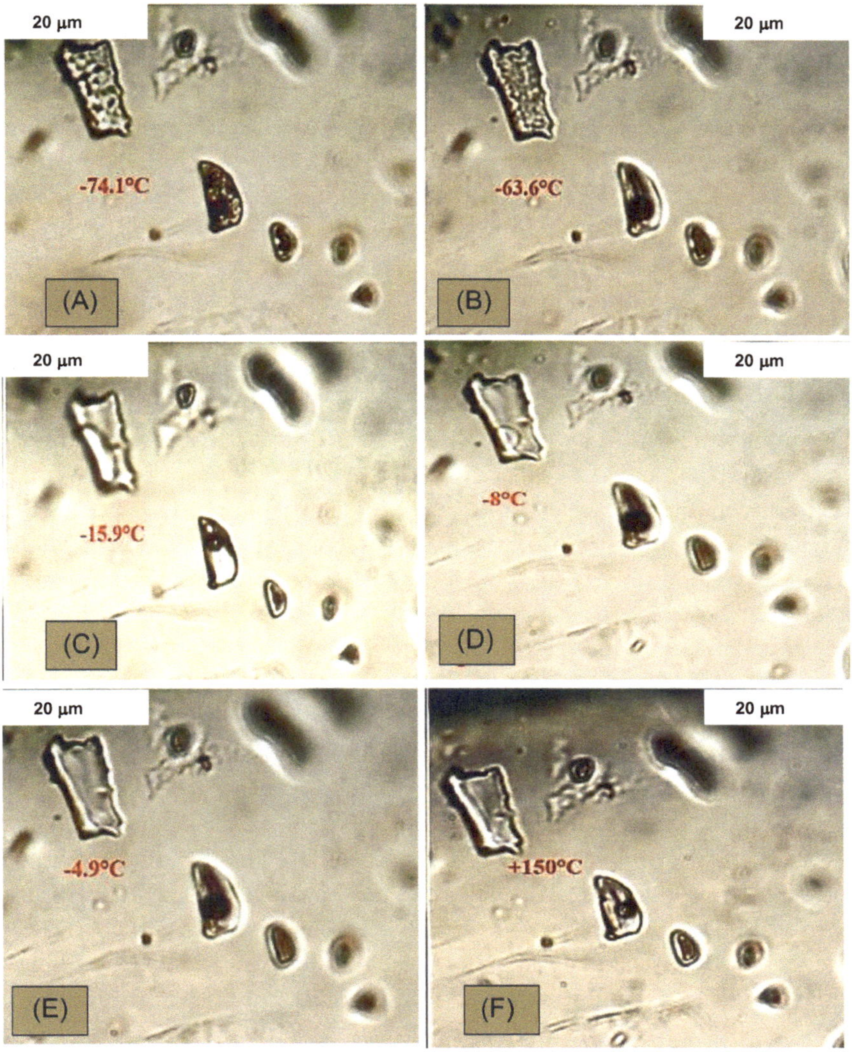

Fig. 5.14 Phase changes observed on fluid inclusions when heating-freezing runs during microthermometry.

The most useful phase transition for biphase petroleum and aqueous with liquid and vapor phase inclusions is homogenization. Liquid and vapor homogenizations are common, and critical homogenization may also occur. For vapor homogenization (L + V to V), the homogenization temperature represents an absolute maximum for the critical point. Likewise, for liquid homogenization (L + V to L), the homogenization temperature represents an absolute minimum for the critical point. In order to use the homogenization temperature quantitatively for calculation of isochores and estimation of trapping conditions, additional information on petroleum fluid composition and microthermometry from coexisting aqueous fluids are necessary.

The fluorescence behavior distinguishes petroleum inclusions from other inclusions; if fluorescence microscopy is not available, the low-temperature behavior can be used to distinguish aqueous and petroleum inclusions. Aqueous inclusions freeze into ice and hydrates and consist only of solids and a vapor phase at low temperature. The liquid appears again at the eutectic point during heating, typically around −50 °C. Another phase transition that may occur in petroleum inclusions is the melting of clathrate. It will always be possible to observe this phase transition if an aqueous phase is visible in the inclusions and may occasionally be observed in inclusions without a visible aqueous phase (Munz, 2001). Homogenization temperatures of biphase liquid–vapor petroleum inclusions give us minimum trapping temperatures. If we have the composition and density data, we can calculate the pressure-volume-temperature (PVT) properties of petroleum inclusions and estimate the actual trapping temperatures (Aplin et al., 1999; Burruss, 1989; Narr & Burrus, 1984; Thiery et al., 2002). Equations of state are used to model PVT properties to calculate phase boundaries (bubble point curve, critical point, dew point curve, and solid-phase coexistence) and isochores. The ultimate objective of microthermometry and PVT interpretation of petroleum inclusions is the estimation of both trapping temperature and pressure. Interpretation of microthermometry of petroleum inclusions is difficult compared to the phase behavior of aqueous (Bodnar, 2003) or gas inclusions (Diamond, 2003). The difficulty is because petroleum is a complex mixture of components and there are not enough phases to yield unambiguous interpretations of composition and density. Therefore, coeval biphase aqueous inclusions are measured for the temperatures of liquid–vapor homogenization yield estimates of the minimum trapping temperature of the inclusions.

The measured temperatures of liquid–vapor homogenization yield estimates of minimum trapping temperature of the inclusions. The phase at which homogenization occurs is important to note. Homogenization to liquid indicates a fluid with liquid-like densities (steep isochores in the P–T space), whereas homogenization to vapor indicates a vapor-like density and the fluid will have an isochore with a relatively shallow slope. The true (critical) homogenization also occurs, but it is not common, where the meniscus fades and the relative volumes of the liquid and gas do not change significantly (Levine et al., 1991). In petroleum fluids, commonly, near-critical homogenization is observed where the meniscus is clearly fading while the liquid or the vapor is rapidly decreasing in volume (O'Reilly and Parnell, 1999). As the temperature of critical or near-critical phase behavior increases above the critical point for pure methane (−82.5 °C), the amount of higher molecular components in the inclusion must increase (Burruss, 2003). Low-temperature solid–liquid–vapor phase behavior in petroleum inclusions is generally difficult to observe, and if seen, it is difficult to interpret quantitatively. Gas-rich systems give more reproducible results for freezing and melting of petroleum sold phases (Levine et al., 1991; Teinturier et al., 2002).

Another phase transition that may occur in petroleum inclusions is the melting of clathrate. It will always be possible to observe this phase transition if an aqueous phase is visible in the inclusions, and may occasionally be observed in inclusions without a visible aqueous phase. However, the clathrate melting temperatures have so far found little quantitative use (Munz, 2001). Kihle et al. (2012) reported that a minimum temperature of fluid trapping is obtained when observing where the gas bubble completely disappears. The actual temperature of trapping may be slightly higher, depending on fluid composition and trapping pressure. The hydrocarbon fluid inclusions are assumed to be in equilibrium with methane; hence, the potential discrepancy between minimum trapping temperature and true temperature of trapping is assumed to be negligible in this context.

The microthermometry technique is invaluable for discovering the temperatures at which minerals form, the thermal history a rock has experienced, and the compositions of the fluids that traversed a rock in its history. Microthermometry is based on the principles of phase equilibria. An inclusion represents a chemically closed (isoplethic) system from the time of entrapment in which the volume of an inclusion remains constant (isochoric) since entrapment. In order to interpret fluid inclusion microthermometric data, the data should be from many inclusions within a single fluid inclusion assemblage (FIA)—the most finely discriminated, petrographically associated group of inclusions that are collected and recorded together as a group (Goldstein & Reynolds, 1994). The goal of measuring inclusions within FIAs is to account for any variabilities in their characteristics. The variability of data within a FIA will alert the researcher to infer about re-equilibration, necking or immiscibility must have affected the constituent inclusions in the assemblage. However, the consistency of data among inclusions of different sizes and shapes within a single FIA indicates that homogeneous fluids were trapped within the inclusions, and are sustaining as chemically closed systems and volumes throughout their histories. All inclusions defining a single healed microfracture and primary inclusions in a single, most finely discernible and petrographically distinguishable growth zone can be the best examples of FIA.

5.4.1 Measuring homogenization temperatures

The measurement of the homogenization temperature (Th) of fluid inclusions requires recording the temperature at which a bubble disappears on heating. In some inclusions, the bubbles will be in constant motion within the vacuole. As explained by Roedder (1984), a bubble is probably moving because of the extremely rapid flow at the bubble's surface due to surface-tensional contrasts caused by infinitesimally small thermal gradients. For smaller inclusions, such movement can greatly enhance the ability to obtain Ths. However, some bubbles will remain stationary through a heating run (they may be wedged between inclusion walls), or may never change position for unknown reasons, and many inclusions are so small that even bubble movement is difficult to discern. Our experience shows that most of the biphase HCFIs show a bubble movement in a sedimentary environment.

5.4.2 Low-temperature phase changes

After inclusions have been studied for high-temperature phase behavior and all of the Ths have been recorded, the inclusions are ready to be studied at temperatures below 0 °C. For aqueous inclusions, microthermometry at low temperatures provides useful information on the composition of the fluid inclusion, which may include the major ions present as well as their concentration in the fluid. Not only are such compositional data useful for understanding the diagenetic environment, but also they provide the basis for determining which phase diagrams to use in developing a further understanding of the Th behavior. For inclusions containing gases such as CH_4 or a more complex mixtures of gases, these analyses may help to approximate gas compositions and define the appropriate phase equilibria, so that entrapment of P–T conditions may be better. Most sedimentary brines are so complex and fluid inclusions are so small that, in general, it is not possible to observe many of the phase changes that take place in aqueous inclusions upon freezing; nor can the solid phases that exist during freezing runs always be identified. One parameter that can be measured relatively accurately during freezing runs is the final melting temperature of ice (Tm ice). This measurement is used to interpret the bulk salinity of aqueous fluid inclusions, but only within certain limits because of the unknown complex composition of many aqueous fluids. In order to use Tm ice to determine bulk salinity, one must assume a model composition, as the true compositions of natural brines enclosed within inclusions are unknown. When a heating-freezing stage runs, after that an inclusion has frozen, the temperature of first melting is the next phase change to observe.

Eutectic melting in a fluid inclusion is the temperature at which one solid-phase melts completely. In practice, that solid phase must be present in a quantity sufficient for the melting to be observed under the microscope. But it is quite impossible to visualize eutectic melting in the small inclusions typically present in diagenetic phases. Therefore, the observation of eutectic temperature (Te) requires either that a studied inclusion be of high enough salinity that a phase can be observed to melt at Te, or that the inclusion is very large.

Paleotemperature, or the temperature determinations on past geologic events, is significant for mining and oil exploration companies for understanding the processes that lie behind the formation of their deposits or new exploration sites. Oil companies are interested in the temperatures of inclusions in core samples from wells to determine whether the rocks have been hot enough to generate oil or overheated such that any oil may have been destroyed. Geothermal companies use these temperature values to guide drilling and to determine the potential of a field. Analysis of the phase behavior and chemical composition of individual inclusions is used to define the PVT properties of the trapped fluids. The homogenization temperature measurement and the effects of changing the molar volume of the trapped fluid can determine the trapping temperature (Tt) from the homogenization temperature. The paleotemperature is determined by considering the PVT properties of both the hydrocarbon fluids and the aqueous fluids.

5.4.3 H_2O-petroleum multicomponent system

There are numerous examples of the coexistence of petroleum with an immiscible aqueous phase in sedimentary basins throughout the world. As a first approximation, the phase behavior of the H_2O-petroleum system is analogous to the H_2O-CH_4 system. When the two phases coexist, there will be a certain amount of one component dissolved in the other. If the aqueous phase is enclosed within an inclusion, this phase will be saturated with hydrocarbon components, so upon cooling, a fraction of these components will instantly separate from the solution. Similarly, if the petroleum phase is entrapped, it will be saturated with H_2O, and upon cooling, H_2O may separate from the solution instantly. Therefore, for a H_2O-CH_4 system, the Th of either type of inclusion should be the true temperature of entrapment (as long as only a single phase is initially entrapped). However, in the low mutual solubility between water and hydrocarbons, the Th may not be visually detectable. In fact, it is predictable that it would be very unlikely for one to be able to observe the tiny amount of H_2O that separates from a petroleum phase in a fluid inclusion (Burruss, 1992).

5.4.4 Gas-petroleum multicomponent system

Some researchers have chosen to treat petroleum phase relations without the H_2O component, ignoring the fact that the inclusions were probably trapped by the growth of a mineral precipitating from an aqueous phase in equilibrium with the petroleum. Burruss (1992) used the Peng and Robinson (1976) equation of state to construct phase diagrams for petroleum of various types and compositions. First note that there is a tremendous range in the position in the P–T space of the boundaries between the one-phase petroleum fields and the two-phase petroleum fields for petroleum of different types: obviously, the composition of petroleum is extremely important for defining phase relationships. The petroleum inclusions of different compositions will yield a great range of homogenization temperatures (Fig. 5.15). Also, since the slopes of the isochores for most petroleum fluids are much shallower than those of aqueous fluids, the apparent Th of petroleum inclusions commonly do not provide a closer estimate of minimum entrapment conditions as do aqueous inclusions (unless they were entrapped in equilibrium with a gas phase). Since the petroleum fluid isochores are commonly equal to or shallower than thermobaric gradients present in sedimentary basins, if a petroleum inclusion is buried after entrapment, the pressure within the inclusion is not likely to rise significantly above the external

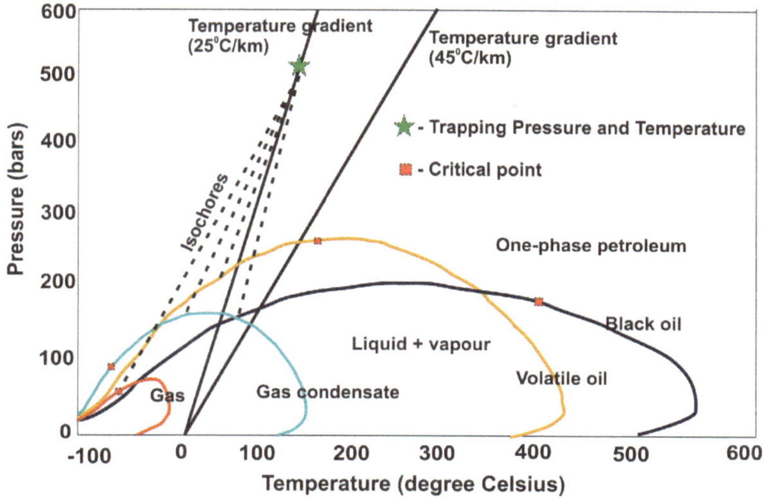

Fig. 5.15 Pressure temperature diagram of four types of reservoir fluids-natural gas, gas condensate, and oil (black oil and volatile oil). *(Modified after Burruss, R.C., 2003. Petroleum fluid inclusions: An introduction. In. I. Samson, A. Anderson & D. Marshall eds. Fluid Inclusions: Analysis and Interpretation. Mineral. Assoc. Canada, Short Course Ser. Vol. 32, 159–174.)*

pore pressure, which is exactly the opposite of the effect in aqueous inclusions. This may have important implications in the interpretation of the petrography of fluid inclusions found in diagenetic phases. The bubble point curves for the various petroleum fluids occur at elevated pressures in comparison to aqueous fluid without gases. Immiscibility between petroleum and gas is predictable to exist over a wide range of conditions in natural diagenetic environments. In addition, if homogeneous entrapment of either phase from an immiscible gas-petroleum fluid system occurred, then the Ths of the petroleum and gas inclusions are the entrapment temperatures. True homogenization of the gas inclusions may be difficult to see, but the Th of a petroleum inclusion trapped when saturated with gas will be easily visible (Goldstein and Reynold, 1994; Burruss, 2003).

Oils (black oil and volatile oil), gas condensate (retrograde gas), wet gas, and dry gas are the main types of reservoir fluids. Black oils are made up of a variety of components including large, heavy, and nonvolatile hydrocarbons. Black oils are dark in color, indicating the presence of heavy hydrocarbons. Volatile oils contain fewer heavy molecules and more intermediate components (ethane through hexane) than black oils. Volatile oils generally have 40° API or higher. The color is generally lighter than black oil—brown, orange, or green. Gas associated with volatile oils tends to be very rich and similar to retrograde condensate gas. A critical point is the P–T condition at which the densities of the liquid and vapor are equal and the phases are indistinguishable. The critical point separates the bubble point (low-temperature side) curve from the dew point (high-temperature side) curve. Homogenization to liquid occurs on the bubble point curve and homogenization to vapor on the dew point curve. The classification of black oil and volatile oil is based on the characteristics of the stock-tank-oil content of their equilibrium gases. Equilibrium gases liberated from volatile *oils* contain an appreciable stock-tank of condensable liquids, whereas the gases from *black oils* contain negligible stock-tank liquids. The higher gas content of the volatile oil shifts the liquid vapor phase boundary to higher pressure than the boundary for the black oil. This difference causes the Th of the volatile oil inclusions to be higher than the Th of the black oil inclusions. The trapping pressure and temperature approaches the P–T conditions of the liquid–vapor phase boundary, whereas the Th approaches the trapping temperature. If the trapping conditions are within the liquid- vapor field of the petroleum, then heterogeneous trapping will occur. Condensate gas is very similar to volatile oils in terms of the color (green, orange, brown, even clear) and gravity (40°–60° API) of the produced oil. However, the reservoir temperature of a condensate gas reservoir is greater than the critical temperature of the fluid, and so where a volatile oil is a liquid at the original reservoir pressure and temperature, a condensate gas is a gas. As pressure is reduced in a condensate gas reservoir, the fluid will pass through the dew point and large volumes of liquid will condense in the reservoir. Natural gas that contains significant heavy hydrocarbons such as propane, butane, and other liquid hydrocarbons is known as wet gas or rich gas. If the gas contains less methane (>85% methane) and more ethane, and other more complex hydrocarbons, it is labeled as wet gas. Wet gas exists solely as a gas in the reservoir throughout the reduction in reservoir pressure. The entire phase diagram of a wet gas will lie below the reservoir temperature. Natural gas that occurs in the absence of condensate or liquid hydrocarbons, or gas that had condensable hydrocarbons removed, is termed dry gas.

5.5 Significance of oil window in petroleum fluid inclusions: A microthermometric approach

Transformation of kerogen to oil is mainly temperature driven and takes place at a specific temperature range known as the "oil window." The oil window is often in the 60–140 °C interval (approx. 2–4 km depth). The temperature of oil generation is also studied using several methods such as vitrinite reflectance values and bottom-hole temperature measurements that provide petroleum geologists with some insight into the probability to find oil in a sedimentary basin. Oil window temperature along with the understanding of a structural or stratigraphic relationship can elucidate oil migration and pooling. When the microcavities within the minerals are sealed, they entrap the surrounding fluids at the prevailing pressure-volume-temperature (PVT) conditions. These PVT data can be used for modeling the fluid-phase behavior and give information on the "oil window" at which oil formation occurs.

To study phase transitions in fluid inclusions, a doubly polished thin section (wafer) of the host rock is prepared, petrographically examined, fluid inclusion assemblages (FIA)—a group of fluid inclusions that were trapped at the same time—are identified and mounted on a heating and freezing stage. These chips are then observed while the stage is either heated or cooled, and the phase changes which take place during a heating or cooling cycle are carefully recorded along with the temperatures at which such changes occur. The best way to interpret fluid inclusion microthermometric data is considering inclusions within a single fluid inclusion assemblage (FIA). An FIA thus defines the most finely discriminated fluid inclusion-trapping event that can be identified based on petrography (Goldstein, 2003). Determination of homogenization temperature (T_h) using a microthermometric experiment by analyzing biphase, secondary fluid inclusion assemblages in the Mumbai offshore basin, India, described as an example, yielded useful information regarding the oil window.

The fluid inclusion samples were selected from six horizons of the RV-1 well of the Panna formation (2900–3500 m), Ratnagiri basin, Mumbai Offshore, India where HCFIs are observed. The Panna formation is mainly arenaceous (sand and siltstone) and is both a producing as well as a reservoir horizon in the Mumbai and Ratnagiri Basins. A number of 110 biphase aqueous secondary inclusions are selected for microthermometric observation using a heating-freezing stage. Temperature lowered up to freezing point where all phases both liquid and gaseous phase was frozen out (gas phase was condensed into liquid phase). When temperature is increased, the ice started melting and the temperature of first melting (T_{FM}) was noted and on continuous heating, the last crystal of ice also melts and the temperature (T_{LM}) was noted, where in the gaseous phase starts its appearance. After heating started, the Brownian movement of gas increased rapidly and ultimately the gaseous phase dissolved into liquid, and this homogenization to liquid gives the homogenization temperature or minimum entrapping temperature.

Microthermometry is a technique for determining the homogenization temperature (Th) of fluid inclusions. As a typical example, microthermometric analysis on 102 biphase, secondary inclusion assemblages from six horizons of the RV-1 well provided a Th range 80–135 °C, which is in tandem with the oil window temperature, and 8-fluid inclusion assemblages provided a Th range 140–165 °C and shows that there is an episode of high-temperature heating favorable for gas generation. Table 5.2 shows the variation in Th showed by coeval aqueous inclusions at different horizon depths in the RV-1 well, Mumbai offshore, India along with other details such as T_{FM}, T_{LM}, salinity, and phase changes observed during microthermometric analysis.

The Panna formation is mainly arenaceous (sand and silt stone) and is both a producing as well as a reservoir horizon in the Mumbai and Ratnagiri Basins. These results indicate the potential of microthermometric analyses on secondary nonhydrocarbon inclusions for identifying the oil window temperature in sedimentary basins.

5.6 Oil window: An Indicator of hydrocarbon potential in a basin

The fluid inclusion PVT data can provide information on the oil window, the temperature at which transformation of kerogen to oil happens (Munz, 2001; Blamey and Ryder, 2007). A microthermometric study on fluid inclusions was used to generate the homogenization data that indicated the oil window (Roedder, 1984; Shepherd et al., 1985).

Usually, the temperature interval at which oil is generated and expelled from the source rock ranges from 60 to 160 °C and is known as the oil window (Hunt, 1995). The relatively high geothermal gradients across the wells in a basin indicate shallow levels of hydrocarbon occurrence, which causes quite a narrow range of the oil window. The low geothermal gradient causes the first formation of oil to begin at deep subsurface levels, making the oil window quite broad. Below, the oil window, any potential source rock will not have begun generating liquid hydrocarbon and above the oil window, liquid hydrocarbons would have been converted to gaseous hydrocarbons. In general, the oil formation process is more in young source rocks where there is a high geothermal gradient and oil can form early at shallow depths.

Table 5.2 The variation in Th showed by coeval aqueous inclusions at different horizon depths in RV-1 well, Mumbai offshore, India.

Sl. No.	Depth (m)	Lithology	T_{FM} (°C)	T_{LM} (°C)	Th (°C)	Salinity (eq. wt % NaCl)	Phase/phase changes
1	1445–1450	Limestone	−5.2	−4.1	122.8	6.59	Biphase ($L+V \rightarrow L$)
2	1700–1705	Limestone	−77.6	−29.8	96	28.53	Monophase ($L+V \rightarrow L$)
3	2065–2070	Limestone, shale	−6.4	−2.1	104	3.55	Monophase ($L+V \rightarrow L$)
4	2420–2425	Limestone	−29.8	−4.7	128	7.45	Biphase ($L+V \rightarrow L$)
5	2665–2670	Limestone, siltstone, shale	−33.4	−28.9	112.8	27.97	Biphase ($L+V \rightarrow L$)
6	2770–2775	Siltstone	−8.2	−2.1	106.7	3.55	Biphase($L+V \rightarrow L$)
7	2985–2990	Siltyshale	−42.3	−31.8	104.3	29.82	Biphase ($L+V \rightarrow L$)
8	3055–3060	Shale, sandstone, siltstone	−26	−10.6	130	14.56	Biphase ($L+V \rightarrow L$)
10	3065–3070	Sandstone	−19.6	−15.4	108	18.96	Biphase ($L+V \rightarrow L$)
11	3105–3110	Sandstone, siltstone	−19	−8.6	132	12.39	Biphase ($L+V \rightarrow L$)
12	3110–3115	Sandstone, shale, siltstone	−22.7	−2.8	128	4.65	Biphase ($L+V \rightarrow L$)
13	3120–3125	Sandstone, shale	−12.4	−8.6	142	12.39	Biphase ($L+V \rightarrow L$)
14	3185–3190	Sandysiltstone, shale	−10.4	−3.2	122	5.26	Biphase ($L+V \rightarrow L$)
16	3245–3250	Sandstone, shale, siltstone	−28.3	−0.9	99	1.57	Biphase ($L+V \rightarrow L$)
17	3270–3275	Sandstone	−32.9	−8.9	120	12.73	Biphase ($L+V \rightarrow L$)
18	3295–3300	Sandysiltstone, shale	−10.2	−8.1	141	11.81	Biphase ($L+V \rightarrow L$)
19	3345–3350	Sandysiltstone, shale	−20.3	−1.8	140	3.06	Biphase ($L+V \rightarrow L$)
20	3395–3400	Sandysiltstone, shale	−9.4	−3.9	110	6.30	Biphase ($L+V \rightarrow L$)
21	3445–3450	Sandstone, shale	−50.4	−48.4	98	45.76	Biphase ($L+V \rightarrow L$)
25	3495–3500	Sandysiltstone, siltyshale, shale	−9.4	−3.6	108	5.86	Monophase ($L \rightarrow V$)

Low burial rates, also arising from the high geothermal gradients, imply that compaction of sediments in the basin should be efficient. As mentioned in Chapter 3, oil is mainly formed during the catagenetic phase, which is known as the oil window or oil zone. With increasing temperature, more molecular bonds are broken (H/C or O/C) and hydrocarbon molecules and aliphatic chains are formed from the kerogen. When burial rates and temperature increase, light hydrocarbons are generated due to cracking and increase the proportion of source rock hydrocarbons and petroleum (Tissot and Welte, 1978, 1984; Hunt, 1995). Subsequently, the hydrocarbons convert to wet gas with an increased amount of methane. The oil window, along with other geological and geophysical investigation, can suggest favorable conditions for hydrocarbon accumulation and maturation.

5.6.1 Th of fluid inclusions from the Mumbai offshore basin

In the RV-1 well, fluid inclusion samples from 22 depths were studied (minimum five fluid inclusion assemblages in a depth-5 m interval) for determining the homogenization temperature. The Mukta (early Oligocene), Bassein (middle-late Eocene), and Panna (Paleocene-early Eocene) Formations of the RV-1 well, Mumbai offshore basin provided a Th, which mostly falls within the oil window. At the Mukta Formation, fluid inclusions in calcite (1445–1450 m depth) provided a Th with the oil window. At the Bassein Formation, fluid inclusions in calcite lithologies at two different depths (1700–1705 m, 2065–2070 m) provided the oil window temperature. In the Panna Formation (with different lithologies of sandstone, shale, siltstone, and sandy siltstone), some depths 3120–3125 m, 3345–3350 m, 2855–2864 m exhibited a gas window (Th range 142–165 °C) along with the oil window at all other depths ranging from 2420 to 3500 m in quartz and feldspar (Table 5.3). The presence of HCFIs and the oil window indicates prospects for finding crude oil in adjacent areas. The Panna Formation is mainly arenaceous (sand and siltstone) and has several shaley (clay) layers and is a producing horizon. The Panna Formation is also a reservoir in the Mumbai Basin, where most of the HCFIs with a Th within the oil window range were observed.

5.6.2 Th of fluid inclusions from the Kerala-Konkan offshore basin

Fluid inclusion samples from 18 depths were studied in the KK-offshore well (five fluid inclusion assemblages in each depth at 5 m intervals). Most of the samples showed a Th above 100 °C, range from 101 to 140 °C, which falls in

Table 5.3 Homogenization temperature observed for the fluid inclusions from different Formations in the offshore basin, Mumbai, India.

Sl. No.	Geological Formation, Age	Lithology/(host mineral)	Depth (m)	Th range (°C)
1	Mukta, early Oligocene	Limestone, calcite	1445–1450	114–123
2	Bassein, Middle-late Eocene	Limestone, calcite	1700–1705	96–108
3		Limestone, shale	2065–2070	104–142
4		Limestone, siltstone	2420–2425	126–128
5		Limestone, siltstone, shale	2665–2670	113–117
6		Shaly limestone, siltstone	2770–2775	105–108
7		Shaly limestone, silty shale	2985–2990	104–110
8	Panna, Paleocene-early Eocene	Shale, sandstone, siltstone	3055–3060	116–130
9		Sandstone	3065–3070	112–132
10		Sandstone, siltstone, shale	3105–3115	109–134
11		Sandstone, shale	3120–3125	142–158
12		Sandstone, siltstone	3145–3150	85–130
13		Sandysiltstone, shale	3185–3190	120–143
14		Sandstone, shale, siltstone	3245–3250	99–147
15		Sandstone	3270–3275	80–135
16		Sandysiltstone, shale	3295–3300	140–145
17		Sandysiltstone, shale	3345–3350	130–142, 161–164
18		Sandysiltstone, shale	3395–3400	109–149
19		Sandstone, shale	3445–3450	94–134
20		Sandstone, siltyshale	3455–3460	99–133
21		Sandysiltstone, shale	3495–3500	108–117

the oil window but are not as similar as the Th obtained in the RV-1 well of the Mumbai offshore basin (85–142 °C). Sandstone, clay stone, and limestone are important lithologies in this area at depth ranges from 3120 to 6110 m, where samples from two depths (3175–3180 m, 3875–3880 m) showed Th 60–70 °C and 87–94 °C, respectively. The Th range 140–165 °C for some fluid inclusion assemblages in these two wells indicates that there is at least one episode of high-temperature heating favorable for gas generation in these basins (Table 5.4). Samples at some depths 3120–3125 m, 3140–3145 m, 3175–3180 m, 3215–3220 m, 3650–3655 m, 6110–6115 m show a high Th (140–146 °C) favorable for gas generation in the KK-basin. Based on the observed Th, we expect that in the KK Basin, only the depth horizons of 3000–4200 m are conducive for oil generation.

Table 5.4 Oil window/Th variation observed at different depths in Kerala-Konkan offshore well with lithological details.

Sample No	Depth (m)	Lithology/host mineral	Th (°C)
1	3120–3125	Claystone, Sandstone	135–138, 145–165
2	3065–3070	Sandstone, Claystone	118–120
3	3080–3085	Sandstone, Claystone	132
4	3140–3145	Sandstone, Claystone	120–127, 150–177
5	3175–3180	Sandstone, Claystone	60–70, 143–172
6	3180–3185	Sandstone, Claystone	111–118
7	3190–3195	Sandstone, Claystone	119–128
8	3200–3205	Sandstone, Claystone	105–110
9	3205–3210	Sandstone, Claystone	118–121
10	3215–3220	Sandstone, Limestone, Claystone	143–146
11	3475–3480	Claystone, Sandstone	115–117
12	3485–3490	Claystone, Sandstone	112–115
13	3650–3655	Claystone, Sandstone	140–142
14	3855–3860	Shale, Sandstone	110–117
15	3875–3880	Claystone, Sandstone, Limestone	87–94, 110–117
16	3890–3895	Claystone, Sandstone	100–140
17	4020–4025	Claystone, Sandstone	101–136
18	6110–6115	Siltstone, Sandstone	150

5.7 Summary

Oil and gas (hydrocarbons) are valuable energy resources hidden below the surface of the Earth. Detection of crude oil as fluid inclusions in rocks has driven the application of fluid inclusion studies in petroleum exploration and development. All studies of petroleum inclusions are anchored on a solid understanding of the properties of fluid inclusions such as the physical properties and geochemical processes that control petroleum formation, migration, and accumulation in sedimentary basins. The presence of petroleum inclusions in diagenetic cements or in healed fractures demonstrates that oil was present at the time of cement formation or later or fracture healing. Such simple observations indicate the relative time of oil generation and migration in the context of the diagenetic and tectonic history of the rock. The textural relationship of HCFIs with the host mineral, especially the HCFI presence in the secondary healed microfractures that cut across the minerals, indicates oil migration in the study area and the oil window determination from coeval aqueous inclusions and the HCFI itself are evidence of the conducive environment for finding oil in this region. Therefore, such studies can help the petroleum industry in exploring adjoining areas of the well or a detailed investigation on the migration pathways of the HCFIs in these formations even if a dug well is "dry."

References

Aplin, A.C., Macleod, G., Larter, S.R., Pedersen, K.S., Sørensen, H., Booth, T., 1999. Combined use of confocal laser scanning microscopy and PVT simulation for estimating the composition and physical properties of petroleum in fluid inclusions. Mar. Pet. Geol. 16, 97–110.

Blamey, N.J.F., Ryder, A.G., 2007. Hydrocarbon fluid inclusion fluorescence: A review. In: Reviews in Fluorescence. Springer, New York, pp. 299–334.

Bodnar, R.J., 2003. Interpretation of data from aqueous-electrolyte fluid inclusions. In: Samson, I., Anderson, A., Marshall, D. (Eds.), Fluid Inclusions: Analysis and Interpretation. Mineralogical Association of Canada: Short Couse Series 32, Vancouver, pp. 81–100.

Burruss, R.C., 1981. Hydrocarbon fluid inclusions in studies of sedimentary diagenesis. Mineral. Assoc. Canada Short Course Handbook 6, 138–156.

Burruss, R.C., 1989. Paleo temperatures from fluid inclusions: Advances in theory and technique. In: Naeser, N.D., et al. (Eds.), Thermal History of Sedimentary Basins. Springer-Verlag, New York Inc, pp. 119–131.

Burruss, R.C., 1992. Phase behaviour in petroleum-water (brine): Systems applied to fluid inclusion studies. In: PACROFI IV, Pan American Conference on Research on Fluid Inclusions, Program and Abstracts. Vol. 4, pp. 116–118.

Burruss, R.C., 2003. Petroleum fluid inclusions: An introduction. In: Samson, I., Anderson, A., Marshall, D. (Eds.), Fluid Inclusions: Analysis and Interpretation. Mineral. Assoc. Canada, Short Course Ser, Vol. 32, pp. 159–174.

Diamond, L., 2003. Introduction to gas-bearing aqueous fluid inclusions. In: Samson, I., Anderson, A., Marshall, D. (Eds.), Fluid Inclusions: Analysis and Interpretation. Mineralogical Association of Canada: Short Couse Series 32, Vancouver, pp. 101–158.

Goldstein, R. H., and Reynold, T. J., 1994. Systematics of fluid inclusions in diagenetic minerals, SEPM (Society for Sedimentary Geology) short course 31, USA ISBN #1–56576-00805.

Goldstein, R.H., 2003. Petrographic analysis of fluid inclusions. In: Samson, I., Anderson, A., Marshall, D. (Eds.), Fluid Inclusions: Analysis and Interpretation. Mineralogical Association of Canada, Short Course, 32, pp. 9–53.

Hunt, J.M., 1995. Petroleum Geochemistry and Geology, second ed. W.H. Freeman, New York. xx + 743 pp. ISBN 0 7167 2441 3.

Kihle, J., Hurum, J.H., Liebe, L., 2012. Preliminary results on liquid petroleum occurring as fluid inclusions in intracellular mineral precipitates in the vertebrae of Pliosaurus funkei. Nor. J. Geol. 92, 341–352.

Levine, J.R., Samson, I.M., Hesse, R., 1991. Occurrence of fracture-hosted impsonite and petroleum fluid inclusions, Quebec CityRegion, Canada. American Association of Petroleum GeologistsBulletin 75, 139–155.

Munz, I.A., 2001. Petroleum inclusions in sedimentary basins: systematics, analytical methods and applications. Lithos 55 (1–4), 195–212.

Munz, I.A., Iden, K., Johansen, H., Vagle, K., 1998. The fluid regime during fracturing of the Embla field, central trough, North Sea. Mar. Pet. Geol. 15 (8), 751–768.

Narr, W.M., Burrus, R.C., 1984. Origin of reservoir fractures in little knife field, North Dakota. Am. Assoc. Pet. Geol. Bull. 68, 1087–1100.

Nedkvitne, T., Karlsen, D.A., Bjorlykke, K., Larter, S.R., 1993. Relationship between reservoir diagenetic evolution and petroleum emplacement in the Ula field, North Sea. Mar. Pet. Geol. 10 (3), 255–270. https://doi.org/10.1016/0264-8172(93)90108-5.

O'Reilly, C., Parnell, J., 1999. Fluid flow and thermal histories for Cambrian–Ordovician platform deposits, New York: evidence from fluid inclusion studies. GSA Bull. 111 (12), 1884–1896.

Oxtoby, N.H., Mitchell, A.W., Gluyas, J.G., 1995. The filling and emptying of the Ula oil-field: Fluid inclusion constraints. In: Cubitt, J.M., England, W.A. (Eds.), The Geochemistry of Reservoirs: Special Publication. Vol. 86. Geological Society, London, pp. 141–157.

Peng, D.Y., Robinson, D.B., 1976. A new two-constant equation of state. Ind. Eng. Chem. Fundam. 15, 59–64.

Roedder, E., 1979. Fluid inclusions as samples of ore fluids. In: Barnes, H.L. (Ed.), Geochemistry of Hydrothermal Ore Deposits, second ed. Wiley, New York, pp. 684–737.

Roedder, I.E., 1984. Fluid inclusions. Mineralogical society of America. Rev. Mineral. 12, 1–644.

Shepherd, T.J., Rankin, A.H., Alderton, D.H.M., 1985. A Practical Guide to Fluid Inclusion Studies. Blackie and Son, Glasgow.

Sorby, H.C., 1858. On the microscopic structure of crystals, indicating the origin of minerals and rocks. Q. J. Geol. Soc. Lond. 14, 453–500.

Teinturier, S., Pironon, J., Walgenwitz, F., 2002. Fluid inclusions and PVTX modelling: examples from the Garn Formation in well 6507/2–2, Haltenbanken, Mid-Norway, June 2002. Mar. Pet. Geol. 19 (6), 755–765.

Thiery, R., Pironon, J., Walgenwitz, F., Montel, F., 2002. Individual characterization of petroleum fluid inclusions (composition and P-T trapping conditions): by microthermometry and confocal laser scanning microscopy: inferences from applied thermodynamics of oils. Mar. Petr. Geol. 19, 847–859.
Tissot, B.P., Welte, D.H., 1978. Petroleum Formation and Occurrence: A New Approach to Oil and Gas Exploration. United States.
Tissot, B.P., Welte, D.H., 1984. Petroleum Formation and Occurrence—A New Approach to Oil and Gas Exploration. Springer-Verlag, New York.
Van den Kerkhof, A.M., Hein, U.F., 2001. Fluid inclusion petrography. Lithos 55 (1–4), 27–47.

CHAPTER 6

Fluorescence spectroscopy in hydrocarbon fluid inclusions

V. Nandakumar[a] **and J.L. Jayanthi**[b]
[a]Scientist - G, National Facility for Geofluids Research and Raman Analysis, National Centre for Earth Science Studies (NCESS), Thiruvananthapuram, Kerala, India
[b]Project Scientist - C, National Facility for Geofluids Research and Raman Analysis, National Centre for Earth Science Studies (NCESS), Thiruvananthapuram, Kerala, India

6.1 Fluorescence spectroscopy

Spectroscopic methods for the analysis of hydrocarbon compounds are based on the phenomena of emission, absorption, fluorescence, and/or scattering. A Jablonski diagram explains the electronic state of a molecule and most of the molecules occupy the lowest vibrational level of the ground electronic state at room temperature, and on absorption of light, they are elevated to higher energy excited states. Excitation can make the molecule reach any of the vibrational sublevels associated with each electronic state, and the excited electronic state is usually the first excited singlet state. From this excited state of the molecule, relaxation can occur via several processes. After excitation, the nuclei adjust their positions to the new excited environment, so the interatomic distances become equal to the equilibrium distances belonging to the excited state. Usually, there is no emission during this kind of relaxation. Fluorescence emission occurs as a result of radiative electronic transition in which an electron jumps from a higher energy state to a lower one and the difference in energy is being released as a photon (Fig. 6.1). The system can return to the ground state spontaneously under emission of radiation from the lowest level of the excited state. The stated emission occurs at an energy lower than the absorbed energy due to the relaxation process and is observed at longer wavelengths than that of the excitation light. Fluorescence has a short lifetime ($\sim 10^{-8}$ s), and the wavelength of the light emitted depends on the energy gap between the ground state and the singlet excited state.

Fluorescence is characterized by quantum yield and lifetime. Quantum yield is the ratio of the number of photons emitted to the number of photons absorbed. The lifetime is the average time that a molecule spends in the

Hydrocarbon Fluid Inclusions in Petroliferous Basins
https://doi.org/10.1016/B978-0-12-817416-6.00008-3

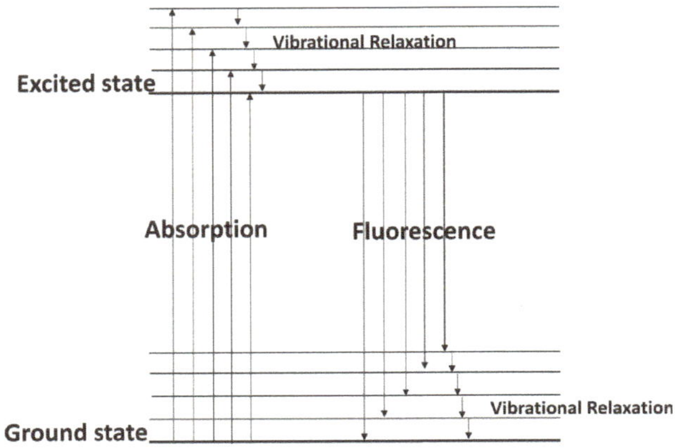

Fig. 6.1 Fluorescence energy level diagram.

excited state before its return to the ground state. A fluorescence emission spectrum represents a plot of emission against wavelength for any given excitation wavelength. If the wavelength of the exciting light is changed and the emission from the sample plotted against the wavelength of exciting light, it is known as the excitation spectrum. The emission spectrum provides information for both qualitative and quantitative analysis of hydrocarbon compounds.

6.2 Petroleum oil fluorescence

The use of fluorescence for the analysis of crude oils is being used for the past 70 years, particularly for mud logging where UV light is used to detect the presence of oil in drilling mud. Fluorescence is also used in the analysis of core samples, again to identify the presence of oils. The fluorescence of crude petroleum oils is derived from the aromatic hydrocarbon fraction, and this fluorescence emission is strongly influenced by the chemical composition (such as fluorophore and quencher concentrations) and physical characteristics (such as viscosity and optical density) of the oil. The excitation wavelength selected for fluorescence spectroscopy of crude oils should enable efficient excitation of all crude oil types. From the previous studies reported by Ryder in 2005, it is evident that visible excitation (>~450 nm) is not suitable for the light crude oils and/or condensates.

The American Petroleum Institute's (API) scale of "lightness" or "heaviness" of crude oil and other liquid hydrocarbons is linked to the market value of crude oil, which depends on the density of a petroleum-based liquid versus that of water.

$$\text{API gravity} = ((141.5/\text{specificgravity at } 15.6^\circ\text{C}) - 131.5)$$

The scale of API gravity indicates the crude's relative density but inversely (i.e., the lighter the crude, the higher the APIG) and vice versa. In this scheme, oil with an API value of >30° is termed light, between 22° and 30° termed medium, <22° heavy, and <10° very heavy. API gravity is one of the simplest and most wide-ranging parameters used for describing crude oils.

Specific gravities of oils are in the 0.73–1.0 range, with paraffin-type oils being lighter than asphalt-base oils. In general, light oils (high API gravity) have relatively narrow, strong fluorescence emission bands with smaller Stoke shifts towards blue than that found for heavier oils (lower API gravity) where the emission tends to be weaker, broader, and red-shifted. These differences in emissions are attributed to a higher concentration of fluorophores and quenchers present in the heavier oils, which leads to a higher rate of energy transfer and fluorescence quenching, producing the broader, weaker, red-shifted emission (Downare and Mullins, 1995; Ralston et al., 1996).

The quantum yields of crude oils are highly dependent on the excitation wavelength. UV excitation is more efficient than visible or infrared excitations because of the reduction in optical absorption and the increase in non-radiative decay pathways (internal conversion) with increasing excitation wavelength. The relative change in quantum yield is accounted for by the energy dependence of internal conversion and was also demonstrated to be very much similar for a wide range of light to heavy oils. The quantum yield or quantum efficiency for fluorescence is the ratio of the number of molecules that luminesce to the total number of excited molecules. Dilution of crude oils also increases the quantum yield by reducing the quenching rate. Therefore, diluted heavy oils have lower quantum yields than diluted light oils due to their higher concentrations of larger (red absorbing) chromophores, which are more likely to undergo nonradiative internal conversion. The balance between collisional energy transfer and quenching is the dominant feature affecting crude oil fluorescence (Ryder, 2005).

There is a significant consideration of excitation sources for petroleum fluorescence applications. The influence of the selection of excitation wavelength on the photophysics of crude oil emission is studied by Downare and Mullins (1995) while plotting the ratio of energy transfer to total emission against excitation wavelengths for the emission spectra of three representative oils. For several studies in the 1970s and 1980s, which employ 316 or 337 nm excitation sources, emission arises largely (>75%) from energy transfer, while other studies employing semiconductor light sources (380 nm

LED or 405 nm violet laser diodes) produce emission that has a lower energy transfer contribution (75%–60%). The changes observed in the fluorescence emission of crude oils with increasing excitation wavelength are; a narrowing of the emission band (Downare and Mullins, 1995; Ryder et al., 2002a,b) and reductions in the Stokes shift, quantum yield (Ralston et al., 1996), and fluorescence lifetime (Ryder et al., 2002a,b). This decrease is caused by the complex interaction between energy transfer and quenching processes. At short excitation wavelengths, energy transfer processes dominate since most of the absorbing fluorophores have large bandgaps and can transfer energy to a large numbers of smaller bandgap molecules. At a longer excitation wavelength, the excited fluorophores have small bandgaps and there are fewer molecules with smaller bandgaps for energy transfer, so most collisions result in quenching, with the subsequent reduction in fluorescence lifetime. Also, as the bandgaps of the excited fluorophores decrease, there is an increased rate of internal conversion, which also contributes to the reduction in lifetime (Zhu and Mullins, 1992; Downare and Mullins, 1995; Nicodem et al., 2000; Ryder et al., 2002a,b).

Excitation by UV light causes petroleum fluids to emit light (or fluoresce) in the visible range (400–700 nm). The fluorescence emission of organic molecules is due to electronic transitions involving carbon–carbon double bonds. The part of a molecule which can be excited by incident light is called a chromophore. Burruss (1991) has described the analytical setup of fluorescence microscopy in detail. A mercury lamp is often used as an excitation source. Monochromatic light can be generated by use of bandpass filters, and the most commonly used excitation wavelength for petroleum is 365 nm. The monochromatic UV light is directed onto the sample and the emitted light is transmitted from the microscope into a spectrograph and detector system. Different combinations of optical components (e.g., filters, lenses, beam splitters, fiber optics) can be used (Burruss, 1991; Guilhaumou et al., 1990; Kihle, 1995). The efficiency of transmission is highly dependent on these components, and absorption in the light path may lead to significantly enhanced detection limits. Optical components may also emit fluorescence, leading to undesired contamination. Therefore, fluorescence studies on crude oils/petroleum require a thorough knowledge not only of the specific optical components used in the light path, but also of calibration of both wavelengths and intensity.

The relationship between fluorescence and chemical composition of petroleum is highly complex. The main fluorescing components in petroleum are aromatic hydrocarbons (Hagemann and Hollerbach, 1986;

Khorasani, 1987). Nitrogen-sulfur and oxygen (NSO) compounds show a less intensive fluorescence at higher wavelengths than the aromatic compounds in the red range (Khorasani, 1987). Various parameters have been used to quantitatively describe the fluorescence emission spectrum, such as the wavelength of maximum intensity, and emission ratio at particular wavelengths. During fluorescence emission, a number of processes may occur; high concentrations of chromophores may lead to absorption of emitted fluorescence; other matrix molecules may absorb both exciting and emitting fluorescence and the emission spectra may be displaced due to molecular interactions (Bertrand et al., 1986). Thus, the fluorescence emission represents a very important fingerprint on petroleum composition. Fluorescence spectroscopy is a valuable tool used to distinguish different generations of inclusions (Guilhaumou et al., 1990). The heavy oils show fluorescence emission at higher wavelengths than lighter oils. The potential of fluorescence emission spectroscopy for the determination of unknown API gravity of crude oils/oils in HCFIs/oil condensates is discussed in this chapter.

6.3 Steady-state and time-resolved fluorescence techniques

Optical (absorption and emission) spectroscopy relies on the creation of excited states and reactive intermediates through the absorption of light, which leads to two fundamentally different approaches. In steady-state spectroscopy, the samples are continuously irradiated with a constant intensity light beam; excited states are continuously created and eventually a steady state is reached. This facilitates the measurement of weak signal levels at the expense of losing kinetic information. Steady-state (continuous wave-CW) spectroscopies are best applied to the measurement of absorption and emission spectra. On the other hand, time-resolved spectroscopies rely on the irradiation of the sample with a light source whose intensity fluctuates as a function of time (a pulsed laser emits flashes of light). Each flash creates a burst of excited states, whose evolution can be monitored as a function of time. Time-resolved spectroscopy provides kinetic information at the expense of a lower sensitivity compared to steady-state techniques.

The use of fluorescence methods for the qualitative and quantitative analyses of crude oils include the study of microscopic petroleum fluid inclusions within rocks. The correlation of different fluorescence parameters with chemical/physical characteristics of bulk crude oils has been used to develop qualitative or quantitative models for industrial or research applications. The selection of the best fluorescence parameters for quantitative crude oil analysis is a complex issue and has not yet been fully resolved. One of the key challenges, however, is to develop a method that is suitable for any crude oil type (heavy/light/condensate) from different geological settings. Earlier studies on the steady-state and time-resolved fluorescence parameters for the analysis of crude oils are discussed here.

6.3.1 Steady-state emission

Ryder (2005) reported the emission spectra obtained from a series of five crude oils, using 337 nm excitation, and the fluorescence emission band differences in light oils and heavy oils have been discussed. Similar results were reported by Rayner and Szabo (1978), also using 337 nm excitation. The changes in fluorescence emission reported (Ryder, 2005) are due to the higher concentration of fluorophores and quenchers present in the heavier oils, which leads to a higher rate of energy transfer and fluorescence quenching, producing the broader, weaker, and red-shifted emission (Downare and Mullins, 1995; Ralston et al., 1996). Energy transfer processes result in red-shifted and spectrally broadened emission, while the quenching processes reduce the emission intensity. This effect was diminished by increasing the excitation wavelength with near-IR excitation, where the profiles of the emission spectra of light, heavy, and diluted samples of each are very nearly identical (Ryder, 2005). At these NIR excitation wavelengths, nearly no energy transfer is observed. Dilution of crude oils with solvents such as benzene, cyclohexane, etc. results in an increase in fluorescence intensity and a blue shift. The Stern-Volmer plots constructed from highly diluted crude oils imply that the quenching and energy transfer processes depend linearly on crude oil concentration (Wang and Mullins, 1994). Analysis of the wavelength dependence of the Stern-Volmer plots indicates that the collisional decay constants are greater at shorter emission wavelengths and excitation wavelengths. The quantum yields of crude oils are highly dependent on the excitation wavelength, with excitation in the visible or red being much less efficient than UV excitation. This is largely due to the reduction in optical absorption and the increase in nonradiative decay pathways (internal conversion) with increasing excitation wavelength. The relative change in quantum yield was also demonstrated to be largely similar for a wide range of light to heavy oils, which can be accounted for by the energy dependence of internal conversion. Dilution of crude oils also increases the quantum yield by reducing the quenching rate (Wang and

Mullins, 1994). Diluted heavy oils, however, have lower quantum yields than diluted light oils due to their higher concentrations of larger (red absorbing) chromophores, which are more likely to undergo nonradiative internal conversion.

The balance between collisional energy transfer and quenching is the dominant feature affecting crude oil fluorescence. From emission spectra for three representative oils, Downare and Mullins (1995) calculated and plotted the ratio of energy transfer to total emission against excitation wavelengths and showed how influential the selection of excitation wavelength is on the photo-physics of crude oil emission. This has glaring consequences for the consideration of excitation sources for petroleum fluorescence applications. For several studies in the 1970s and 1980s, which employ 316 nm or 337 nm excitation sources, emission arises largely (>75%) from energy transfer, while studies employing semiconductor light sources (380 nm LED or 405 nm violet laser diodes) produce emission that has a lower energy transfer contribution (75%–60%). Therefore, the selection of excitation sources is to be specially taken care of. The changes observed in the fluorescence emission of crude oils with an increasing excitation wavelength are: a narrowing of the emission band and reductions in the Stokes shift, quantum yield, and fluorescence lifetime (Ryder et al., 2002a,b; Downare and Mullins, 1995; Wang and Mullins, 1994). This decrease is caused by the complex interaction between energy transfer and quenching processes. At short excitation wavelengths, energy transfer processes dominate since most of the absorbing fluorophores have large bandgaps and can transfer energy to the large numbers of smaller bandgap molecules. At a longer excitation wavelength, the excited fluorophores have small bandgaps and there are fewer molecules with smaller bandgaps for energy transfer, so most collisions result in quenching, with the subsequent reduction in fluorescence lifetime. Moreover, as the bandgaps of the excited fluorophores decrease, there is an increased rate of internal conversion, which also contributes to the reduction in lifetime.

Three of the most common steady-state parameters used by the geological science community for oil analysis are (a) fluorescence parameter (emission wavelength, in particular the point of maximum emission), (b) the red-green ratio (shows a better correlation with oil viscosity, which is linearly related to oil density), and (c) the QFT-II method developed by Texaco in the 1990s for oil well logging. The method is designed for on-site analysis and only yields an estimate for API gravity; it is unsuited for heavy oils (API gravity <15).

6.3.2 Time-resolved fluorescence

The fluorescence lifetime of crude oils is very sensitive to composition, with heavy oils having shorter lifetimes than light oils. The change in fluorescence lifetime is a result of the interplay between energy transfer and quenching; it is also emission wavelength dependent, since each emission wavelength represents a different population of emitting fluorophores. The numerical value of the fluorescence lifetime is, however, very dependent on the method by which it is calculated like via a single exponential convolution process or a deconvolution method (Rayner and Szabo, 1978; Ryder, 2005). However, no attempt was made to correlate the lifetime data with chemical composition. The complex composition of crude oils ensures that the measured fluorescence decay curve is the sum of a distribution of individual decay curves; therefore, interpreting fluorescence lifetimes is to be done very cautiously. In general, the wavelength at which the maximum average lifetime is measured occurs at a longer wavelength for heavy crude oils than it does for light oils. The variation in lifetime ranges from 450 to ~600 nm emission wavelength and is more pronounced for the lighter oils; a maximum value for the average lifetime (τ max) is recorded at a wavelength λτmax, which is dependent on the chemical characteristics of each oil. As discussed by Ryder (2005), a noticeable decrease in the average lifetime is observed at emission wavelengths greater than ~600 nm. This wavelength dependence is due to the complex interplay between collisional energy transfer and quenching processes (Ryder et al., 2002a,b; Ryder, 2002). At short emission wavelengths, fluorophores have large bandgaps and can readily undergo collisional energy transfer with larger, aromatic, small bandgap molecules. The rate of energy transfer is dependent on the bandgap between the two energy level populations, and so the value of τ relative to the maximum lifetime (τ max) is shortened the most at short emission wavelengths where the bandgap is greatest. As the emission wavelength increases, the energy gap between the donor and acceptor decreases, resulting in an increase in the energy transfer rate. In addition, as the emission wavelength increases, the concentration of large aromatic acceptor molecules decreases, which leads to a decrease in the rate, as there are fewer acceptor molecules. There is a net overall decrease in the rate of energy transfer leading to an increase in τ until τ max is reached at a wavelength λτmax. Λτmax is red-shifted as the API gravity of the oils decreases.

At emission wavelengths longer than λτmax, collisional quenching takes over as the dominant effect because as the emission wavelength increases, the bandgap between the excited fluorophores and acceptor molecule gets smaller. Eventually, the bandgap is so small that the acceptor molecules act as quenchers since they can de-excite themselves (internal conversion), and so collisional quenching dominates, resulting in a decrease in τ relative to τ max. The excited fluorophores can also undergo internal conversion, the rate of which is inversely proportional to the bandgap, and this leads to a decrease in τ. The greater degree of lifetime change associated with lighter oils is due to the comparatively dilute concentration of fluorophores, so the average lifetime is more sensitive to collisional and internal conversion processes. With heavy oils, the concentration of fluorophores and quenchers is so high that small changes in the concentration of donor or quencher species will have very little effect on the average lifetime. The decrease in lifetime attributable to internal conversion is far outweighed by collisional quenching for the heavy oils.

Time-resolved fluorescence measurements are largely insensitive to the negative factors that affect steady-state measurements and are more easily referenced, making sensing applications more robust. Time-resolved fluorescence techniques, employed for characterization of petroleum products, are not only largely free of these common artifacts, but in addition contain information that is lost in the time averaging process inherent in steady-state methods. In the case of crude oils, which are complex mixtures of fluorophores, time-resolved fluorescence measurements offer the best approach for fully revealing the influence of quenching and energy transfer processes on fluorescence behavior. The use of an intensity average fluorescence lifetime for the characterization of crude petroleum oils and to correlate fluorescence lifetime data with physical characteristics (API gravity) and compositional factors such as alkane, aromatic, polar, and sulfur concentration is very interesting. Using a longer excitation wavelength (405 nm) resulted in a slightly better correlation for both corrected alkane and polar concentrations. (In each case, however, there is still considerable scatter about the best-fit line, preventing the development of accurate, quantitative models.)

The wide diversity of sources from which the test oils were derived almost certainly contributes to the large scatter in the plots. The use of simple lifetime models is another potential source of error, since they cannot fully account for the complex decay kinetics of crude oils. Further investigations are underway to try to resolve this issue and will be reported in the near future. The use of fluorescence lifetime ratios for quantitative analysis was also investigated, but these studies used a restricted set of oils from the North Sea. The application of lifetime ratios to the more diverse oil sample sets does not result in accurate correlations suitable for analytical use.

6.3.3 Multidimensional techniques

Multidimensional techniques that involve the collection of multiple fluorescence parameters offer a more detailed method for studying petroleum photo-physics or providing a spectroscopic fingerprint for identification purposes. Excitation-emission matrix (EEM) methods are a convenient way of mapping the fluorescence space that complex oils occupy. They can be used in the laboratory to determine the optimum excitation-emission parameters for simpler measurements/instrumentation, or as a method for the discrimination of oil types. The main drawbacks of EEM include time-consuming data collection and the requirement for complex instrumentation. Synchronous fluorescence spectroscopy (SFS) has been applied to the study of petroleum-based materials since the mid-1970s. Constant wavelength mode SFS has been used to study crude oils, motor oils, and asphaltenes. SFS has also been applied to the study of shale oils, fluid inclusions, and reservoir compartmentalization. Constant energy SFS was used for the analysis of both crude oil and gasoline. Three-dimensional fluorescence spectra have been used to discriminate oils into condensate, light oil, and heavy oils. In addition, Total Synchronous Fluorescence Scan Spectroscopy (TSFS) has been used to discriminate different refined and crude petroleum liquids. Generally, as a crude oil matures, the aromatic fraction is gradually reduced, and therefore it would seem possible to assess the maturity of the oils by measuring the changes in the TSFS profile. Unfortunately, this is probably only applicable to oils from a single source.

6.3.4 Fluorescence studies-applications in HCFIs

The fluids entrapped within HCFI are fossilized samples of oils, gases, aqueous solutions, and solids whose composition might not have changed since the trapping event. The accurate determination of the chemical composition of entrapped fluids can provide (together with microthermometric studies) essential information about the crystal growth, temperature, and timing of fluid migration. This information is of significant importance to the petroleum exploration industry, particularly with regard to the study of petroleum reservoirs. The analysis of HCFI is done by either crushing bulk rock samples or extracting the entrapped fluid for analysis, or by the analysis of single fluid inclusions. The oil composition data obtained from bulk fluid inclusion analysis (by crushing) suffer from a variety of problems such as sample destruction, mixing of fluids from heterogeneous fluid inclusions, and contamination from materials within the rock sample itself as discussed.

Fluorescence-based methods are widely used for studying HCFI, and the most common identification method is by observing their fluorescence under UV illumination. The use of visually determined fluorescence color is widely used as a qualitative guide for assessing the maturity of oil in HCFI. Unfortunately, the use of fluorescence color is intrinsically prone to error; it does not yield quantitative results. In addition, there are issues with instrumental variation (excitation wavelength, emission filters, etc.) and reproducibility.

The red-green quotient Q (Q = intensity 650 nm/intensity 500 nm) and the wavelength of maximum fluorescence emission intensity (max) were found to correlate with API gravity and gross chemical composition (%w saturates, aromatics, polars, and asphaltenes) for a set of HCFIs synthesized using a sample set of Canadian crude oils (Stasiuk and Snowdon, 1997). The apparent fluorescence intensity and spectral distribution can get adversely influenced by the physical properties of the sample such as geometry, opacity, sample turbidity, and the scattering properties of both the actual inclusion and host mineral. These things are to be carefully considered while studying HCFIs where there is a large variation in sample type. Photobleaching of synthetic and natural HCFI has been shown to cause changes in the fluorescence emission intensity. In 1987, McLimans outlined the potential for using fluorescence lifetime measurements for HCFI analysis. He showed that fluorescence lifetime (355 nm excitation) increased from condensate, to very mature, to moderately mature, to immature oils. There was also a noticeable increase in lifetime for all oils with emission wavelength over the 400–600 nm range. However, no data were provided on the method for calculating fluorescence lifetime or about the chemical composition/source of the oil samples; nor were any HCFI data presented. The utility of lifetime-based measurements for HCFI characterization, though still in its infancy, does offer considerable benefits when compared to the more traditional color, intensity, and spectral measurements. One novel application of fluorescence to inclusion studies involves using confocal microscopy to measure the volume of oil and gas within fluid inclusions by utilizing the spatial resolution of modern confocal microscopes.

The use of fluorescence spectroscopy in a wide variety of methodologies for the characterization and analysis of crude petroleum products is well established. Yet there are considerable opportunities for the development of robust, accurate, and quantitative fluorescence methods for crude oil analysis. The rapid miniaturization and increasing capability of fluorescence instrumentation is further expanding the range of applications, particularly with regard to on-site and in situ measurements. This diversity of excitation sources, each of which excites different fluorophore populations within the crude oils, makes it increasingly difficult to compare quantitatively results from different studies. This, combined with the fact that most research laboratories work on widely different crude oils, further fragments this economically important research field. Therefore, a set of crude oil standards (either natural or synthesized in the laboratory) needs to be developed, freely available to all, with which new methods and instrumentation can be calibrated, to enable results from different laboratories to be more easily compared.

6.4 Fluorescence spectroscopy for hydrocarbon fluid inclusions

Fluid inclusions (FIs) are microsamples of fluids trapped and sealed in mineral cavities during phases of mineral growth and/or as in subsequent modifications. FI technology is greatly diverse in its application in oil exploration and invaluable in the reconstruction of fluid flow histories in reservoir rocks (Roedder, 1984; Middleton et al., 2001), indirectly yielding information about the temperature, salinity, and composition of fluids that migrated through the basin in the geological past. Hence, FIs represent samples of reservoir fluids of certain geologic time bands. In petroliferous basins, FIs, mostly in the size range of 5–20 μm, are monophase, biphase, or multiphase (containing oil, gas, solid, or water, or any combination), and the oil-bearing ones typify microsamples of petroleum that once migrated through the rocks prior to trapping (Munz, 2001). Oil inclusions are easily identified in thin sections using ultraviolet (UV) light-induced fluorescence due to the presence of aromatic and polar fractions in oil (McLimans, 1987). According to Goldstein, FIs are one of the most sensitive and critical indicators of diagenetic environments and are an important toolset in oil exploration and development (Goldstein, 2001).

Precise characterization of the petroleum fluids trapped in inclusions is burdened with considerable challenges. Usually, HCFIs are obtained from core and cutting samples from exploratory wells, and considering the tedious procedure adopted for wafer preparation to spot HCFIs, they are to be carefully handled; so it requires noncontact, nondestructive, analytical methods and procedures. The composition of petroleum in HCFIs is obtained typically by gas chromatography and mass spectrometry of an extract of petroleum after crushing (a destructive process) a small quantity of the HCFI-bearing sample material, but this results in mixing of fluids from multiple sources (i.e., multiple inclusions) (George et al., 2001; Kihle, 1995; Anderson et al., 1998; Parnell et al., 1998, 2001; Blamey and Ryder, 2007). The study of HCFIs through optical methods is wholly devoid of such mixing; therefore, by using nondestructive optical methods, more precise control on discriminating individual inclusions can be achieved.

Optical-spectroscopic methods provide valuable data on the profile (composition and properties) of petroleum inclusions and entrapped fluids and are representative of the hydrocarbon fluids that once existed. Most importantly, optical procedures (a nondestructive/noncontact analysis) are of reasonable sensitivity, facilitating the analysis of individual micron-sized inclusions (Blamey and Ryder, 2007; Guilhaumou et al., 1990; Przyjalgowski et al., 2005). In this chapter, a combination of microscopy and fluorescence-based approaches are presented for the analysis of hydrocarbon fluid inclusions that ensure reliable data/results with high sensitivity and diagnostic potential.

Fluorescence and infrared spectroscopy are widely employed as nondestructive methods for the characterization of crude oils and hydrocarbon fluid inclusions (Blamey and Ryder, 2007; Stasiuk and Snowdon, 1997; Bourdet et al., 2012, 2014). Quick information on API gravity (APIG) is unquestionably an important step in managing the crude oil extraction in any oil field. A methodology has been reported to deduce the APIG of oil right from HCFIs (Nandakumar and Jayanthi, 2016) in samples (single mineral grain) of well cuttings from lithologies (source, reservoir, or carrier beds) encountered during drilling, having the probability to act as oil-producing zones. This approach involves a nondestructive and noninvasive procedure to capture the fluorescence emission spectra of HCFIs trapped in the specially prepared sample "wafers." The ratio of emission spectral intensity is used to interpolate the API gravity from a "standard" line-fitted scatterogram of known API values versus emission spectral intensity ratios.

This chapter elaborates on fluorescence emission data of pure petroleum oils of known APIG and proposes an empirical graphical tool for predicting the APIG of oils in micron-sized HCFIs. It also documents the application of this empirical tool for inferring the APIG in HCFIs by considering the samples from the RV-1 well, which is a nonproducing dry well from a basin with proven commercial productivity from the Mumbai offshore basin, India.

6.5 Fluorescence emission of oils and its relation to API gravity

Stasiuk and Snowdon (1997) used fluorescence spectra to estimate API gravity using correlation equations calculated from a calibration set of crude oils. These correlations are based on a red/green intensity ratio (Q = intensity 650 nm/intensity 500 nm) and the wavelength at maximum spectrum intensity (Lmax). The application of this correlation to light oil or condensate inclusions (as in the case of Phuong Dong Field) proved to be difficult because the fluorescence signal at 650 nm is often weak with a low signal-to-noise ratio, making the Q values imprecise. Later on, Bourdet et al. (2012) calculated a green-blue fluorescence intensity ratio ($Q_{G/B}$ ratio) from the fluorescence spectra and used it to estimate the density of oil, expressed as API gravity, by comparison to a calibration set of crude oils. In 2014, Bourdet et al. derived correlations based on a new set of crude oils (Australian crude oils and one Middle East crude oil) with API between 30.3 and 59 and a different definition of Q (intensities at 510 and 430 nm), but it could not achieve a perfect correlation.

Determination of the APIG of oils in HCFIs is important since it offers a unique insight into the quality of oils that are present in the different Formations of petroliferous basins. Detection of API gravity of oil in HCFIs from RV-1 well cuttings serves as an example to demonstrate the relation between emission intensity ratio (at F620/F560) and known API gravity (of crude oils). Fluorescence spectral data (on excitation with a 405-nm diode laser) of 13 crude oil samples of known API gravities are used to develop an empirical relation between the API gravity and emission spectral ratios.

6.5.1 Fluorescence analysis of crude oil samples

Thirteen fresh and pure oil samples of known APIG (collected from the chemical analytical facility of the Regional Geological Laboratory (RGL) of ONGC at Panvel, Mumbai) are used in the study as a reference set of standards for the derivation of emission spectra. These oil samples are with different APIGs, ranging from 20.49 to 64.88, out of which two are condensates with APIGs 52.5 and 64.88 (Tapti condensate). All the crude oil samples are from the Mumbai offshore basin, India.

One drop of oil (0.05–0.1 mL) placed in the sample holder is viewed through a 5 × objective lens of 0.12 numerical aperture (NA) at a working distance of 14 mm. A laser beam of 1.0 mW output power is directed to a spot size of 4 μm on the oil sample to induce fluorescence and the emitted fluorescence recorded on a laser Raman microspectrometer (Renishaw inVia, UK) with 2400 L/mm (UV) grating providing a 1-nm spectral resolution. The entire spectrum is captured in the wavelength range of 406–750 nm in one acquisition window, with an acquisition time of 10 s.

Normalized emission spectra of oil samples of known APIG and a mean spectra of each group (API = below 30, 30–40, 40–50, and above 50) are given in Fig. 6.2A. Normalization of peak emission intensity on each spectrum reduces the effect of variation in excitation laser light intensity. Medium (APIG = 30–40) and heavy (APIG <30) oils show an emission peak, each around 620 and 650 nm, respectively. Oil samples of APIG in the range of 40–50 and APIG >50 (i.e., lighter oils) show an emission peak around 560 nm with two shoulder peaks around 450 and 460 nm. Such peaks are characteristic of lighter oil. It is evident from Fig. 6.2A that light oils (of higher APIG) have relatively narrow, intense fluorescence emission bands with small blue shifts, while heavy oils (of lower APIG) have emissions that tend to be broader, weaker, and red-shifted. Spectral emission from the 13 crude oil samples is shown (Fig. 6.2B). The emission spectra from 13 crude oil samples are grouped into four categories based on the known APIG values.

Fluorescence spectral data (of 13 oil samples of known APIG) are normalized to the maximum emission intensity using the Renishaw WiRE software of the spectrometer system. Gaussian curve fitting for each spectrum (using Origin 7.0) located the exact emission peaks around 500, 560, and 620 nm. Mean spectral emission intensity is calculated for these wavelengths with a standard deviation of ±30 nm. Table 6.1 lists APIG data of 13 oil samples along with the corresponding F620/F560 ratios. Spectral intensity ratio at F620/F560 is then calculated for each oil sample since the spectral profile shows a clear distinction of blue shift in light oils around 560 nm and a red shift around 620 nm for heavy oils.

6.5.2 Fluorescence analysis of HCFIs

In order to predict the APIG of oils in HCFIs, samples of RV-1 well cuttings (a nonproducing well in the Ratnagiri block of the offshore Mumbai basin, India) have been collected from the core library of RGL (Oil and Natural Gas Corporation), Panvel, Mumbai, India. The samples were crushed, washed, dried, sieved, and then securely stored prior to preparation of "specialized" wafers (wafer preparation is detailed in Chapter 2). Fluid

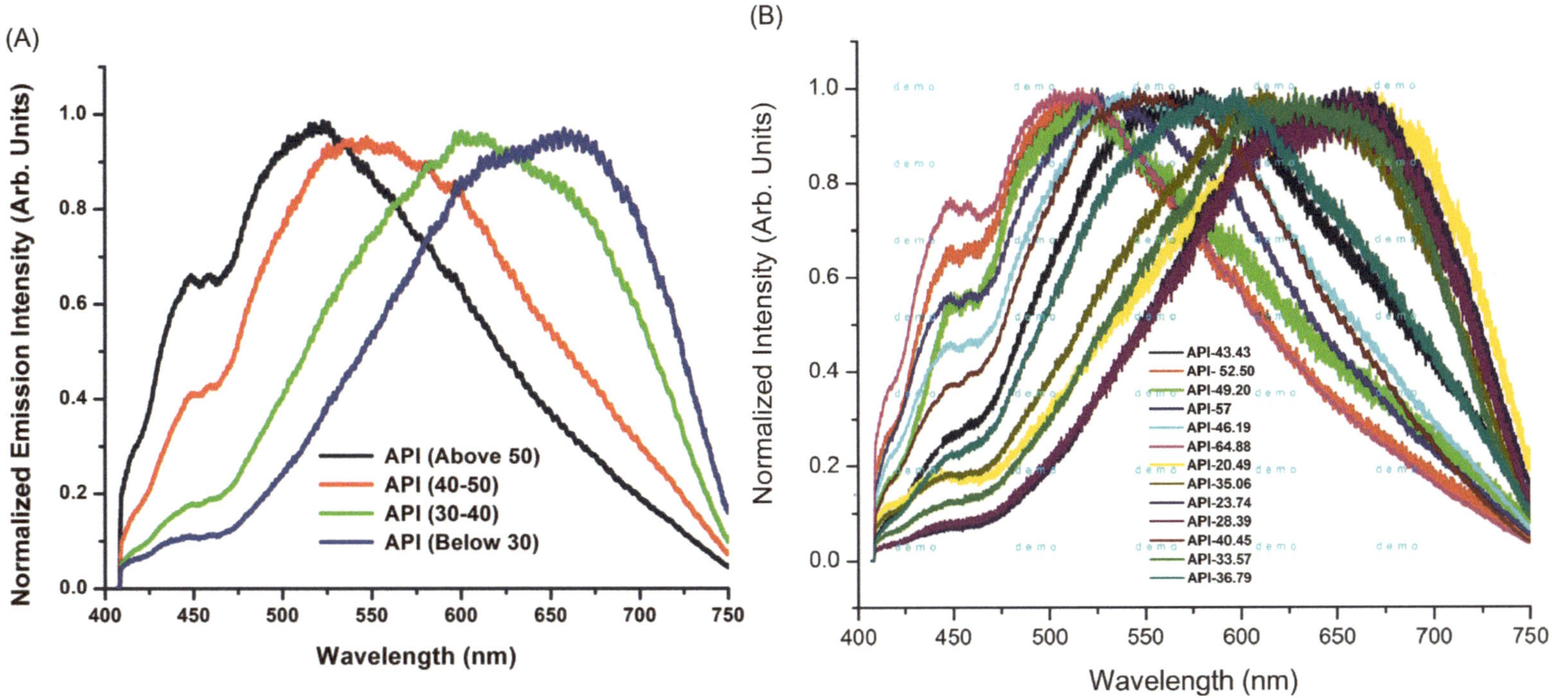

Fig. 6.2 (A) Normalized fluorescence spectra of petroleum oils of known API gravity at 405 nm excitation. The emission spectra are grouped into four categories according to API ranges as shown. (B) Fluorescence emission spectra from the 13 crude oil samples with different API gravities.

Table 6.1 APIG and corresponding emission ratios of crude oil samples with their location.

Sample no.	APIG	Oil type	Emission ratio, F620/F560	Oil location	Well no.
1	20.49	Heavy oil	1.650772	Mumbai basin, India	SB-13
2	23.74	Heavy oil	1.554814	Mumbai basin, India	BH 134
3	28.39	Heavy oil	1.508974	Mumbai basin, India	B-15A-1
4	33.57	Medium oil	1.359441	Mumbai basin, India	B-22-5
5	35.06	Medium oil	1.237669	Mumbai basin, India	Hy-8
6	36.79	Medium oil	0.96095	Mumbai basin, India	B-179 A
7	40.45	Medium oil	0.746964	Mumbai basin, India	VSEA-16
8	43.43	Volatile oil	0.703765	Mumbai basin, India	BH-68
9	46.19	Volatile oil	0.673091	Mumbai basin, India	DRB-1
10	49.2	Volatile oil	0.62868	Mumbai basin, India	C-39
11	52.5	Gas condensate	0.584046	Mumbai basin, India	N_2O C-series
12	57	Light oil	0.475672	Mumbai basin, India	B-149
13	64.88	Gas condensate	0.406672	Mumbai basin, India	Tapti

inclusions and HCFIs in grains are identified using UV excitation (with a Mercury arc lamp) under a fluorescence microscope. In RV-1, HCFIs are reported from quartz and feldspar grains, especially in the sandstone-siltstone lithologies at well depths ranging from 3070 to 3690 m. Minor quantities of hydrocarbon fluid inclusions are also noticed in the 1345–1350 m horizon (i.e., the Mukta formation comprising limestone-shale lithologies). HCFIs occurring in secondary trails showed good fluorescence under UV illumination (fluorescence microscopy) and are mostly biphase inclusions (i.e., 80%–85% liquid and 20%–15% gas). A dark spot in the inclusion is the vapor bubble, which does not fluoresce under UV excitation (Fig. 6.3).

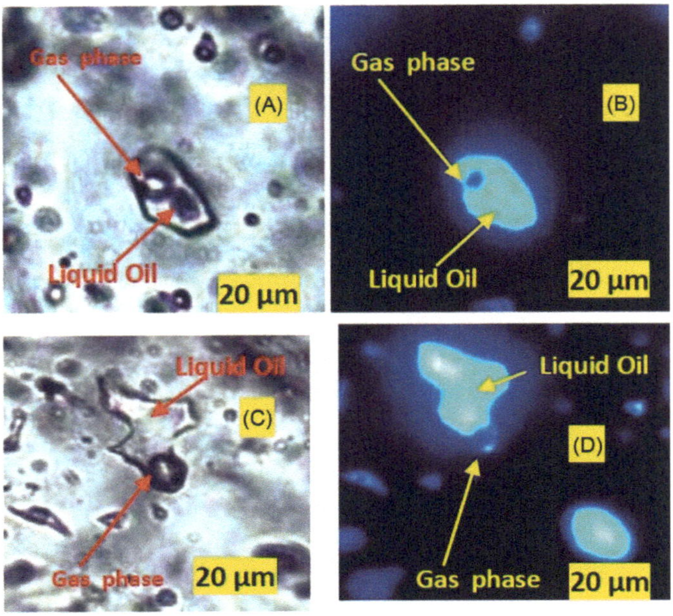

Fig. 6.3 Paired-transmitted light and UV photomicrographs of fluorescing biphase oil inclusions from (A – B) secondary fractures in feldspar, depth 3495–3500 m, and (C – D) secondary annealed fractures in quartz, depth 3445–3450 m.

Once HCFI occurrences are located, the fluorescence emission studies are carried out using an inVia Raman microscope (Renishaw, UK, Model no. M-9836-3991-01-A) attached to an inVia Reflex Raman Spectrometer offering automated alignment via the Renishaw WiRE software. A spectrum stabilized laser module at 405 nm excitation (20 mW, CW, Model. no. I0405SR0050B, Class 3B laser product, Innovative Photonic Solutions, USA) is used for the study. The system is fully automated and self-validating with auto aligning and optimization of the input laser.

The presence of HCFIs is explored on specially prepared representative wafers of 11 well cuttings of RV-1 from six different horizons (depths) following the same procedure and settings used in respect of samples of oil of known APIG (i.e., fluorescence emission on UV excitation of the wafers) but at 100 × magnification, whereas a 5 × objective was used for crude oil samples. Laser output power was the same as for the HCFI study with a working distance of 0.33 mm, 0.85 NA, and a laser spot size 0.6 μm. Fluorescence emission was recorded for the same wavelength region (406–750 nm) for the 11 HCFI samples from six different horizons (Fig. 6.4). A minimum of 20 individual inclusions were analyzed from each sample and averaged. HCFIs from

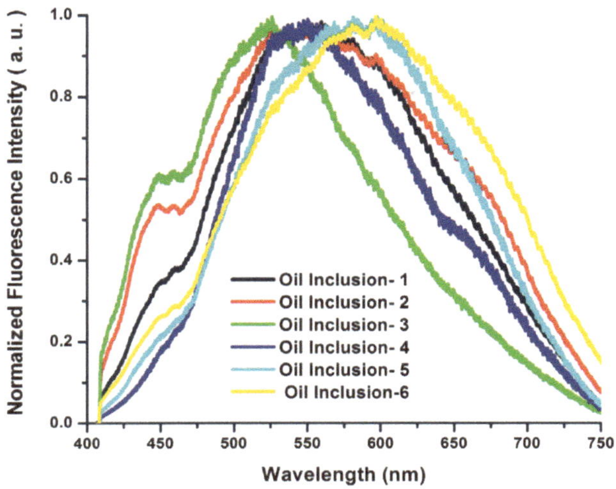

Fig. 6.4 Normalized fluorescence spectra of oil trapped as HCFIs at 405 nm excitation.

the same depth show a similar fluorescence emission pattern. Therefore, from the 11 HCFI samples, six HCFIs in different horizon depths were retained for further analysis.

6.5.3 Fluorescence emission ratio and API gravity

A bivariate scatter plot of ratio of emission intensity (on the X axis) vs APIG (on the Y axis) is plotted for all 13 oil samples, followed by a smooth line fitting. This plot shows a fall in spectral ratio (F620/F560) with rising API gravity values, a distinct inverse trend (Fig. 6.5).

F620/F560 emission ratios in the scattergram show an inverse relation with APIG (Fig. 6.5). An empirical algebraic expression is derived based on the correlation of API gravity of crude oils with intensity emission ratio (F620/F560) and is as follows:

$$y = y_0 (x_0 - x)^t \pm 1 \tag{6.1}$$

where y_0, x_0, and t are constants with values, $y_0 = 23$, $x_0 = 2.55$, and $t = 1.4$, and x is the emission ratio value at F620/F560.

In order to predict the APIG of oils in HCFIs, the fluorescence emission data from the six HCFI samples were smoothed and normalized using the WiRE software and the intensity ratio F620/F560 for each sample was calculated. The F620/F560 emission ratio for HCFl inclusion in RV-1 samples is used to read off the predicted APIG from the scatterplot. Table 6.2 shows the details of HCFIs in the study along with the lithological details and

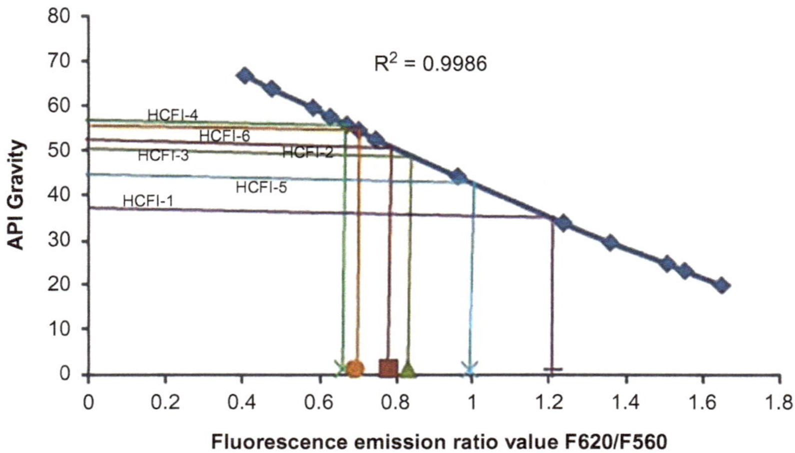

Fig. 6.5 Scatterogram designed to predict API gravity for known emission ratios.

spectral emission ratios (or F620/F560) of oil inclusions. From the scattergram of crude oils (Fig. 6.5), the API gravity of HCFIs can be inferred by extrapolating their emission intensity ratios. The derived APIGs of HCFI samples in the RV-1 well are an outcome of the algebraic expression and are also shown in Table 6.2. Oil inclusion from the Mukta Formation showed the highest emission ratio and an API gravity of 34.7 (1345–1350 m depth). Inclusions from the Panna Formation show varying APIGs. At 3495–3500 m depth, the APIG is 42.8; at 3065–3070 m depth, the APIG is 51.1; at a depth of 3445–3450 m, the APIG is 56; at a depth of 3685–3690, the APIG is 54.6; and at 3270–3275 m depth, the APIG is 49.1 (Table 6.2).

Here the fluorescence emission from 13 crude oils (from the Mumbai offshore basin, India) with known APIGs is used to develop a reference database. A detailed analysis of the relation between emission ratios at F620/F560 and corresponding APIGs for the 13 crude oil samples provided an empirical correlation based on the emission ratio. A bivariate plot of this empirical equation fitted well to a straight line with a coefficient of correlation, $R^2 = 0.9982$ with equation $y = -37.686 \times (\text{F620/F560}) + 81.136$. In order to study the effectiveness of this empirical tool for inferring the APIG of oils in HCFIs, samples from the RV-1 well were used. Fluorescence emission spectra from the HCFIs of the RV-1 well were recorded and the ratio F620/F560 was calculated. By extrapolating the emission ratio value of HCFIs in the bivariate plot of the crude oil reference standard data set, the APIGs were determined.

Table 6.2 Partial well log of RV1, emission ratios and characteristics of hydrocarbon fluid inclusions in sample wafers.

Sample no.	Depth (m)	F620/F560 ratio	Age	Lithology	HCFIs characteristics	Derived APIG
1	1345–1350	1.207741	Mukta formation/ early oligocene	Limestone and shale	Biphase secondary inclusions in cements. Size between 5 and 12 μm, shaped as rectangular, irregular, and elongate. Extent of fill 5%–10%	34.7
2	3065–3070	0.781682	Panna formation/ palaeocene-early eocene	Sandstone	Biphase inclusions occurring along secondary fractures in quartz grains of 2–5 μm and rounded, eye shaped, elongated and stretched outlines. Extent of fill 1%–5%	51.1
3	3270–3275	0.831704	Panna formation/ palaeocene-early eocene	Sandstone	Biphase inclusions along secondary annealed fractures of quartz grains. Size varies from 5 to 9 μm and shapes are elongated, irregular and rounded. Extent of fill 5%–10%	49.1
4	3445–3450	0.662772	Panna formation/ palaeocene-early eocene	Sandstone and shale	Biphase secondary inclusions along secondary annealed fractures in quartz grains of size range of 3–12 μm and shaped as irregular, oval and elongate. Extent of fill 1%–5%	56
5	3495–3500	0.991743	Panna formation/ palaeocene-early eocene	Sandy-siltstone and shale	Biphase secondary inclusions concentrated along secondary fractures in quartz and feldspar grains of size of 5–17 μm and shaped as irregular and elongate. Extent of fill 10%–20%	42.8
6	3685–3690	0.695561	Panna formation/ palaeocene-early eocene	Shaley-siltstone	Biphase secondary inclusions in cements, size 3–8 μm, shaped as rectangular and elongate. Extent of fill 10%–20%	54.6

6.6 Previous studies on fluorescence emission and API gravity of crude oils/HCFIs

HCFIs are isolated microscopic cavities within mineral grains of host rock that are filled with oil, gas, or mixtures of both, the composition of which, we assume, might not have changed since the trapping event. Petroleum fluids entrapped in HCFIs fluoresce because of the presence of polycyclic aromatic hydrocarbons. However, the complex chemical composition of the entrapped oils determines the precise absorption and fluorescence characteristics of HCFIs. Therefore, fluorescence-based methods are widely used for analysis of crude oil as they offer robust, noncontact, and nondestructive measurement options (Burruss, 2003; Downare and Mullins, 1995; Ralston et al., 1996).

The excitation wavelength plays a critical role in the emission properties of petroleum (Ryder, 2002; Wang and Mullins, 1994; Hagemann and Hollerbach, 1986). Fluorescence studies in crude petroleum oils that correlated steady-state emission properties to oil composition have been reported (Hagemann and Hollerbach, 1986; Stasiuk and Snowdon, 1997) using an excitation source of 365 ± 30 nm, wherein the wavelength of maximum fluorescence emission (λ_{max}) was shown to correlate with API gravity and aromatic composition of crude oils. Similar to the $Q_{R/G}$ correlated to viscosity proposed by Stasiuk and Snowdon (1997), a parameter Q-535 based on the ratio between 535 and 750 nm flux and 430–535 nm flux also was used to characterize crude oils and HCFIs, but no quantitative correlation to petroleum composition was described (Guilhaumou et al., 1990; Pradier et al., 1990). Fluorescence-based techniques are widely used in the study of HCFIs, and combined fluorescence microscopy and spectroscopy have been in use in the characterization and analysis of HCFIs (Blamey and Ryder, 2007; Stasiuk and Snowdon, 1997; Ryder, 2004). McLimans (1987) proposed the use of fluorescence spectroscopy for the analysis of HCFIs. Earlier studies reported that the red-green quotient (Q = intensity 650 nm/intensity 500 nm) and the wavelength of maximum fluorescence emission intensity (λmax) correlated with API gravity and gross chemical composition for a set of synthesized HCFIs using a sample set of Canadian crude oils (Ryder, 2004). There exist several studies using fluorescence spectral data to estimate the API gravity of the entrapped fluids. However, there are only a few studies that compared HCFIs to crude oil based on fluorescence characteristics due to the variation in fluorescence lifetime in HCFIs and bulk oils (Pradier et al., 1990; Ryder et al., 2001, 2002a,b; Blamey et al., 2009; Dearing et al., 2011).

Ryder (2005) reviewed the fluorescence techniques in the analysis of crude oils with a specific focus on HCFIs and suggested the need for a standard methodology to map the capability of fluorescence-based techniques to analyze and compare results emerging from various studies. Ryder et al. (2001) successfully applied fluorescence lifetime imaging microscopy (FLIM) with 380 nm excitation to determine APIG and aromatic content in 12 crude petroleum oil samples from the North Sea. The emission spectra recorded using a standard fluorometer showed a strong correlation between APIG and the ratio of the fluorescence intensities at 650 and 450 nm (Wang and Mullins, 1994; Hagemann and Hollerbach, 1986). The FLIM method depends on the average fluorescence lifetime of a molecule, which again depends on specific excitation and emission wavelengths. Qualitative correlations of fluorescence lifetime data with compositional parameters such as API gravity have been reported by Ryder et al. (2001), although accurate quantitative correlations have not been possible with fluorescence lifetimes particularly when trapped in HCFIs.

Bourdet et al. (2012) used the intensities at 510 and 430 nm (intensity 510 nm/intensity 430 nm) to define a new parameter called the green/blue (QG/B) ratio, since the Q ratio (Q = intensity 650 nm/intensity 500 nm) from Stasiuk and Snowdon (1997) was found unsuitable for characterizing inclusions in this part of the visible spectrum. The spectroscopic measurements were made with an Ocean Optics USB2000 FL spectrometer with a bandwidth between 350 and 1000 nm. Bourdet et al. (2014) used an empirical relationship between fluorescence of oils and API gravity and applied it to petroleum inclusions. Fluorescence spectra were measured in the range of 400–900 nm with an Avantes spectrometer (Anglia Instrument, model Avaspec-2048TEC) equipped with a thermoelectric cooled Sony CCD detector. The correlation equation ($\mathrm{API} = -24.824 \times Q_{G/B} + 66.434$, $R^2 = 0.84$) is used to calculate the API of inclusion oils from their measured QG/B values with 90% confidence interval curves, which results in a precision of ±6 API for a QG/B of 0.4 and ±2 API for a $Q_{G/B}$ of 1.8. The calculated API gravities of petroleum inclusions from the Phuong Dong Field are presented for each of the oil inclusion assemblages in both gas and oil zones.

6.7 Observations and highlights of the HCFI fluorescence emission study

The visual fluorescence colors of each inclusion in RV-1 are clearly obse-rved under UV illumination during petrographic observations. The high emission ratio (1.20) of the Mukta sample distinguishes it from the rest of the Panna samples (0.78, 0.83, 0.66, 0.99, and 0.69) (Table 6.2). Mukta inclusions appeared more whitish under UV compared to the Panna Formation samples (whitish blue to blue). The Mukta Formation samples are at shallow levels (1700 m above the other samples from the Panna Formation), and potentially the oil here could be of a different generation or could be from a different source or has been altered. The Panna Formation oils are found to be more commercially viable compared to the Mukta oil samples. Evidence of entrapment of light oils in a dry well could be a useful indicator in petroleum exploration. Microthermometric studies of 102 non-HCFI biphase-secondary-coeval aqueous inclusions experimentally attribute an oil window (80–135°C) to these Formations, which is very conducive to oil generation and hence provides a possibility for intense exploratory activities in the adjoining areas in the basin.

It has been reported that the Mumbai offshore is the most prolific oil- and gas-producing basin in the Western offshore of India. The major source rocks in the Mumbai basin lie in the Paleocene-early Eocene Panna Formation having total organic carbon in the range between 0.5% and 20% that has contributed mostly to the generation and migration of hydrocarbons (Bastia and Radhakrishnan, 2012). The HCFIs from the Panna Formation (Paleocene-early Eocene) indicate the presence of light oil when compared to those from the Mukta Formation (Early Oligocene) samples. HCFI samples from the RV-1 well (nonproducing well in a proven basin) was the right choice for analyzing the potential application of the empirical correlation between emission of crude oils and APIG. Crude oil samples for the calibration data set were also from the Mumbai basin, India, which emphasizes the accuracy of the APIGs presented. Fresh oil samples have probably lost their gas component (i.e., they are degassed). However, HCFIs retain their gas component and there is a slight difference in their bulk densities. This might have caused the minor variations in API gravity of HCFIs compared to crude oils (Fig. 6.5).

The scattergram of emission ratio (F620/F560) of crude oils versus API gravity is an accurate graphical tool to predict or determine the API gravity of oils in HCFIs. The APIG of unknown samples (HCFIs or bulk oils) that can be inferred from the derived algebraic expression linking emission spectra to APIG for known crude oil samples is the important outcome of the study. The inferred APIGs of HCFIs from the RV-1 well are the result of applying the equation derived from the emission ratio (F620/F560) of crude oils versus API gravity to an unknown sample. This underscores and highlights the measurement of the fluorescence emission of the HCFIs as a valuable signature for quick detection of APIG. The presence of commercially viable oil in the HCFIs from the Panna Formation of the RV-1 well, Mumbai offshore basin, India, seems to be a useful indicator for a detailed exploration activity in the area.

Determination of APIG of oils in HCFIs is a challenging task for the geologists in the petroleum exploration industry. A methodology to determine the APIG of oil samples in a nondestructive and noninvasive way using an algebraic expression linking emission spectra to APIG for known crude oil samples (whether they are HCFIs or otherwise) is a breakthrough. A combination of our methodology along with the relative chronology of fluid inclusions (both aqueous and HCFs), as well as the composition and temperature–pressure-volume characteristics of individual inclusions, could provide valuable information about the thermal history and migration properties of hydrocarbons in petroleum basins. The present methodology could therefore help the petroleum industry to relook into the oil exploration activities in the adjoining areas of the well and investigate the migration pathways of the HCFIs.

6.8 Determination of the API gravity of oils in HCFIs from the Kerala-Konkan basin using the fluorescence emission technique

Fluorescence emission of oils in HCFIs from the Kerala-Konkan basin (see Chapter 2 for more details) is studied (Fig. 6.6), and using the emission ratios, the unknown API gravities are determined from the derived equation and scattergram. This is a significant step in the petroleum exploration industry. Secondary hydrocarbon fluid inclusion trails and some occluded bitumen fluorescence are observed at different depths in the KK-well at 3065–3070 m, 3075–3080 m and 3080–3085 m, and 3115–3120 m depths with sandstone, clay stone lithology. The secondary fluid inclusion trail observed at 3065–3070 m has a migrating trend (Fig. 6.7). Occluded bitumen is observed with dim fluorescence at a depth of 5900–5905 m and fluorescence of diffused bitumen-like samples is observed at a depth of 3220–3225 m. With the standard empirical equation developed, we report an API gravity of 31 for migrating oil (KK − 1), while for the bitumen-like oil (KK − 2) the API gravity was 22 (Fig. 6.8).

Based on the microthermometric analysis, we postulate that in the Kerala-Konkan basin only the horizon depth ranging from 3200 to 4200 m is conducive to oil generation.

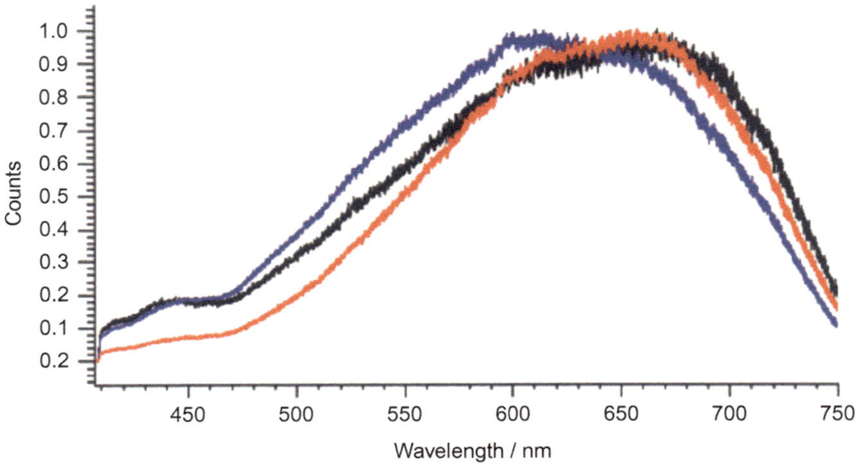

Fig. 6.6 Fluorescence emission profile for oils in HCFIs from the KK-basin.

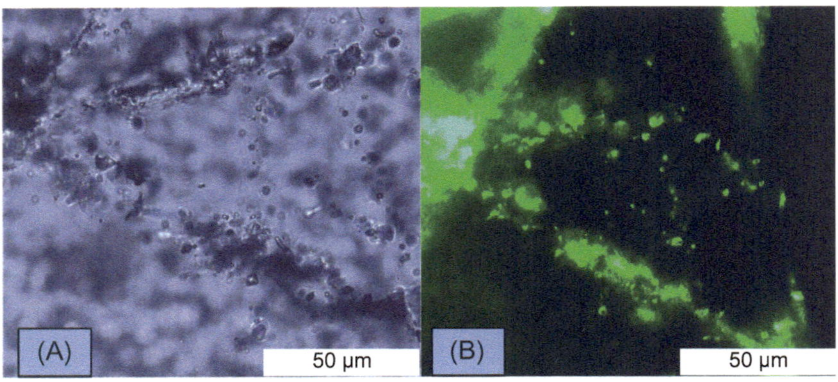

Fig. 6.7 Hydrocarbon fluid inclusion trails in the KK-4C-A-1 well in the Kerala-Konkan basin at a depth of 3065–3070 m. (A) Bright field image and (B) fluorescence image.

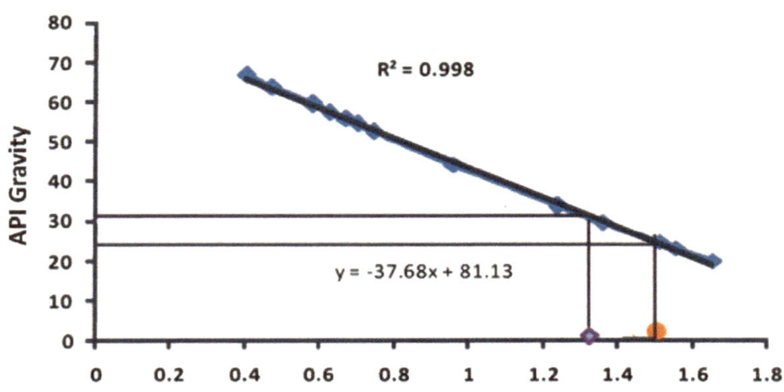

Fig. 6.8 The unknown API gravity of oils in HCFIs from the Kerala offshore basin determined using the standard bivariate plot based on the empirical equation $y = y_0\ (x_0\text{-}x)^t \pm 1$.

6.9 Comparison of API gravity of oils in HCFIs from dry wells in a proven and nonproven basin

The quality of oils in terms of American Petroleum Institute's (API) gravity is significant in the petroleum industry since it determines the market value of crude oil. Fluid inclusion study is gaining importance in the petroleum exploration industry since hydrocarbon fluid inclusions (HCFIs) are direct samples of oil that have migrated through various stratigraphic successions during the past. The determination of the quality of petroleum oils in HCFIs using nondestructive methods is advantageous in many respects. Here we present the utility of the relation between fluorescence emission ratios and API gravity of petroleum oils to determine the unknown API gravity of oils in HCFIs from two dry wells in the Mumbai offshore basin and Kerala-Konkan basin, India, respectively. To this end, an empirical equation was developed and discussed in this chapter (Eq. 6.1) that relates API gravity and emission ratio at 620/560, which determines the unknown API gravity of oil/oils in HCFIs/oil condensates. A comparison of the API gravity from these two dry wells in a proven and nonproven basin is also discussed. This comparative observation signifies the potential of the nondestructive fluorescence emission method to find out the unknown API gravity of oils in HCFIs. It is a significant step for the petroleum exploration industry to find the API gravity of petroleum oils from micron-sized HCFIs nondestructively, where conventional methods cannot be used.

Crude petroleum oils are complex mixtures of different compounds obtained from an extensive range of different geological sources and are mainly organic. They are a mixture of hydrocarbons that existed in the liquid phase in natural underground reservoirs and remain liquid at atmospheric pressure after passing through surface separating facilities (Definition by American Petroleum Institute, American Association of Petroleum Geologists, and Society of Petroleum Engineers). Their physical appearance can vary from solid black tars to almost transparent liquids. Crude oils can also contain varying amounts of gases such as carbon dioxide, methane, etc. in their natural state within an oil field reservoir or entrapped within hydrocarbon bearing fluid inclusions (HCFIs). Noncontact, nondestructive, quantitative analysis of crude petroleum oils is a highly desirable pursuit for both research (study of microscopic HCFI) and industry (real-time assessment of oil production). Optical methods with their high sensitivity, good diagnostic potential, relatively simple instrumentation, and suitability to both microscopy and portable instrumentation and fluorescence spectroscopy are the best available technique for studying micron-sized fluid inclusions (Blamey and Ryder, 2007). API gravity is one of the simplest and most wide-ranging parameters used for describing crude oils. It is inversely related to density by the formula: API gravity = [(141.5/specific gravity at 15.6°C)–131.5]. Specific gravities of oils tend to be in the 0.73–1.0 range, with paraffin-type oils being lighter than asphalt-base oils. In general, the higher the API gravity value, the lighter the oil. The fluorescence of crude petroleum oils derives from the aromatic hydrocarbon fraction (Pradier et al., 1990), and this fluorescence emission is strongly influenced by the chemical composition (such as fluorophore and quencher concentrations) and physical characteristics (such as viscosity and optical density) of the oil.

Hunt and North provided a detailed understanding on the basics of petroleum composition and the geological factors that influence changes in oil composition (Hunt, 1995; North, 1985). The primary compositional parameters reported in fluorescence studies by Ryder (2005)as obtained from column chromatography are alkane (or paraffin, or aliphatic) fraction, an aromatic fraction, a polar (or resin) fraction, and an asphaltene fraction. The asphaltene fraction is the portion of the oil which is insoluble in pentane or heptane and comprises high molecular weight species, depending on the chromatographic conditions used and whether the light, volatile fraction has been removed prior to chromatographic analysis. Further, the results obtained for chemical composition will differ from laboratory to laboratory. Therefore, comparing fluorescence data and results from different crude oil studies needs much attention. For accurate characterization of crude oil fluorescence, information on the source rock, its density (API gravity), and the gross chemical composition (alkane, aromatic, polar, and asphaltene concentration) is needed.

The present study attempts to establish the utility of the relation between fluorescence emission ratios and API gravity of petroleum oils (Nandakumar and Jayanthi, 2016; Indian Patent No. 315456, 2019) to determine the unknown API gravity of oils in HCFIs from two dry wells as discussed previously. One dry well RV-1 is from the Mumbai offshore basin, which is a proven basin, and another dry well is the KK-well from the Kerala-Konkan basin, which is a nonproven basin. To this end, the fluorescence emission from the oils in HCFIs at some depths was studied for both the wells using laser excitation at 405 nm. The emission ratios at F620/F560 were determined from the emission profiles and these values were used in the empirical equation previously developed by our group that relates API gravity and emission ratio at 620/560, which determines the unknown API gravity of oil/oils in HCFIs/oil condensates. A comparison of the API gravities from these two dry wells is also discussed in this chapter.

6.9.1 The two basins

Mumbai offshore is a pericratonic rift basin situated on the western continental margin of India. It is located on the western continental shelf of India between the Saurashtra basin in NNW and Kerala-Konkan in the south. The basin falls under category- 1 (NDR-DGH, Govt. of India), which implies that the basin has proven commercial productivity. The age of the basin ranges from late Cretaceous to Holocene with thick sedimentary fill ranging from 1100 to 5000 m. HCFI samples from a dry well RV-1 were used here and are from the Panna Formation, which is a synrift tectonic clastic sequence and comprises alternations of shale, siltstone, sandstone with occasional coal and limestone streaks (see Fig. 4.1. in Chapter 4).

The Kerala-Konkan basin is situated south of the east–west trending Vengurla Arch and extends up to Cape Comorin in the south of the Indian subcontinent in the western offshore basin (see Fig. 4.1 in Chapter 4). The sedimentary sequence is composed of Mesozoic and Tertiary sediments. The basin evolved through early rift and post rift phases and contains more than 5 km of Cretaceous to Recent sediments. Postrift mature sediments with sufficient organic carbon content are present in the basin. Drilling results and adsorbed gas anomalies confirm generation of hydrocarbons in the basin. This basin covers an area of around 580,000 km^2. This basin comes under category III, which has oil prospects (NDR-DGH, Govt. of India). There are 15 exploratory wells drilled so far but none of them are producing, although some wells show faint oil fluorescence. Hydrocarbon has been discovered all through the coastal areas of India, except the Kerala-Konkan basin. In the Kerala-Konkan basin, the Tertiary and Mesozoic sediments are separated by a thick Basaltic layer. Imaging below basalts has always been a problem in oil exploration. The structural style of the Kerala-Konkan

Basin is similar to that of the Bombay Offshore Basin. However, the horst and graben features on the continental shelf are a comparatively less pronounced area. The Kerala-Konkan basin is an opportunity and further value addition could be possible with a paradigm shift in exploration strategy in terms of concepts, technology, and personnel expertise. Here we are discussing the application of fluid inclusion techniques to find the oil resource in the Kerala-Konkan basin, India.

6.9.2 Sample details

Sidewall cuttings for the RV-1 well in the Ratnagiri Block of the Mumbai Offshore Basin and the Kerala-Konkan Basin, India were collected from the core library of the Regional Geological Laboratory (RGL), Oil and Natural Gas Corporation (ONGC), Government of India, Mumbai. The samples were washed, dried, and sieved and specialized wafers were prepared for the petrographic and fluorescence study with the wafer preparation technique explained in Chapter 2. In RV-1, HCFIs were observed in quartz, calcite, feldspar grains, and cements in the sedimentary rock cuttings, especially from the sandstone, siltstone lithologies towards the lower portion of the well depth ranging from 3270 to 3690 m. Using the lithostratigraphic nomenclature of the Mumbai Offshore Basin, the hydrocarbon inclusions are mainly observed in the Panna Formation (Paleocene-early Eocene age), which is both a producing as well as a reservoir horizon. Petrographic studies confirmed the presence of hydrocarbon inclusions that are showing fluorescence when observed under 365 nm (mercury arc lamp) UV excitation. Fluorescence spectral studies were carried out (using the same analytical procedure explained in Chapter 5) for HCFI assemblages at four different depths from the RV-1 well (and a minimum of 15 individual inclusions from each assemblage) in order to detail any hydrocarbon species present. Lithostratigraphic details of the study samples are given in Table 6.3.

In the Kerala-Konkan basin, HCFIs are mostly present in sandstone, siltstone, and clay stone lithologies within the depths of 3000–4000 m. The HCFIs observed are mostly in secondary trails and are showing a bituminous nature without phase distinction. The Lithostratigraphic details of the study samples are given in Table 6.4.

A comparative observation of the fluorescence spectral features of the HCFIs from both the basins is given in Figs. 6.9 and 6.10 and shows a clear-cut distinction between the fluorescence emissions of oils in HCFIs from these two basins. Fluorescence emission features from the RV-1 well

Table 6.3 Derived API gravity of the oils in HCFIs studied from the RV-1 well, Mumbai offshore Basin, India with lithological details and HCFIs characteristics.

Sl. no.	Horizon (m)	Lithology	HCFIs description	Derived API gravity
1	3055–3060	Sandstone, siltstone, and shale	Monophase liquid and biphase inclusions along the secondary trails and some monophase inclusions as primary also; size varies from 1 to 4 μm; shapes of HCFIs are elongated and irregular	47.7
2	3445–3450	Sandstone and silty shale	Primary biphase inclusions are seen and biphase inclusions in secondary trails are observed; size varies from 1 to 6 μm and shapes are irregular	54.1
3	3455–3460	Sandstone and silty shale	Biphase secondary inclusions seen along the secondary fluid inclusion trails of quartz; monophase bitumen-like inclusions also observed; size of HCFIs varies from 0.5 to 8 μm, and shapes are elongated and irregular	53.6
4	3495–3500	Sandy siltstone, shale	Biphase inclusions in secondary trail trapped along the micro fractures of quartz grains and cements; size varies from 0.5 to 12 μm and shapes are subrounded, elongated, irregular, etc.	42.5

of the Mumbai offshore basin show emission patterns (Fig. 6.9) that are explained in this chapter and in Nandakumar and Jayanthi (2016). The fluorescence emission from the HCFIs in the KK-basin is showing a different pattern with a sharp shoulder peak around 450 nm for samples from two depths (Fig. 6.10). Such shoulder peaks were observed for HCFI samples in the RV-1 well, which shows very high API gravities. In the KK-well samples, the shoulder peaks are of very less intensity compared to those in the RV-1 well. The red and blue shifts at 600 nm and 525 nm are clearly visible in the emission features of the RV-1 well, but the emission profiles of HCFIs from the KK-basin have a broad peak emission around 560 nm and

Table 6.4 Derived API gravity of the oils in HCFIs studied from the KK-well, Kerala-Konkan offshore Basin, India with lithological details and HCFIs appearance.

Sl. no.	Horizon (m)	Lithology	HCFIs description	Derived API gravity
1	3475–3480	Sandstone and claystone	Monophase secondary inclusions seen along he secondary trails. Crosscutting relationship is observed. Size varies from 0.5 to 4 μm. Shapes of HCFIs are rounded	31.6
2	3890–3895	Sandstone and claystone	Monophase primary inclusions are seen. Occluded bitumen-like florescence is observed. Size varies from 1 to 10 μm and shapes are irregular	22.5
3	3180–3185	Sandstone and claystone	Monophase secondary inclusions seen along the secondary fluid inclusion trails of quartz. Size of HCFIs varies from 1 to 4 μm, and shape is elongated, fusiform and irregular	25.3
4	3340–3345	Limestone	Monophase inclusions in secondary trail trapped along the micro fractures of calcite grains. Size varies from 0.5 to 3 μm and shapes are subrounded, elongated, irregular, etc.	32.8

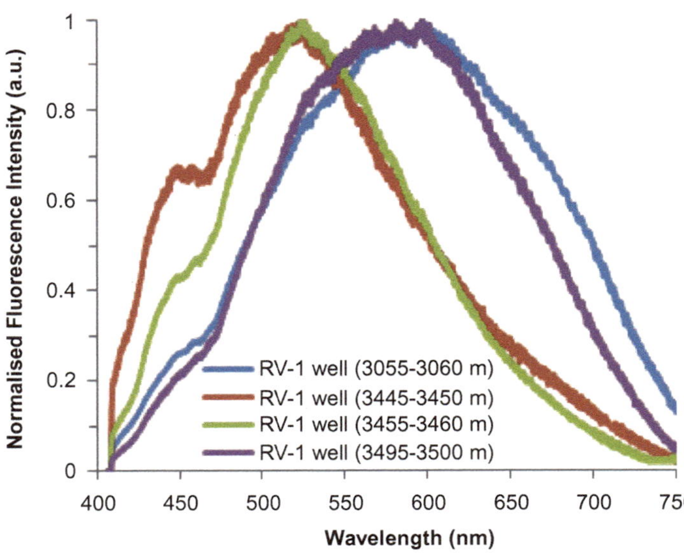

Fig. 6.9 Normalized fluorescence emission of oils in HCFIs from RV-1 well, Mumbai offshore basin, India at 405 nm laser excitation.

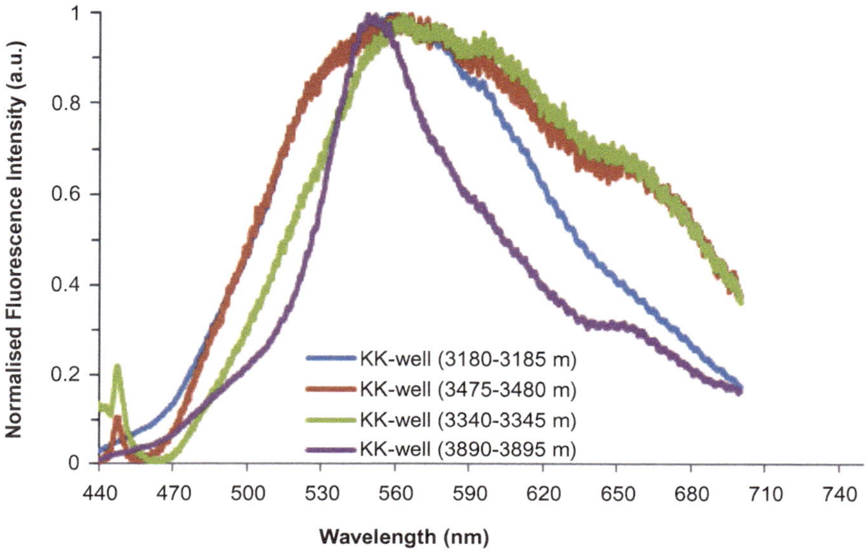

Fig. 6.10 Normalized fluorescence emission of oils in HCFIs from KK- well, Kerala-Konkan offshore basin, India at 405 nm laser excitation.

are not showing any shift towards the red side of the spectrum like the 600 nm shift in Fig. 6.10 for the samples that are having a shoulder peak around 450 nm. A minor shift around 600 nm is shown by the samples without a shoulder peak emission around 450 nm in their emission profile (Fig. 6.9).

The emission profile variations/similarities are related to the API values that are derived using the empirical equation shown in this chapter. HCFI samples with identical emission profiles are showing almost similar API gravities. You can see the emission profiles at depths 3475–3480 m with sandstone, clay stone lithology and at 3340–3345 m with limestone lithology. Both are showing a shoulder peak around 450 nm and not showing any shifts around 600 nm (Fig. 6.10). When applying the emission ratio values in Eq. (6.1), samples from these two depths show that the API gravities are around 31.6 and 32.8, respectively (Table 6.4). Samples at two other depths 3180–3185 m and 3890–3895 m with sandstone, clay stone lithology are showing an almost similar emission profile picture without a shoulder peak around 450 nm and broad peak around 560 nm with a small red shift around 620 nm (Fig. 6.10). These samples show an API gravity around 25.3 and 21.5, respectively.

Fluorescence spectroscopy, where the radiation of one wavelength that is absorbed by a molecule with the appropriate molecular structure can be remitted at a higher wavelength (lower energy), is increasingly used in petroleum related research because of its sensitivity and selectivity. The increase in oil concentration shows a decrease in the emission intensity due to the difficulty in light penetration through the sample. Crude oil contains fluorescent molecules. Ryder (2005) discussed fluorescence for the different HCFI samples excited at 450 nm that shows an emission band around 560 nm and observed a shift to red and decrease with the increase in the oil concentration. Ryder (2005) also showed the fluorescence for the different HCFI samples excited at 532 nm. For this excitation, the emissions appear around 660 nm and shift to red and decrease with the increase in oil concentration. Energy transfer is the process responsible for red shift and broadening in the emission spectral band and at the same time a quenching process reduces the emission intensity. Other authors using solvents such as benzene, cyclohexane, and n-heptane observed the same effect. Ryder (2005) again shows the fluorescence spectra of 350, 450, and 532 nm for different oil concentrations and observed that in the samples when the oil concentration increases, a displacement occurs in the wavelength possibly because of the higher oil concentration.

In the petroleum industry, the characterization of crude oil is a key factor for improved optimization of the refining process. Spectroscopic techniques have gained relevance for this purpose because of their potential to give rapid responses containing valuable information related to the intrinsic chemical characteristics of each analyzed sample. Petroleum oils are mixtures of aliphatic, aromatic, and high molecular weight organic compounds. Due to this heterogeneity, chemical analysis is generally complex. Petroleum oils are typically characterized using liquid chromatography, separating into saturates, aromatics, resins, and asphaltenes. Each of these components can be further characterized by the use of gas chromatography–mass spectrometry (GC–MS), which provides unambiguous identification of individual components. Infrared (IR, mid-IR, MIR, and near- IR, NIR) spectroscopies are most widely used in spectroscopic techniques for crude oil analysis. The simplicity of sample handling and the rapidity of the analysis have facilitated the development of new ways to determine the physical chemical properties of petroleum products (gasoline, kerosene, diesel, etc.). In spite of its good signal to noise ratio, rapid response, simplicity, and low cost, IR still presents several limitations, such us the overlapping of absorption bands and saturation of the signal due to the high absorbance of the dense sample such as crude petroleum.

Fluorescence spectroscopy is gaining importance in the petrochemical field due to its sensitivity and selectivity. Fluorescence has been extensively used in the petroleum industry for the analysis and classification of different petroleum samples, and reviews describing the basis of the technique and its inherent advantages and disadvantages can be found in the literature. Factors such as the specific chemical composition (concentration of fluorophores and quenching species) and physical properties (viscosity and optical density) influence emission intensity and wavelength. In the case of heavy oils, fluorescence emission is generally broad, very weak, and has short lifetimes, whereas lighter oils have narrower, more intense emission bands, and longer lifetimes. The nature of the emission is governed by the complex interplay between energy transfer and quenching caused by the high concentrations of fluorophores and quenchers in petroleum oils. The complexity of crude oils usually prevents the resolution of any specific chemical component in terms of individual emission parameters. Fluorescence emission techniques can show the components that are responsible for emission around a particular wavelength, and its concentration difference or absence is the reason for the same profile that cannot be observed in another basin/area of observation. Therefore, while studying the emission features of characteristic components in HCFIs, one can easily find out which component is absent or whose concentration makes the changes in the fluorescence emission profile.

Constituent characterization is very important while considering HCFIs, and how it is possible with Laser Raman spectroscopy is discussed in the next chapter.

6.10 Conclusions

The effectiveness of using fluorescence emission from crude oils/HCFIs in determining its API gravity is well demonstrated from the fluorescence study of hydrocarbon compounds recorded with laser excitation at 405 nm. Such a nondestructive method for determining API gravity is a valuable tool in the petroleum exploration industry. The changes in the emission ratio at F620/F560 are significant while considering the 405 nm laser-induced fluorescence emission. With this technique of microscope-based fluorescence spectroscopy, which is a nondestructive optical analytical technique, we could characterize different generation/differently matured oils as found in the Panna and Mukta Formations of the Mumbai offshore basin, India. Thus, fluorescence emission of oils provides an understanding of different generation/different source rocks for oils, and the methodology can be used to determine the API gravity of crude oils/oils in HCFIs/oil condensates.

References

Anderson, B., Munz, I.A., Johansen, H., Johansen, I., 1998. Abstracts of Papers, 215th National Meeting of the American Chemical Society, Dallas, TX, Mar 29 – Apr 2, 1998. American Chemical Society, Washington, DC.

Bastia, R., Radhakrishnan, M., 2012. Basin evolution and petroleum prospectivity of the continental margins of India. In: Developments in Petroleum Science. vol. 59. Elsevier, USA, pp. 1–417. 2012.

Bertrand, P., Pittion, J.L., Bernaud, C., 1986. Fluorescence of sedimentary organic matter in relation to its chemical composition org. Geochemistry 10, 641–647.

Blamey, N.J.F., Ryder, A.G., 2007. Hydrocarbon fluid inclusion fluorescence: a review. In: Geddes, C.D. (Ed.), Reviews in Fluorescence. Springer, New York, pp. 299–334. 2012.

Blamey, N.J.F., Conliffe, J., Parnell, J., Ryder, A.G., Feely, M., 2009. Application of fluorescence lifetime measurements on single petroleum-bearing fluid inclusions to demonstrate multicharge history in petroleum reservoirs. Geofluids 2009 (9), 330–337.

Bourdet, J., Eadington, P.J., Volk, H., George, S.C., Pironon, J., Kempton, R.H., 2012. Chemical ahnges of fluid inclusion oil trapped during the evolution of an oil reservoir. Mar. Pet. Geol. 2012 (36), 118–139.

Bourdet, J., Burruss, R.C., Chou, I.M., Kempton, R., Liu, K., Hung, N.V., 2014. Evidence for a palaeo-oil column and alteration of residual oil in a gas-condensate field: integrated oil inclusion and experimental results. Geochim. Cosmochim. Acta 142, 362–385.

Burruss, R.C., 1991. In: Barker, C.E., Kopp, O.C. (Eds.), Practical aspects of fluorescence microscopy of petroleum fluid inclusions. Luminescence microscopy and spectroscopy: qualitative and quantitative applications. 25. SEPM Short Course, pp. 1–7.

Burruss, R.C., 2003. Petroleum fluid inclusions, an introduction. In fluid inclusions: analysis and interpretation, mineralogical Association of Canada. Vancouver 2003, 159–169.

Dearing, T.I., Thompson, W.J., Rechsteiner Jr., C.E., Marquardt, B.J., 2011. Characterization of crude oil products using data fusion of process Raman, infrared, and nuclear magnetic resonance (NMR) spectra. Appl. Spectrosc. 65 (2), 181–186.

Downare, T.D., Mullins, O.C., 1995. Visible and near infrared fluorescence of crude oils. Appl. Spectrosc. 49 (6), 754–764.

George, S.C., Ruble, T.E., Dutkiewicz, A., Eadington, P.J., 2001. Assessing the maturity of oil trapped in fluid inclusions using molecular geochemistry data and visually determined fluorescence colours. Appl. Geochem. 16 (4), 451–473.

Goldstein, R.H., 2001. Fluid inclusions in sedimentary and diagenetic systems. Lithos 55, 159–193.

Guilhaumou, N., Szydlowskii, N., Pradier, B., 1990. Characterization of hydrocarbon fluid inclusions by infra-red and fluorescence micro spectrometry. Mineral Mag. 54 (375), 311–324.

Hagemann, H.W., Hollerbach, A., 1986. The fluorescence behavior of crude oils with respect to their thermal maturation and degradation. Org. Geochem. 10 (1–3), 473–480.

Hunt, J.M., 1995. Petroleum Geochemistry and Geology, second ed. W.H. Freeman, New York. xx + 743 pp. ISBN 0 7167 2441 3.

Khorasani, K., 1987. Novel development in fluorescence microscopy of complex organic mixtures: application in petroleum geochemistry. Org. Geochem. 11 (3), 157–168.

Kihle, J., 1995. Adaptation of excitation-emission micro-spectroscopy for characterization of single hydrocarbon fluid inclusions. Org. Geochem. 23, 1029–1042.

McLimans, R.K., 1987. The application of fluid inclusions to migration of oil and diagenesis in petroleum reservoirs. Appl. Geochem. 2 (5–6), 585–603.

Middleton, D.W.J., Parnell, J., Green, P.F., Xu, G., McSherry, M., 2001. Hot fluid flow events in Atlantic margin basins: an example from the Rathlin Basin. Geological society, London, special publications. Geol. Soc. Spec. Publ. 188, 91–105.

Munz, I.A., 2001. Petroleum inclusions in sedimentary basins: systematics, analytical methods and applications. Lithos 55 (1 – 4), 195–212.

Nandakumar, V., Jayanthi, J.L., 2016. Hydrocarbon fluid inclusions, API gravity of oil, signature fluorescence emissions and emission ratios: An example from Mumbai Offshore, India. Energy & Fuels 30, 3776–3782.

Indian Patent No. 315456 dated 03/07/2019, 2019. Method of detecting API gravity of oil present in hydrocarbon bearing fluid inclusions. Patentee: V. Nandakumar, JL Jayanthi.

Nicodem, D.E., Da Cunha, M.F.V., Guedes, C.L.B., 2000. Time-resolved single photon counting study of the quenching of fluorescent probes by petroleum: probing the energy distribution of the nonaliphatic components. Appl. Spectrosc. 54 (9), 1409–1411.

North, F.K., 1985. Petroleum Geology. Allen & Unwin Inc., USA, ISBN: 0-04-553003-3.

Parnell, J., Carey, P., Duncan, W., 1998. History of hydrocarbon charge on the Atlantic margin: evidence from fluid-inclusion studies, West of Shetland. Geology 26 (9), 807–810.

Parnell, J., Middleton, D., Honghan, C., Hall, D., 2001. The use of integrated fluid inclusion studies in constraining oil charge history and reservoir compartmentation: examples from the Jeanne d'Arc Basin, offshore. Newfoundland Mar. Pet. Geol. 18 (5), 535–549.

Pradier, B., Largeau, C., Derenne, L., Bertrand, P., Pouet, Y., Martinez, L., 1990. Chemical basis of fluorescence alteration of crude oils and Kerogens-II. Fluorescence and infrared micro-spectrometric analysis of vitrinite and liptinite. Org. Geochem. 16 (1–3), 451–460.

Przyjalgowski, M.A., Ryder, A.G., Feely, M., Glynn, T.J., 2005. Analysis of hydrocarbon-bearing fluid inclusions (HCFI) using time resolved fluorescence spectroscopy. Proc. SPIE 5826, 173–184.

Ralston, C.Y., Wu, X., Mullins, O.C., 1996. Quantum yield of crude oils. Appl. Spectrosc. 50 (12), 1563–1568.

Rayner, D.M., Szabo, A.G., 1978. Time-resolved laser fluorosensors: A laboratory study of their potential in the remote characterization of oil. Appl. Opt. 17 (10), 1624–1630.

Roedder, E., 1984. Fluid inclusions. Rev. Mineral 12, 1–644.

Ryder, A.G., 2002. Quantitative analysis of crude oils by fluorescence lifetime and steady state measurements using 380 nm excitation. Appl. Spectrosc. 2002 (56), 107–116.

Ryder, A.G., 2004. Time resolved fluorescence spectroscopic study of crude petroleum oils: influence of chemical composition. Appl. Spectrosc. 58 (5), 613–623.

Ryder, A.G., 2005. Analysis of crude petroleum oils using fluorescence spectroscopy. In: Geddes, C.D., Lakowicz, J.R. (Eds.), Reviews in Fluorescence. Springer, New York, pp. 169–198.

Ryder, A.G., Power, S., Glynn, T.J., Morrison, J.J., 2001. Time domain measurement of fluorescence lifetime variation with pH. Proc. SPIE 4259 (2001), 102–109.

Ryder, A.G., Glynn, T.J., Feely, M., Barwise, A.J.G., 2002a. Characterization of crude oils using fluorescence data. Spectrochim. Acta Part A 58 (5), 1025–1038.

Ryder, A.G., Glynn, T.J., Przyjalgowski, M., Szczupak, B., 2002b. A compact violet diode laser based fluorescence lifetime microscope. J. Fluoresc. 12 (2), 177–180.

Stasiuk, L.D., Snowdon, L.R., 1997. Fluorescence microspectrometry of synergic and natural hydrocarbon fluid inclusions: crude oil chemistry, density and application in petroleum migration. Appl. Geochem. 12 (3), 229–241.

Wang, X., Mullins, O.C., 1994. Fluorescence life time studies of crude oils. Appl. Spectrosc. 48 (8), 977–984.

Zhu, Y., Mullins, O.C., 1992. Temperature dependence of fluorescence of crude oils and related compounds. Energy Fuel 6 (5), 545–552.

CHAPTER 7

Raman spectroscopic studies of hydrocarbon fluid inclusions

V. Nandakumar[a] and J.L. Jayanthi[b]
[a]Scientist - G, National Facility for Geofluids Research and Raman Analysis, National Centre for Earth Science Studies (NCESS), Thiruvananthapuram, Kerala, India
[b]Project Scientist - C, National Facility for Geofluids Research and Raman Analysis, National Centre for Earth Science Studies (NCESS), Thiruvananthapuram, Kerala, India

7.1 Introduction

Raman spectroscopy is a nondestructive analytical technique that characterizes liquid and gaseous compounds, solid phases, and solute species in fluid inclusions. The main advantages of Raman spectroscopy is that it allows chemical and structural characterization of samples as small as 1 μm in diameter, a resolution not possible by conventional petrography, microthermometry, or other spectroscopic methods such as infrared spectroscopy. Raman spectroscopy has become a conventional method in fluid inclusion research starting from the 1970s (Burke and Lustenhouwer, 1987; Dhamelincourt et al., 1979; Dubessy et al., 1982, 1989; Guilhaumou, 1982; Pasteris et al., 1986, 1988; Rosasco et al., 1975; Seitz et al., 1987) and the continuing interest and the importance of Raman spectroscopic technique is demonstrated by a number of publications and review papers (Burke, 1994, 2001; Burruss, 2003a,b; McMillan et al., 1996; Nasdala et al., 2004). Frezzotti et al. (2012) introduced Raman spectroscopy for the analysis of geological fluids trapped as inclusions.

Raman spectroscopy is based on the inelastic scattering of light by matter in its solid, liquid, or gas state. Monochromatic light scattered by matter contains radiations with frequencies different from the exciting light. This effect was predicted by Smekal in 1923 and was demonstrated by CV Raman, an Indian physicist, in 1928 (Smekal, 1923).

In several liquids, CV Raman observed scattered light, which had energy greater than the incoming light (Raman anti-Stokes). For the discovery of this optical scattering phenomenon he was awarded the Nobel Prize in physics in 1930 (Raman, 1928; Raman and Krishnan, 1928). The fundamentals of Raman spectroscopic techniques are detailed in Chapter 2. Raman

Hydrocarbon Fluid Inclusions in Petroliferous Basins
https://doi.org/10.1016/B978-0-12-817416-6.00002-2

spectroscopy is the measurement of the photons arising from inelastic (Raman) scattering of light. A Raman spectrum is the plot of light intensity expressed as arbitrary units, or counts, vs the frequency of scattered light in units of wavenumbers.

Raman shift occurs because photons (particles of light) exchange part of their energy with molecular vibrations in the material. When illuminating a sample with a monochromatic light source, such as a laser, most of the light that scatters off is unchanged in energy ("Rayleigh scattered"). A minute fraction (perhaps 1 part in 10 million) has lost or gained energy ("Raman scattered"). The change in energy depends on the frequency of vibration of the molecule. If it is very fast/high frequency (light atoms held together with strong bonds) the energy change is significant. If it is very slow/low frequency (heavy atoms held together with weak bonds), the energy change is small. The energy-loss process is the most probable because always there will be more molecules in lower states than in higher states and usually only the more intense Stokes side of the spectrum is analyzed. Where energy is lost the Raman scattering is designated as "Stokes"; where energy is gained the Raman scattering is designated as "anti-Stokes" (Fig. 7.1.). We rarely use anti-Stokes Raman light as it is less intense than the Stokes; however, it does represent equivalent vibrational information of the molecule (Panczer et al., 2012; Dubessy et al., 2012). A Stokes Raman spectrum is a plot of the intensity of the scattering vs the energy loss, expressed in wavenumbers relative to the source ($\Delta \upsilon$ cm^{-1}, the Raman shift, i.e., the changes in wavenumber

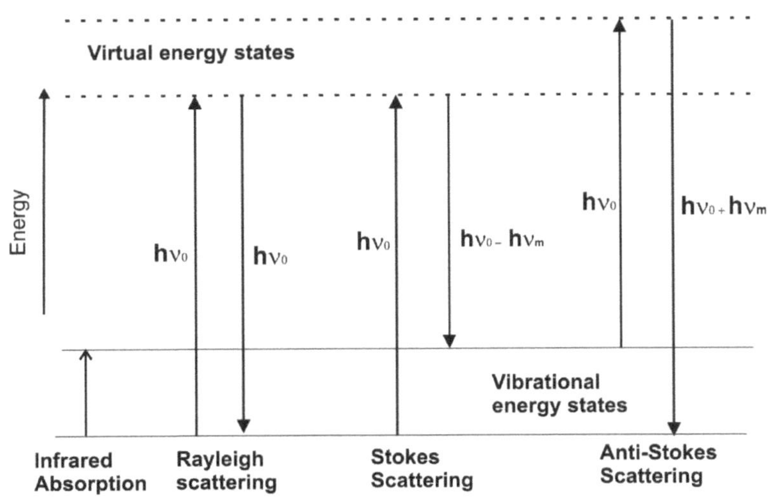

Fig. 7.1 Jablonski diagram showing the scattering events of an excited photon.

compared to the incident light). The vibrational frequency (υ_m) as a shift from incident light beam frequency (υ_0) is measured in Raman spectroscopy.

7.2 Raman spectrometer

Raman spectrometer systems consist of the following: monochromatic light sources (usually lasers), lenses (to focus the light onto the sample and to collect the scattered light), filters (to filter out the reflected and scattered light so that only the Raman light is collected), a diffraction grating or prism; a very sensitive detector (to detect the weak light); and a computer to control the whole system that integrates with the software and displays the spectrum (Fig. 7.2).

Raman spectroscopy reveals the chemical and structural composition of samples. Raman scattering occurs when light interacts with molecular vibrations. Molecular groups have a number of vibrational modes corresponding to specific energy states of the molecule. If a molecule has a centre of symmetry, then Raman active vibrations are infrared inactive and vice versa. If there is no centre of symmetry then some (but not necessarily all) vibrations may be both Raman and infrared active. In order to be Raman active a molecular rotation or vibration must cause some change in a component of the molecular polarizability. Symmetric vibrations give rise to very intense Raman lines where as non-symmetric ones are usually weak and some times unobservable. The peaks in the spectrum correspond to the energies of the vibrational modes of the different species in the sample. The theory of Raman scattering and of its application in the earth sciences are given in the reviews by McMillan (1985, 1989), McMillan and Hofmeister (1998), Roberts and Beattie (1995) and a handbook for practical Raman spectroscopy by Gardiner and Graves (1989).

7.3 Challenges with Raman analysis

Fluorescence and overlapping bands from the host mineral are possible competing effects during Raman analysis, since they often subdue and conceal the weak Raman features from the fluid inclusions. Fluorescence appears as a very broad background and is more intense than the Raman scattering. Background fluorescence arises from the epoxy used to embed the fragmented piece of rocks for fluid inclusion studies. Fluorescence can also be emitted by fluids contained in inclusions (hydrocarbons) or by the surrounding host mineral (e.g., calcite). Increasing the wavelength of the excited light source is a way of overcoming fluorescence (red or near infrared lasers

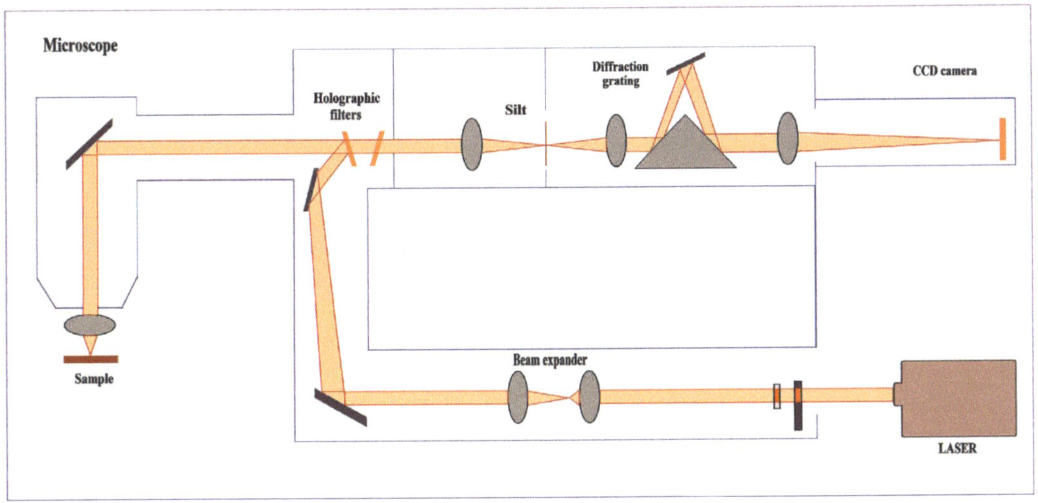

Fig. 7.2 Schematic diagram of a Laser Raman micro spectrometer.

($\lambda = 630$–$1060\,\text{nm}$) that should not give rise to fluorescence (Carey, 1999; Dubessy et al., 2012). Another method to mitigate background fluorescence is repeating spectral accumulations for several times in order to bleach out this effect by continuous exposure to laser light (photo bleaching).

7.4 Choice of laser excitation

A Raman system depends on the choice of exciting radiations. The selection of laser depends on: (1) sample type and its fluorescence properties; (2) the λ^4 law (Rull, 2012); (3) the efficiency of the whole Raman system; and (4) the type of application (in the field or in the laboratory). As discussed, fluorescence may make it difficult to obtain a good-quality Raman signal (Panczer et al., 2012). Therefore, the choice of excitation wavelength is critical and is possible if the spectral range of the fluorescence of the majority of the sample is known. There are two ways to eliminate or significantly reduce the fluorescence intensities; one is not by exciting the fluorescence and it requires a laser line that has insufficient energy to reach the excited electronic state of the sample and therefore the excited wavelength should be in the red and in the very near IR spectral region. The second one is by exciting the fluorescence in a spectral range at higher wavelengths than the spectral range of the Raman spectrum and this is achieved with UV lasers that are not commonly used in Earth sciences research (Dubessy et al., 2012). For a laboratory study, the main selection criteria for the laser excitation wavelength depends on the samples which you are studying (taking into account their fluorescence spectrum, their photosensitivity, and thermal degradation under laser irradiation).

7.5 Importance of Raman studies in fluid inclusions

Fluid inclusions can provide vital information that is fundamental to the petroleum exploration and production industry. The organic fluids present in rock include hydrocarbons ranging from methane to high molecular-weight hydrocarbons and these fluids can undergo complex chemical changes depending on the conditions of their formation. When sediments undergo diagenesis, the organic matter present in sediments is transformed into organic liquids and gases. Hydrocarbon fluid inclusions (HCFIs) can provide information on the composition and the conditions under which the fluids were entrapped (Roedder, 1984). Hydrocarbon inclusion's compositions are sensitive to the complex interplay between source, transport, and deposition and are therefore of interest. The variation in composition between present-day crude oils and inclusions can provide information on the genesis of the fluids (Orange et al., 1996). Since the composition of petroleum fluids is complex, its species-wise characterization is fundamental in the geological interpretation of the source and maturity of oils (Goldstein, 2001; Munz, 2001).

The rapid expansion of research on petroleum inclusions is being observed recently because petroleum inclusions are "hidden petroleum shows" (Lisk et al., 2002). Analysis of petroleum inclusions and its application to petroleum exploration and reservoir assessment have undergone rapid development during the last decade. Primary and secondary petroleum inclusion trails in fractures and cements can provide evidence for the presence of petroleum undergoing migration (Karlsen et al., 1993, 2004; Munz et al., 1998; Munz, 2001; Karlsen and Skeie, 2006). Also, petroleum migration can be inferred from compositional differences between inclusions in successive growth zones of fractures and in cement samples (Bodnar, 1990) or from the presence of different populations of inclusions in different generations of cements (Burruss, 2003a).

Raman spectroscopy is widely applied successfully to aqueous and dry gas inclusions. Burruss (2003a,b) detailed the importance of petroleum inclusion studies in petroleum exploration and development and suggested Raman spectroscopy as one method of getting detailed compositional data suitable for the analysis of petroleum inclusions if suitable methods to eliminate fluorescence interference are employed (Kihle et al., 2012). Therefore, as discussed the main challenge is to eliminate the fluorescence while studying Raman peaks in petroleum inclusions. And the next is selection of an optimum excitation wavelength since many previous studies noticed the problem of fluorescence from the fluids contained in the inclusions (hydrocarbons) or by the surrounding host mineral with shorter excitation wavelengths around 260 nm used for Raman analysis of HCFIs (Orange et al., 1996). Increasing the wavelength of the excitation source as discussed—red or near-infrared lasers—is the way to overcome fluorescence, since longer wavelengths excite less fluorescence from fluorophores that have visible wavelength excitation bands (Ahmadjian and Brown, 1976; Orange et al., 1996; McCreery, 2000; Burke, 2001; Bourdet et al., 2011; Frezzotti et al., 2012). Detection of hydrocarbon components in fluids or solids is useful for investigating the nature of fluids that interact with various geological formations, as well as the quality of oil in a petroliferous basin.

7.6 Raman analysis of HCFIs

Fluid inclusions are useful indicators of the migration pathways of hydrocarbons. Oil constituents may differ depending on the source rocks, maturity, organic matter trapped, and the pressure–temperature conditions of entrapment. Fluid inclusions can delineate the evolution of the chemistry of hydrocarbons; thus, they remain an important tool for understanding the thermal history of basins and related fluid migration events in the evolution of reservoir systems. HCFIs generally contain complex mixtures of organic compounds that depend on their source. Accurate analysis of the chemical composition of the entrapped hydrocarbons in HCFIs yields vital information about the history and evolution and migration of petroleum fluids, and can thus yield crucial data for the petroleum exploration industry (Goldstein, 2001; Munz, 2001; Frezzotti et al., 2012).

Raman spectroscopy provides useful inputs for the better understanding of the structure and maturity of both HCFIs and crude oils (Orange et al., 1996; McCreery, 2000; Burruss, 2003b; Bakker, 2004; Dearing et al., 2011). Rosasco et al. (1975) published the first results of Raman analyses of natural fluid inclusions. Bény et al. (1982) and Touray et al. (1985) published more comprehensive studies on fluid systems. Raman spectra of natural HCFIs are less common (Touray et al., 1985; Frezzotti et al., 2012; Lopponow et al., 2012). The reasons for the lack of feasibility of such studies in natural fluid inclusions are: (1) the presence of laser-induced fluorescence from both natural hydrocarbons and many host minerals can mask the Raman scattering signal (Guilhaumou, 1982; Guilhaumou et al., 1988; Wopenka et al., 1990), (2) laser-induced photodegradation may alter the chemical composition of the inclusion (Guilhaumou, 1982; Touray et al., 1985), and (3) the low laser power required for studies of these inclusions again make the Raman scattering intensity correspondingly low (Berlman, 1965; Touray et al., 1985). However, Guilhaumou (1982), Guilhaumou et al. (1988) and Piranon and Pradier (1992) have shown that Raman studies on HCFIs are possible, in spite of the obvious problems and difficulties, by increasing the wavelength of excitation light to red or near infrared (630–1060 nm) and thereby mitigating background fluorescence. Orange et al. (1996) opined that avoiding the absorption bands of hydrocarbons is crucial when choosing a Raman excitation line since both photodegradation and fluorescence are critically dependent on the absorption spectrum of the inclusion or host minerals.

The presence of petroleum inclusions in diagenetic cements or in healed fractures demonstrates that oil was present at the time of cement formation or during fracture healing. This observation can provide a lead on the relative time of oil generation and migration in the context of the diagenetic and tectonic history of the rock. Most of the petroleum inclusions occur as secondary inclusions within detrital grains (e.g., quartz, calcite, and feldspars). Accurate determination of the chemical composition of the hydrocarbon fluids can provide (together with microthermometric studies) information about the temperature, timing of fluid migration, and entrapment. By analyzing inclusions on a grain-by-grain basis, it is possible to differentiate inclusions from different source rocks. Such information is significant to the petroleum exploration industry, particularly concerning the study of petroleum reservoirs, where it may be necessary to chart small changes in fluid composition (Karlsen et al., 1993; Ryder et al., 2004).

We made an effort to characterize petroleum fluid inclusions from Mumbai and Kerala offshore, India, using a laser exciting at a wavelength of 785 nm, pioneering work in this field, and it highlights the utility of Raman spectroscopy as a powerful tool for characterizing hydrocarbons and for identifying the major functional groups in natural HCFI samples. The choice of excitation wavelength, laser power, and the fluorescence-quenching techniques employed during wafer preparation yielded the best conducive conditions for obtaining adequate Raman signals from the HCFIs studied.

Sidewall cuttings and core samples of the RV-1 well in the Ratnagiri Block of Mumbai Offshore Basin and KK-4C-A-1 well in the Kerala-Konkan offshore basin, India, were collected from the Regional Geological Laboratory (RGL), Oil and Natural Gas Corporation (ONGC), Government of India, Panvel, Mumbai. The samples were washed, dried, and sieved, and were securely stored for specialized wafer preparation in the Lapidary Facility at the National Centre for Earth Science Studies (NCESS). Sieved samples were impregnated with a suitable epoxy resin mixture, doped with a fluorescent-quenching dye, and then made into doubly polished wafers for fluid inclusion studies. These specially prepared wafers were used for petrographical, microthermometric, and Raman studies. Wafer preparation details are given in Chapter 2.

In RV-1, HCFIs were observed in quartz, calcite, feldspar grains, and cements in the sedimentary rock cuttings, especially from the sandstone-siltstone lithologies towards the lower portion of the well depth ranging from 3270 to 3690 m. Using the lithostratigraphic nomenclature of the Mumbai Offshore Basin, the hydrocarbon inclusions are mainly observed in the Panna Formation (Paleocene-Early Eocene age), which is both a producing as well as a reservoir horizon. Minor quantities of HCFIs are also found at the 1345–1360 m depth, which falls within the Mukta Formation (Early Oligocene) comprising limestone-shale lithologies. Petrographical studies confirmed the presence of CO_2-rich inclusions and aqueous inclusions, as well as hydrocarbon inclusions, in RV-1 well. Hydrocarbon inclusions showing fluorescence observed under 365 nm (mercury arc lamp) UV excitation during petrographical studies were selected for Raman studies (Figs. 7.3 and 7.4). Lithostratigraphic details of the HCFI samples are given in Table 7.1. The Mukta Formation samples are at shallow levels (1700 m above the other samples from Panna Formation), and potentially, the oil

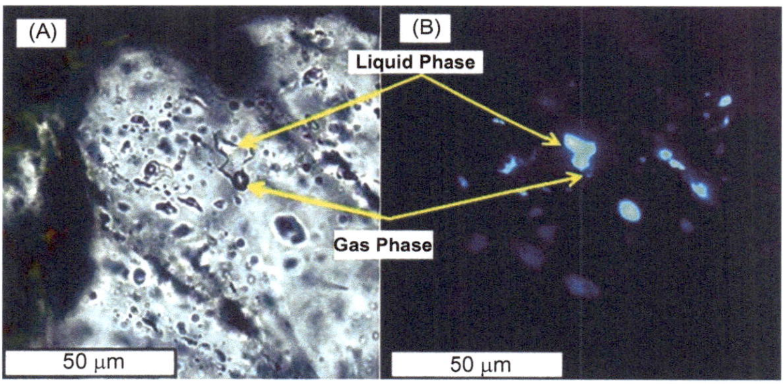

Fig. 7.3 Bright field (A) and fluorescence image (B) of HCFIs at 3495–3500 m depth sandy siltstone and shale lithology from Panna Formation, Mumbai Offshore Basin, India. HCFIs are fluorescing; where the vapor phase is not fluorescing.

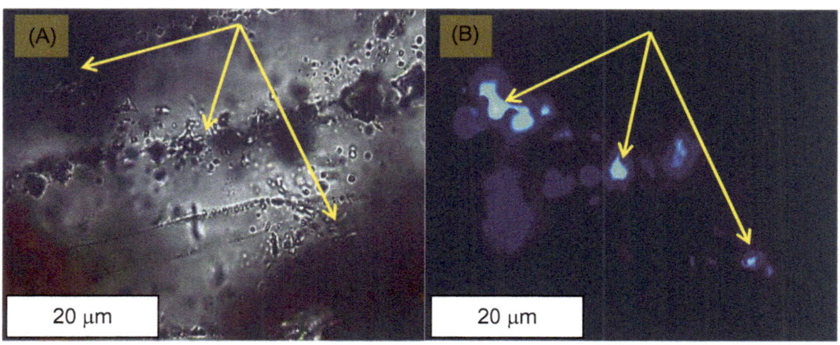

Fig. 7.4 Biphase HCFIs (marked with yellow arrows) at a 3055–3060 m depth with shale, sandstone and siltstone lithology in Panna Formation, Mumbai Offshore Basin, India. (A) Bright field and (B) fluorescent images.

Table 7.1 Details of the hydrocarbon fluid inclusions (HCFIs) studied from RV-1 well along with lithostratigraphic nomenclature, Mumbai offshore basin, India.

Sl. no.	Horizon (m)	Formation/age	Lithology	HCFIs description
1	3055–3060	Panna Formation/ Paleocene-Early Eocene	Sandstone and siltstone	Biphase secondary inclusions concentrated along the annealed fractures within quartz grains. Size varies from 2 to 11 μm and shapes are rectangular, fusiform, and elongated. Degree of fill 10%–20%
2	3065–3070	Panna Formation/ Paleocene-Early Eocene	Sandstone	Biphase inclusions concentrated along secondary fractures in quartz grains. Size varies from 2 to 5 μm and shapes are rounded, eye shaped, elongated, and stretched. Degree of fill 1%–5%
3	3270–3275	Panna Formation/ Paleocene-Early Eocene	Sandstone	Biphase inclusions concentrated along the secondary annealed fractures in quartz grains. Size varies from 5 to 9 μm and shapes are elongated, irregular and rounded. Degree of fill 5%–10%
4	3445–3450	Panna formation/ Paleocene-early Eocene	Sandstone and shale	Biphase secondary inclusions concentrated along the secondary annealed fractures in quartz grains. Size varies from 3 to 12 μm and shapes are irregular, oval and elongated. Degree of fill 1%–5%
5	3455–3460	Panna Formation/ Paleocene-Early Eocene	Sandstone and silty shale	Monophase secondary inclusions seen along the secondary trails. Size varies from 2 to 6 μm and shape is rectangular, fusiform, and elongated
6	3685–3690	Panna Formation/ Paleocene-Early Eocene	Shaley siltstone	Biphase secondary inclusions in cements. Size varies from 3 to 8 μm and shapes are rectangular and elongated. Degree of fill 10%–20%
7	3495–3500	Panna Formation/ Paleocene-Early Eocene	Sandy siltstone and shale	Biphase secondary inclusions concentrated along the secondary fractures in quartz and feldspar grains. Size varies from 5 to 17 μm and shapes are irregular and elongated. Degree of fill 10%–20%
8	1345–1350	Mukta Formation/ Early Oligocene	Limestone and shale	Biphase secondary inclusions in cements. Size varies from 5 to 12 μm and shapes are rectangular, irregular and elongated. Degree of fill 5%–10%
9	1355–1360	Mukta Formation/ Early Oligocene	Limestone	Biphase secondary inclusions concentrated along the annealed fractures in sporadic feldspar grains as well as in cements. Size varies from 2 to 5 μm and shape is irregular and elongated. Degree of fill 2%–5%

here could be of a different generation, or could be from a different source, or has been altered. Panna Formation oils are found to be more commercially viable compared to Mukta oil samples based on its API gravity value and is described in Chapter 6. Evidence of entrapment of light oils in the study area could be a useful indicator in petroleum exploration. In the KK basin, HCFIs are mainly observed at depths ranging from 3050 to 4000 m to 5025–5700 m with sandstone, limestone, siltstone, and claystone lithology, where bitumen such as heavy oil (low API gravity, <25) and sluggishly moving oil with migrating trend (medium heavy, API gravity <35) are observed (Figs. 7.5 and 7.6).

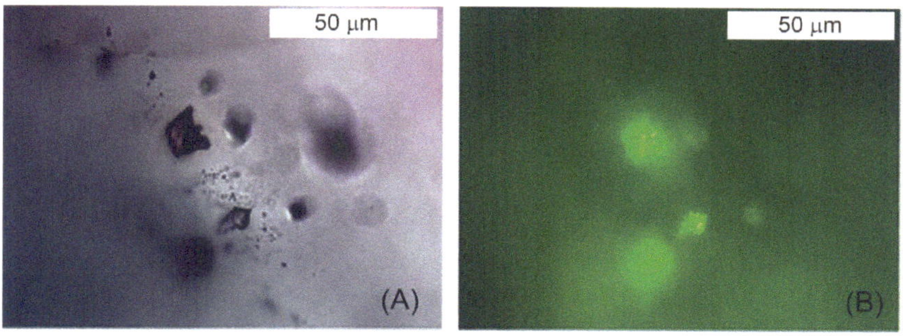

Fig. 7.5 Bright field (A) and fluorescent (B) images from the oil inclusion in the KK-4C-A-1 well (depth 3890–3895 m) with claystone, sandstone lithology.

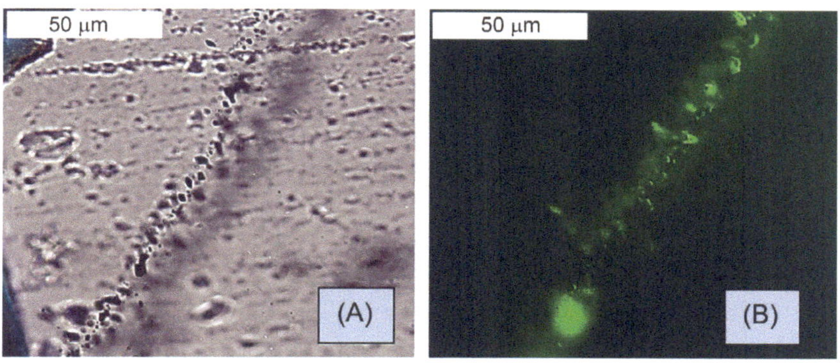

Fig. 7.6 Hydrocarbon fluid inclusion trail in the KK-4C-A-1 well in the Kerala-Konkan basin at a depth of 3065–3070 m with sandstone, and claystone lithology showing migrating oils. Bright field (A) and fluorescent (B) images.

Raman spectral analyses were carried out in nine HCFI assemblages from the RV-1 well (and a minimum of 15 individual inclusions from each assemblage) in order to detail any hydrocarbon species present. During petrography, HCFIs were identified, and then Raman studies were carried out at the National Facility for Geofluids Research and Raman Analysis of the NCESS, Thiruvananthapuram, using an inVia Raman Microscope (Renishaw, UK, Model No. M-9836-3991-01-A) attached to an inVia Reflex Raman Spectrometer offering automated alignment through the Renishaw WiRE 3.4 software. A diode laser with an exciting wavelength of 785 nm (300 mW, CW, Model. No. HPNIR 92E371, Class 3B laser product, Renishaw plc, UK) is used for Raman studies of fluid inclusions. Instrument calibration performed by measuring the Raman spectra of the internal and external silicon samples shows a sharp peak at a full-width half-maximum (FWHM) of $520 \pm 0.5\,cm^{-1}$. The Raman probe that we used has built-in neutral density filters that can reduce the beam intensity by several orders of magnitude, and the beam can be positioned precisely in the area intended for analysis.

The spectral range of the equipment is from 50 to $4000\,cm^{-1}$ shift from the laser line, accomplished with an edge filter. The Raman scattered light is dispersed with a grating of 1200 lines/mm. The detection is done by a Peltier-cooled charge-coupled device (CCD) detector with 576×384 pixels, with a spectral resolution of $1\,cm^{-1}$.

The HCFIs are viewed through the Renishaw inVia Raman Microscope, and are focused using a $100\times$ magnification objective lens at a working distance of 0.33 mm and 0.85 numerical aperture (NA). The Raman analysis was performed using a laser power around 1.5 mW with an exposure time of 20 s for three accumulations for each sample. Raman shifts were recorded, and the spectra were baseline corrected and smoothed using the Renishaw WiRE software supplied with the spectrometer system.

7.7 Earlier studies on the Raman spectra of HCFIs

Only a limited number of studies are reported in the analysis of HCFIs using the Raman technique, especially in natural fluid inclusions (Orange et al., 1996; Nai et al., 2007; Bourdet et al., 2011; Verma et al., 2012). Orange et al. (1996) have carried out a Raman feasibility study on natural hydrocarbon inclusions by combining an appropriate selection of the laser excitation line (the 633 nm light of a He—Ne laser) with a minimization of the excitation volume, low laser power, and long integration times. However, this technique remains qualitative because of complications associated with the quantification of the Raman cross-sections. Bourdet et al. (2011) reported that an excitation wavelength of 244 nm for Raman analysis caused some photochemical alterations in the inclusions and the Raman bands of hydrocarbon molecules in oil inclusions are masked by carbon bands related to molecules containing aromatic rings. This necessitated an optimum excitation wavelength for Raman spectral studies. Burke (2001) and Frezzotti et al. (2012) provided reviews on Raman spectroscopy in fluid inclusion analysis and mentioned that the influence of fluorescence might be avoided while considering hydrocarbon inclusions.

7.8 Raman spectra of HCFIs from Mumbai offshore, India

Raman spectroscopy has the advantage of having a better spatial resolution (in our observation, it is $<1\ \mu m$ while using a $\times 100$ objective), a resolution not possible with conventional petrography, microthermometry, or other spectroscopic methods (e.g., infrared spectroscopy). Thus, it is possible to carry out studies on an inclusion-by-inclusion basis in a sample, thereby characterizing multiple phases of fluid interaction. The main components of petroleum are hydrocarbons, nitrogen-sulfur-oxygen-bearing (NSO) compounds, and inorganic gases, such as N_2, CO_2, and H_2S. While analyzing the HCFIs with a 785 nm laser excitation, we observed Raman peaks of benzene at 606 cm^{-1} (ring deformation), 1310 cm^{-1} (ring-stretching vibration), 1486 cm^{-1} (ring stretching + deformation) and a bromobenzene peak at 1580 cm^{-1} (ring stretching). These vibrations, because of their highly symmetrical character, are quite intense and show the presence of benzene and its derivatives in the inclusions we have studied. The 990–1010 cm^{-1} region is of interest since this spectral region is diagnostic of the presence of aromatic rings. These vibrations are quite intense, and we observed peaks at 1010 cm^{-1} corresponding to the breathing vibration of the benzene ring. Peaks corresponding to cyclohexane are observed at 786 cm^{-1} (CH_2 rock vibration) and at 3245 cm^{-1}. A peak at 2055 cm^{-1} is observed, which corresponds to carbonyl sulfide (COS), and another peak observed at 2331 cm^{-1} shows the presence of nitrogen in the inclusions. A peak at 524 cm^{-1} represents the presence of sulfur dioxide and a peak at 2580 cm^{-1} shows the presence of H_2S in liquid form. The presence of carbon monoxide is seen with a peak at 2143 cm^{-1}. Calcium carbonate (calcite) peaks are seen at 710

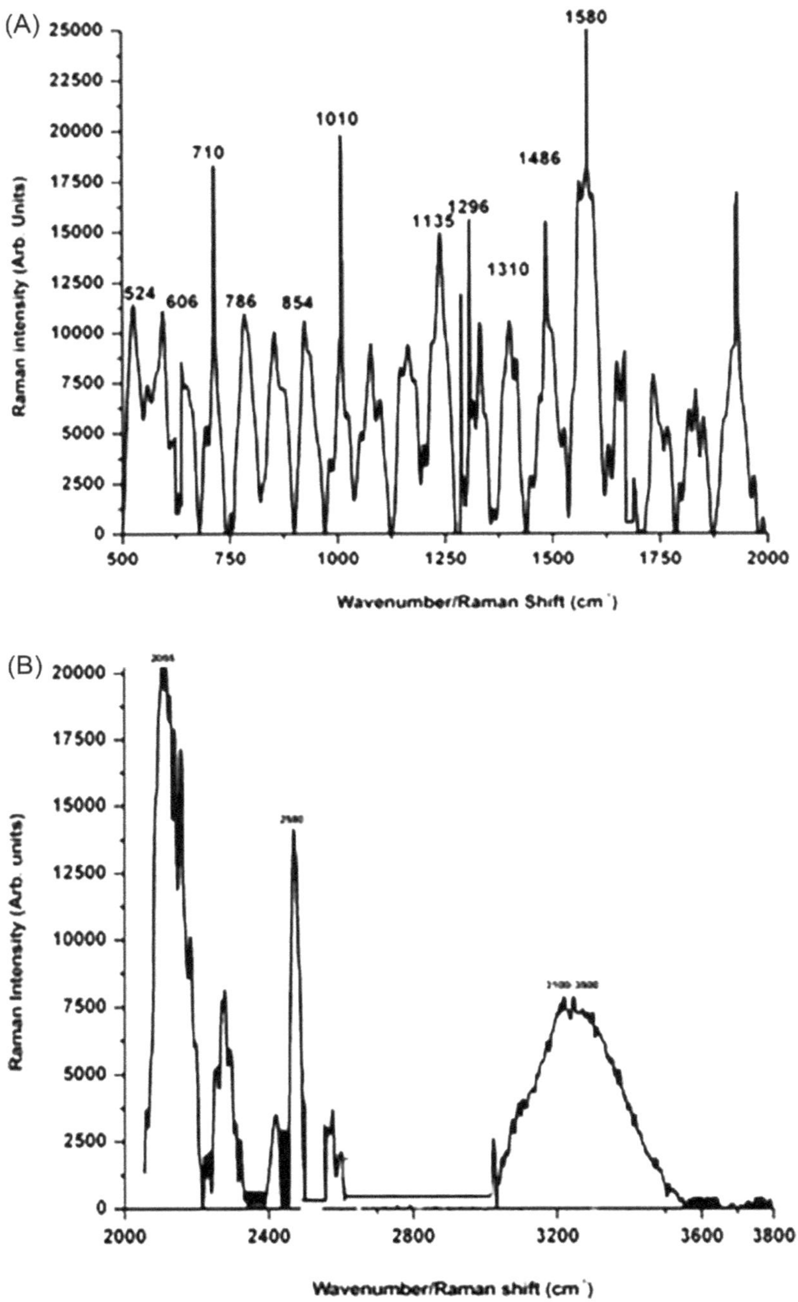

Fig. 7.7 Raman spectrum of natural HCFIs from the Mumbai offshore basin observed at 785 nm laser excitation

Table 7.2 Raman peaks and the major hydrocarbon species identified by Raman analysis for the natural hydrocarbon inclusions from Mumbai offshore basin, India.

Raman peaks/shifts (cm^{-1}) observed	Species/compound identified
606,1010,1310,1486	Benzene
1580	Bromo-benzene
786,3245	Cyclohexane
710,854	Calcium carbonate ($CaCO_3$)
2143	Carbon monoxide (CO)
2331	Nitrogen (N_2)
1135	Calcium sulfate ($CaSO_4$)
524	Sulfur dioxide (SO_2)
1296	Ethylene
2055	Carbonyl sulfide (COS)
2580	Hydrogen sulfide (H_2S)
3100–3500	H_2O (Liquid)

and 854 cm^{-1}, and a calcium sulfate ($CaSO_4$) peak is observed at 1135 cm^{-1}. A broad peak of water in liquid form is also observed in the region 3100–3500 cm^{-1}. Observed peaks shown in Fig. 7.7A and B have been consolidated and are shown in Table 7.2.

We have used a fluorescence quencher with a resin-hardener mix to eradicate the background fluorescence while preparing the wafers. Further, we have selected an excitation wavelength of 785 nm, which is above the absorption wavelength of aromatic hydrocarbon, which again eliminates the fluorescence from the hydrocarbons and, thus, also photodegradation. Advanced Raman spectrometers, such as the Renishaw inVia Raman spectrometer that we used in our studies, avoid the fluorescence effect with the use of notch holographic filters and collect the Raman signal directly from the fluids within the sample. Orange et al. (1996) reported that they were not able to see the presence of benzene and its derivatives in the natural fluid inclusions they studied, whereas they observed the same in oil samples and we observed benzene and its derivatives in natural fluid inclusion samples. The methodology we followed in wafer preparation and the selection of the excitation wavelength at 785 nm helped us to obtain desirable results.

Raman spectroscopy provides a wealth of information about the composition of microscopic mineral and fluid inclusions. Very fast acquisition times and lesser amount of laser energy penetrating into the HCFIs prevent any kind of alteration in the composition of the fluid inclusions and the use of notch filters in the latest Raman spectrometers allows the detection of even very weak signals with high sensitivity. Highly polished standard thin sections and doubly polished wafers are used for this purpose. The near-infrared (NIR) region has special importance for ultra-sensitive Raman spectroscopy at the single-molecule level in which background fluorescence is reduced and photo-induced degradation from the sample is diminished (Orange et al., 1996). High-intensity NIR diode lasers are easily available for low-cost Raman instrumentation. Further, the development of low-noise, high quantum-efficiency multichannel detectors (CCD arrays), combined with high-throughput single-stage spectrographs used in combination with holographic laser rejection filters have led to high-sensitivity Raman spectrometers (Kneipp et al., 1999; Dubessy et al., 2012). Recently, sophisticated data analysis techniques based on multivariate analysis have made it possible to exploit the full information content of Raman spectra (Tarcea and Popp, 2012).

The composition of oils in HCFIs determined using high-resolution Raman spectroscopy can very well be compared with the composition of the bulk oil obtained from the same basin using GCMS. We, therefore have analyzed fresh oil samples from the same basin using GC–MS (Single Quadrupole GC/MS 5977A GC/MSD with 7890B GC and 7693A Automatic Liquid Sampler) with a column HP-5 ms (30 m length × 250-μm internal diameter × 0.25-μm film thickness, working with Agilent Mass Hunter software), which identified the presence of benzene, cyclohexane, ethylene, sulfur, bromine, and nitrogen-based compounds. The chemical constituents in natural HCFIs from the same basin identified using laser Raman spectrometric methods with a 785 nm laser excitation agrees well with the GC–MS results of oil in the same basin, which again supports the utility of the laser Raman technique with 785 nm laser for the chemical constituents' identification in natural HCFIs.

The difference in the constituents of oils in secondary trails varies based on the source, generation, oil maturity, etc. The methodology we are discussing in Chapters 5 and 6, and this chapter is the only nondestructive method that can extract all the information from HCFI/assemblages along with coeval aqueous inclusion such as their PVT condition, salinity, API gravity, chemical constituents, etc. The nondestructive methodology we are introducing can revolutionize data generation that can be useful for petroleum exploration activities. Presently, the petroleum exploration industry is technologically more intensive with several exploration techniques to identify new/unidentified reserves, wherein the hydrocarbon fluid inclusions could be unique because of its capability to provide exclusive information that can significantly contribute to petroleum exploration activities.

7.9 Raman spectra of non-HCFIs

Raman analysis was also carried out for non-HCFIs (fluid inclusions that are not HCFIs, but are coeval aqueous inclusions associated with HCFIs in secondary trails) with the Raman spectrometer system for the fluid inclusion samples from the Kerala-Konkan basin. The Raman spectra observed are shown in Figs. 7.8. and 7.9. Peaks corresponding to species such as CO_3^{2-}(1064 cm^{-1}), calcite (1081 cm^{-1}), SiO_2 (1160 cm^{-1}), CO_2(l) and CO_2(g) (1281, 1385 cm^{-1}), N_2 (2330 cm^{-1}), and CH_4 (2926 cm^{-1}) were present in the fluid inclusion samples. The presence of water in liquid form shows a broad peak around 3000–3500 cm^{-1}. With the Raman spectrometer system, we were able to individually focus the gas phase in a biphase hydrocarbon fluid inclusion and the gas composition was determined as ethane (Fig. 7.10).

7.10 Raman peaks of major constituents in HCFIs

Detection of hydrocarbon components in fluids or solids could be useful in the investigation of the petroleum potential of a basin. FTIR or Raman spectrometry techniques are the two most commonly used tools for the analysis of fluid inclusions. Raman spectroscopy offers a laser spot size of 1–2 μm, which allows the analysis of individual phases in a fluid inclusion. Using visible light excitation, fluorescence of aromatic compounds prevents analysis of petroleum-bearing fluids because fluorescence emission is more intense than Raman scatter. Aromatic compounds present in oil and gas fluoresce in the UV, visible and NIR, depending on the excitation wavelength and the number of aromatic rings. Benzene is the smallest aromatic molecule and its fluorescence emission maximum is at the shortest wavelength of the aromatics at about 280 nm. In order to avoid fluorescence, UV-Raman uses short wavelength excitation from a frequency-doubled argon ion laser at 244 nm, below the fluorescence maximum of benzene. UV-Raman scatter can be detected between the excitation wavelength and the wavelength of fluorescence emission.

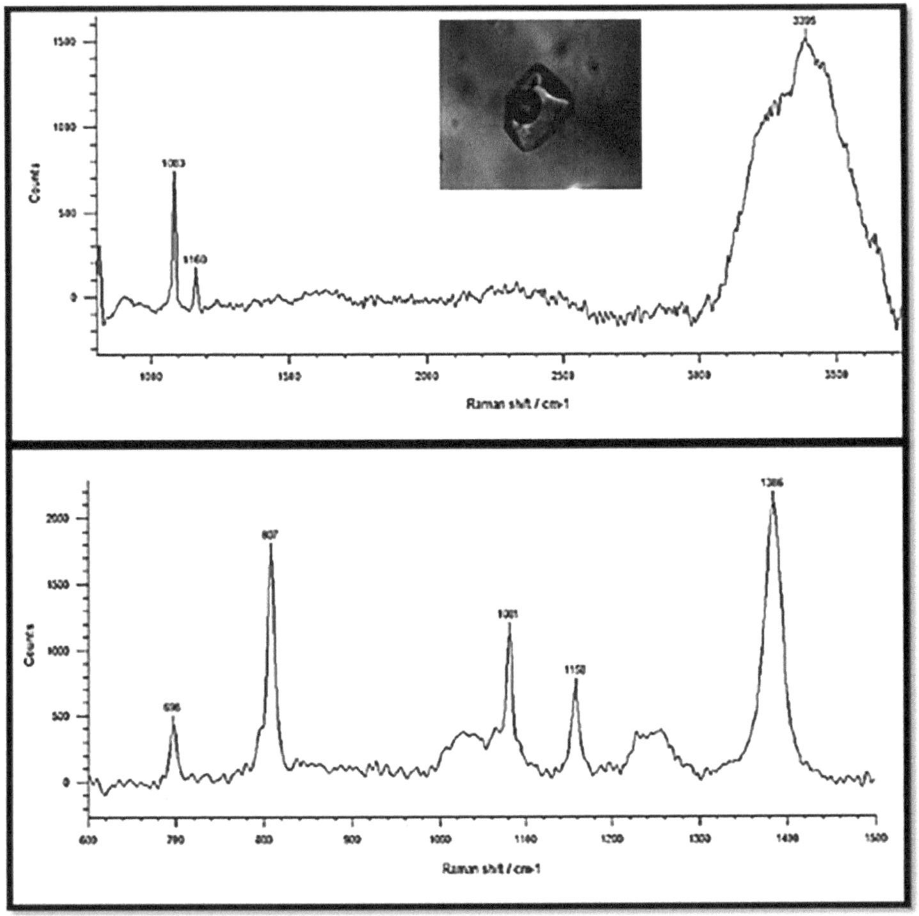

Fig. 7.8 Raman spectra of non-HCFIs from the Mumbai offshore basin showing peaks of H_2O (broad peak 3200–3600 cm^{-1}) and CO_2 (1386 cm^{-1}) along with peaks of quartz and calcite.

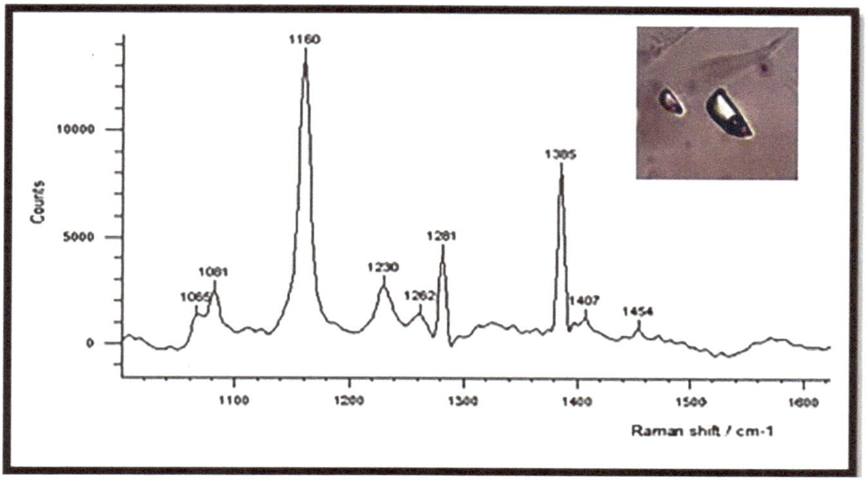

Fig. 7.9 Raman peaks for non-HCFI from the Kerala-Konkan basin at (3940–3945 m) depth showing the peaks of CO_2 at 1281 and 1285 cm^{-1} with Al laser excitation of 785 nm.

No technique other than Raman spectroscopy can analyze liquid, gas, and solid constituents in fluid inclusions. Incorporating Raman techniques with evolving new technologies such as spectral imaging could be a strong analytical tool for geological fluids. Burke (2001), Burruss (2003a,b), Frezzotti et al. (2012) and Vratislav et al. (2015) have provided the details of Raman spectroscopic analysis of fluid inclusions and have recommended further references, even though there are limited illustrations on HCFIs.

Bourdet et al. (2011) presented spectra acquired on pure paraffinic oils and on petroleum inclusions (oil and condensate). Pure paraffinic oil contains exclusively n-alkanes. Its spectrum shows typical C—H bands with two dominant groups, one between 1550 and 750 cm^{-1} and the other between 2600 and 3150 cm^{-1}. They represent C—C stretching, bending, and twisting bands, and C—H stretching bands, respectively, with additional features from Fermi resonance. The spectrum of oil in a paraffin-rich natural oil inclusion, enclosing a solid wax phase, shows a main band at around 1610 cm^{-1}, two bands at 1380 and 1435 cm^{-1}, and a minor band at 1230 cm^{-1}. Two wide bands with low intensities are located at about 2925 and 3210 cm^{-1} (Bourdet et al., 2011). The spectrum of a rim of oil in a gas condensate inclusion is reported (Bourdet et al., 2011), which is very similar to that of the paraffin-rich natural oil inclusion (heavier oil). All the bands observed in the petroleum inclusion spectra, except methane and

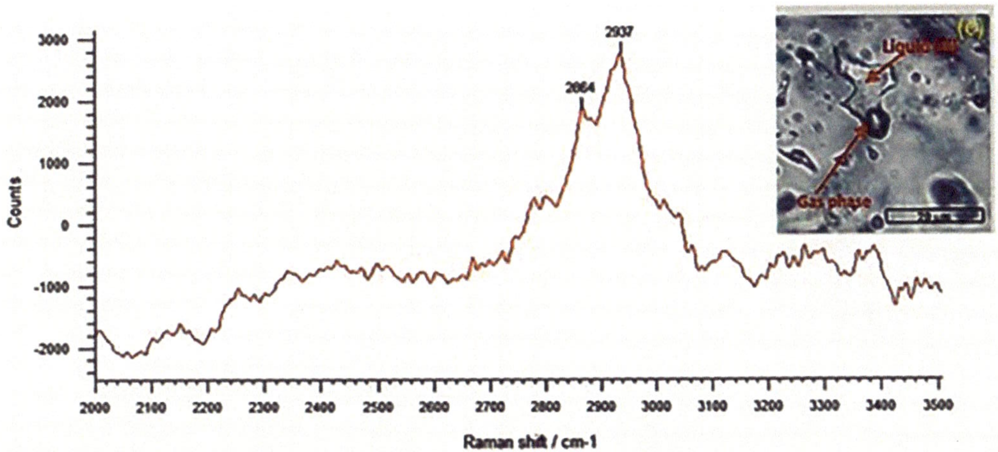

Peak no.	Centre	Height	Width	Area	Absolute inten...	Low edge	High edge
1	2864.48	483.398	14.5619	125670	1871.76	2819.77	2877.4
2	2936.92	1700.32	52.4505	307105	2782.59	2890.53	3006.25

Fig. 7.10 Raman spectrum of individual gas phase (ethane) from natural biphase hydrocarbon fluid inclusions at 3495–3500 m depth from the Mumbai offshore basin observed using a 785 nm laser excitation.

quartz, can be attributed to the vibration of graphite in ordered and disordered forms. It appears that those vibrations mask all CH or C=C bands produced by hydrocarbon molecules. Since the ordered/disordered carbon bands are not present in the pure paraffinic oil, those are interpreted to be related to the aromatic/resin/asphaltene molecules present in natural oil. During the Raman scatter spectra acquisition (D1 filter, 25 s, 2 acc) a brownish spot formed that fluoresces with an orange color (observed subsequently with a 365 nm UV lamp). A slight red shift of the whole inclusion fluorescence color of the heavier oil inclusions has been noticed following laser irradiation. This suggests that irradiation of petroleum inclusions with focused 244-nm laser light causes some photochemical alteration of the inclusions. Detection of Raman bands of hydrocarbon molecules in oil inclusions using a 244-nm laser are masked by carbon bands related to molecules containing aromatic rings (Bourdet et al., 2011).

Crude oil compositions can be divided into saturated hydrocarbon, aromatic hydrocarbon, nonhydrocarbon compound, and bitumen. Several types of hydrocarbons in fluid inclusions were identified using Raman spectroscopy, including saturated hydrocarbons, alkenes, aromatic hydrocarbons, bitumen, and methane. An overview of the characteristic Raman vibrational modes of some selected hydrocarbon compounds such as bromobenzene, PAHs, cycloalkanes, n-alkenes and mono- and polycyclic aromatics is given by Vratislav et al. (2015). Saturated compounds do not contain C=C bonds and other bonds vibrating between 1500 and 1570 cm^{-1}; therefore, they exhibit first-order Raman bands below 1470 cm^{-1}. The strongest Raman band of saturated hydrocarbons is the C—H deformation mode at 1430 cm^{-1} and its overtone at 2860 cm^{-1} (Lopponow et al., 2012). The UV Raman spectra of alkylated monocyclic aromatic hydrocarbons exhibit one or more bands at 1600 cm^{-1}, assigned as a C=C or ring stretch, while most of it exhibits a medium-strong band at this frequency. Petroleum with highly saturated hydrocarbon contents exhibits strongest bands between 2700 and 2970 cm^{-1}, and corresponds to C—H stretching and bending modes (Nai et al., 2007).

Aromatic hydrocarbons containing one benzene ring can be identified by a stable Raman band at 1600 cm^{-1} with two peaks. The combination of the strongest Raman peaks of 1005 and 3060 cm^{-1} is the evidence of aromatic hydrocarbons containing single chains, while the strongest Raman peaks of 1250 and 2910–2920 cm^{-1} are the evidence of aromatic hydrocarbons containing three chains. A stable Raman band at 1600 cm^{-1} showing double peaks can identify aromatic hydrocarbons containing two benzene rings and a strongest Raman peak at 3060 cm^{-1}. All these typical Raman bands can be used to identify alkanes and aromatic hydrocarbons (Chen et al. 2016). Chen et al. (2016) reported compositions of hydrocarbon-bearing inclusions in rock salt (halite) from the Shahejie Formation in the Dongying sag, Bohai Bay Basin, eastern China. The inclusions were analyzed using Raman spectroscopy coupled with a 244 nm UV laser source.

Raman spectroscopy gives a better evidence of C—C vibrations than C—H in aromatic hydrocarbons. Monocyclic aromatic hydrocarbons show the strongest bands at 1020–1200 cm^{-1} and one more band at around 1600 cm^{-1} (Vratislav et al., 2015). The band at 1380–1386 cm^{-1} is the δCH_3 vibration of methyl groups, and the band at 1435–1447 cm^{-1} is the δCH_2 vibration of methylene groups (Bruno et al., 1978; Zhu et al., 1998). The two bands are common in saturated hydrocarbons, especially in long chain hydrocarbons. Their wavenumbers depend on the carbon number of the chain. The band at 1620 cm^{-1} is the vibration of aromatic rings (Asher, 1984; Court et al., 2007). A small difference in the excitation wavelength can have a large effect on the relative intensities of 1400 and 1600 cm^{-1} bands in aromatic hydrocarbons.

By employing suitable filters with both mercury-arc and helium-arc excitation, a number of new lines have been recorded for Raman spectrum of benzene. The complete spectrum is given in frequency differences ($\Delta\nu$) as 606, 849, 992, 1178, 1584, 1603, 2460, 2542, 2597, 2617, 2784, 2928, 2947, 3046, 3060, and 3164 cm^{-1}. Anti-Stokes lines correspond to $\Delta\nu$ values 606, 849, and 992 cm^{-1} for benzene (Wood, 1930). The Raman spectrum of diphenyl consists of five faint diffuse lines and seven strong lines. Their $\Delta\nu$ values are 416, 606, 731, 766, 810 (these are very faint) and 990, 1023, 1280, 1544, 1584, 1603, 3055, and 3170 cm^{-1}. It will be noted that six of the lines are found in the spectrum of benzene (Wood, 1930).

The bands between 835 and 910 cm^{-1} indicate the single hydrogen atom in benzene. If two adjacent peripheral atoms are present, Raman bands occur between 800 and 855 cm^{-1}. Bands between 765 and 800 cm^{-1} indicate three adjacent hydrogen atoms at the edges of molecules. The presence of four adjacent peripheral hydrogen atoms show strong Raman features between 740 and 770 cm^{-1} (Lopponow et al., 2012). A typical phenyl (C_6H_5) band occurs at 988 cm^{-1} in some HCFIs, where weak bands between 2970 and 3200 cm^{-1} are attributed to methyl ($-CH_3$) and methylene ($-CH_2$) groups. The Raman spectra of bitumen usually show two main peaks centered at 1400 and 1620 cm^{-1}, corresponding to the C—C stretching in aromatic hydrocarbons. Additional weak bands between 2970 and 3200 cm^{-1} may be attributed to methyl and methylene groups (Nai et al., 2007).

Polycyclic aromatic hydrocarbons (PAHs) are molecules consisting exclusively or predominantly of carbon and hydrogen arranged in multiple aromatic (phenyl) rings. Structure and background PAHs refer to molecules composed of one or more phenyl (benzene) rings and are composed exclusively or predominantly of carbon and hydrogen. The nature of PAHs present in any environment can provide insights into the origin and evolution of the environments in which they are found. PAHs are widely distributed on the Earth, forming the backbone of a wide range of biomolecules and fossil fuels. For PAHs involved in the various geological processes, Raman spectroscopy is well suited for the analyses of organic matter. It can provide information on the degree of metamorphism to which organic matter has been subjected (Schopf et al., 2005). From this, the biogenic origin of Archean organic matter can be inferred (Schopf et al., 2005), as well as the degree of thermal maturity (Marshall et al., 2010). It is also a powerful technique for characterizing the structure of aromatic-rich synthetic and natural materials (Larouche and Stansfield, 2010). It is finding widespread use in characterizing the degree of structural evolution and metamorphism/graphitization in terrestrial organic-bearing materials such as coal, and for determining the hydrocarbon-generating potential of source rocks (Cloutis et al., 2016).

7.10.1 Raman spectra of bitumen

The Raman spectra of bitumen from sedimentary rocks and fluid inclusions show well-defined C—C stretching modes of PAHs at 1336–1345 and 1582–1596 cm^{-1}. A minor band centered at 1050 cm^{-1} in the bitumen from calcareous breccia could be due to small amounts of monocyclic aromatic hydrocarbons (Vratislav et al., 2015). Solid bitumen usually appears when the hydrocarbon source rock reaches a high maturity (Schoenherr et al., 2007a; Zhou et al., 2013). Solid bitumen has two typical first-order characteristic bands, Raman peaks (D and G) and the D band at 1355 cm^{-1} and G band at 1620 cm^{-1} (Jehlička and Beny, 1999; Shoute et al., 2002; Jehlička, Urban, and Pokorný, 2003; Zhou et al., 2014). The positions of the D and G bands depend on the thermal metamorphic degree of the organic matter (Jehlička and Beny, 1999; Jehlička, Urban, and Pokorný, 2003). Raman spectra of solid bitumen in inclusions indicate a very weak structural organization. The G band appears as a wide band at around 1600 cm^{-1}, which shows Raman bands at 1612 cm^{-1} or at 1614 cm^{-1}, indicating highly mature petroleum. However, the D band is just a weak shoulder band at the left of the G band. Schoenherr et al. (2007b) pointed out that oil inclusions can be converted in situ into solid bitumen, a process that releases gas. However, the inclusions containing bitumen do not show any Raman signal for gas, which indicates that the bitumen was not formed in situ in the inclusions (Chen et al., 2016).

7.10.2 Raman peaks of water

The stretching vibration of water shows a high and wide band at about 3000–3800 cm^{-1} with a sharp peak. The aqueous phase should be saturated with NaCl if the host mineral is halite (Mernagh and Wilde, 1989). The Raman modes of water consist of two main O—H stretching modes at 3657 and 3756 cm^{-1} and one very weak H-O-H bending mode at 1595 cm^{-1}. However, the Raman spectrum of liquid H_2O consists of several large overlapping bands in the OH stretching region from 2750 to 3900 cm^{-1}, and of a weak bending mode at ~1630 cm^{-1} (Frezzotti et al., 2012). It has been widely reported that petroleum inclusions are often filled with an oil–water immiscible mixture. (Pironon et al., 2000; Teinturier and Pironon, 2004; Pironon and Bourdet, 2008; Kettanah, 2013) Water is supposed to be present as a nonvisible liquid film under the microscope, wetting the inclusion walls. However, O—H and C—H bonds from water and oil can be easily detected and differentiated by Raman spectra because of their characteristic stretching and bending vibration regions. Pironon et al. (2000) reported the presence of water in petroleum inclusions as evidenced by Raman and FT-IR results. Our studies show that most hydrocarbon inclusions contain water. The water in the liquid phase shows a strong Raman band, but a weak band or no signal in the gas phase. Although the HC-bearing inclusions emit blue fluorescence, the Raman spectra indicate that these inclusions contain water in the liquid phase and the coexisting gas phase, which was mainly filled with hydrocarbons. The band at 1389 cm^{-1} is the δCH_3 vibration of methyl groups, and the band at 1440 cm^{-1} is the δCH_2 vibration of methylene groups, while the vibration of aromatic rings is shown as a Raman shift at 1620 cm^{-1}. The stretching vibration of water is separated by hydrocarbons showing two bands at 3220 and 3450 cm^{-1} in the gas phase (Chen et al., 2016). The middle and wide bands around 2930 cm^{-1} in the gas phase belonging to stretching vibration of the C—H bond (Zhu et al., 1998), while water shows a strong and wide crossing band at 3000–3800 cm^{-1} in the liquid phase.

A Raman spectrogram of normal paraffin has characteristic peaks of 2870–2873 cm^{-1} and 2934–2936 cm^{-1}. Different normal paraffins have the same Raman spectrogram because their molecules are composed of methyl ($-CH_3$) and methylene ($-CH_2-$). Normal pentane, hexane, and heptane have different numbers of carbon in their molecules. The characteristics of isoparaffin Raman spectrogram have been described by Zhang et al. (2007). A Raman spectrogram of all isoparaffins has a characteristic peak of 749–763 cm^{-1}; and that peak of 2912–2913 cm^{-1} of isoparaffin is stronger than that of normal paraffin. The 2911 cm^{-1} peak is regarded as the characteristic peak of methane. In mixed hydrocarbons or oil inclusions, the 2912–2913 cm^{-1} peak is the characteristic Raman peak of methylene, but not that of methane. In methane and inorganic compounds, the 2905–2921 cm^{-1} Raman spectrum peak is of methane. Different saturated compositions have different Raman spectra. The methyl special Raman peak (2872 cm^{-1}) is the strongest at normal paraffin; the methylene special Raman peak (2911 cm^{-1}) is the strongest for isoparaffin; and the methylene special Raman peak (2857 cm^{-1}) is the strongest for cycloparaffin. The particular peak of branched methyl is 748 cm^{-1}.

The Raman spectrograms have similar features due to the same alkyl group present in different hydrocarbons. The hydrocarbon types cannot be identified due to the complexity of hydrocarbon compositions and comparability of different alkyls. The Raman spectrogram of normal paraffin has very strong peaks of methyl and methylene (from 2700 cm^{-1} to 2970 cm^{-1}) groups; branch methyl has a particular peak of 748 cm^{-1}; phenyl has two particular peaks of 988 cm^{-1} and 3058 cm^{-1} and the 988 cm^{-1} peak is stronger than the 3058 cm^{-1} peak. Hexene has three alkenyl spectrum peaks of 1294 cm^{-1}, 1635 cm^{-1} and 2996 cm^{-1}, with the 1635 cm^{-1} peak being the strongest, showing that the number of carbon atoms in hydrocarbon does not affect its Raman spectrogram, and the hydrocarbon molecular structure and base groups affect its Raman spectrogram (Nai et al., 2007).

7.10.3 Raman peaks of host minerals

Hydrocarbon fluid inclusions are usually observed in limited number of minerals such as quartz, calcite, feldspar, etc. The strongest peak of quartz is at 466 cm^{-1}, relatively strong lines occur at 128, 206, 1082, and 1160 cm^{-1}, and a weak band at 1230 cm^{-1}. Calcite has a very strong line at 1085 cm^{-1}, and lesser peaks at 156, 283, 711 and, 1450 cm^{-1}; the lower peaks are important to distinguish the mineral from aragonite, which has its peaks at 152 and 209 cm^{-1}. Feldspar has two peaks at 475–480 and 510 cm^{-1}. Silicate minerals are critically important to fluid inclusion research: quartz, olivine, pyroxenes, and garnet represent very common host phases for fluid inclusions, and their Raman bands should be fully characterized before analyzing fluid inclusions. Compared to carbonates and sulfates, silicate minerals are weaker Raman scatterers, due to the low polarizability of the Si—O bonds. Carbonates are common phases in fluid inclusions. Raman vibrational modes are dependent on the main carbonate groups, modified by the interactions with the bonded mineral lattice. CO_3^{2-} exhibits three main distinct internal vibrational modes over the range 400–1400 cm^{-1}. Generally, strong Raman modes appear around 1050–1100 cm^{-1} due to the symmetric stretching vibration (ν1) of the carbonate group, while weaker (around 20 times less intense) Raman bands near 700 cm^{-1} and 1400 cm^{-1} are due to the in-plane bending mode (ν4) and the antisymmetric stretch (ν3) of CO_3, respectively.

7.10.4 Raman spectra of alkanes

Spectroscopic properties of alkanes in the homologous series (methane, ethane, propane, and butane) have been a subject of interest for many years. In the case of methane, the Raman spectrum was studied in detail, especially in the Fermi resonance (FR) region (Atamas et al., 2004). The broad range of alkanes in this series (C_nH_{2n+2}) gives the possibility of studying their optical properties. The molecular structure changes from methane to butane through successive addition of the CH_2 group, which is reflected by changes in the Raman spectra. Three Raman bands at 1438, 2890, and 2850 cm^{-1}, can identify the normal alkanes of carbon number >10. Iso-alkanes (like C_8H_{18}) have a stable strong Raman peak at 2875 cm^{-1}. The strongest Raman peak combination of C—C (1450 cm^{-1}) and C—H (2875 cm^{-1}) is the evidence of the isoalkanes containing one chain.

Methane-bearing inclusions are usually observed in sedimentary environments. Methane is a spherical top molecule, with four distinct normal vibrational modes that are Raman active (Magnotti et al., 2015). The dissolved CH_4 peak is visible at 2913 cm^{-1}. Methane has been detected in gas-rich inclusions. The gas phase has a smooth spectrum with wide peaks at about 1400, 1610, 2925, and 3210 cm^{-1}. The vapor CH_4 peak is identified at 2911.5 cm^{-1}. The υ1 mode is a symmetric C—H stretching mode around 2918 cm^{-1}, the υ3 is a triply degenerate C—H stretching mode at 3020 cm^{-1}, the υ2 is a doubly degenerate deformation (bending) mode centered at 1534 cm^{-1}, and the υ4 is a triply degenerate deformation mode centered at 1306 cm^{-1} (Magnotti et al., 2015). Usually, the υ3 and 2υ2 bands cannot be detected with Raman spectroscopy because the concentration of methane is less in most fluid inclusions and the 2υ2 band at 3070 cm^{-1} (Hansen et al., 2002). Chen et al. (2016) have used the 514 nm laser as the exciting light for the Raman studies on methane-bearing inclusions since the fluorescence is weak for methane-bearing inclusions and it gives a stronger response than that of 244 nm. Raman spectra from the methane-bearing gas phase show water and methane, a wide band at 3000–3800 cm^{-1} for the stretching vibration of water while there is a sharp peak at 2918 cm^{-1} for the υ1 band of methane (Chen et al., 2016). The band position of υ1 may show little change for different inclusions, and sometimes it may be at 2917 and 2919 cm^{-1} for inclusions; with Raman spectra, we can easily find out the presence of methane in inclusions.

The spectrum of methane (CH_4) shows a rich rotational fine structure in the low-frequency region (1300–1800 cm^{-1}) with a clear maximum at 1532 cm^{-1}. In the high-frequency region, very intense bands are observed. The one of higher intensity lies at 2919 cm^{-1}, whereas that at 3030 cm^{-1} is of lower intensity. Both bands give overlapping rotational structure in the region of 2750–3350 cm^{-1} and these bands are symmetrical. Another band of low intensity can be found close to the one at 3030 cm^{-1}. This band (with a maximum at 3090 cm^{-1}) has rotational structure on its right shoulder. Similarly to the low-frequency region, the right shoulder is more intense than the left one (Atamas et al., 2004).

Ethane (C_2H_6) has two high intensity bands at 2895 and 2955 cm^{-1} in the spectral region 2600–3250 cm^{-1}. The rotational structure of the shoulders was not observed for ethane. The band intensities were similar to those for methane, except for the second high intensity band with a maximum at 2895 cm^{-1}. Another two bands of low intensity can be found at the low-frequency end. In the low-frequency region, again two symmetrical bands with maxima at 1000 and 1440 cm^{-1} are observed, where the band at 1000 cm^{-1} shows higher intensity.

For propane, a three-component band (at 2883, 2937, and 2973 cm^{-1}) is observed in the high-frequency region 2400–3400 cm^{-1}. The peak at 2883 cm^{-1} is the most intense of these, while the antisymmetric stretching modes give rise to peaks at 2937 and 2973 cm^{-1}. On the left shoulder of the band with a maximum at 2883 cm^{-1}, three weaker bands can be observed. The group of bands in the low-frequency region (500–1600 cm^{-1}) are similar to those observed in methane and ethane. Another complicated band shape is observed for butane. The spectral region 2800–3100 cm^{-1} consists of at least three sub-bands with peaks at 2885, 2927 and 2960 cm^{-1}. The sub-band located at 2885 cm^{-1} is the most intense, whereas the one with a maximum at 2927 cm^{-1} is the weakest. It should be noted that the bands at 2885 and 2960 cm^{-1} each display shoulders at 2865 and 2987 cm^{-1}, respectively. In the low-frequency region (0–1600 cm^{-1}) two groups of bands can be seen. The first group is in the region 0–450 cm^{-1}, and the other in the region 450–1600 cm^{-1}.

Although petroleum contains trace amounts of alkenes (Horsfield and McLimans, 1984), it is easy to identify these compounds using gas chromatography and mass spectrometry (GC–MS), FT-IR, and Raman spectroscopy. The band at 1647–1655 cm^{-1} is the C=C vibration of long chain alkenes. The position of the C=C vibration depends on the chain length of these hydrocarbons, but it only changes within a few wavenumbers (Bruno et al., 1978; Zhu et al., 1998).

7.10.5 Factors affecting the intensity of Raman peaks

Several factors can affect the relative intensities of Stokes and anti-Stokes lines in a Raman spectrum. These include polarizability—the degree of ease with which the electron clouds around nuclei can be distorted in response to an electric field (Smith and Geoffrey, 2005; Carey and Sundberg, 2007). Molecular vibrations that exhibit a strong polarizability result in intense Raman peaks (Smith and Geoffrey, 2005). This occurs in molecular vibrations that involve symmetric motion and lack a dipole moment. These molecular vibrations are commonly described as being Raman active whereas those that do not demonstrate such properties are often referred to as being Raman inactive and IR active (Smith and Geoffrey, 2005). The delocalization of π electrons in a conjugated system in a molecule also enhances polarizability (Mitsui et al., 2007). Conjugated systems are those in which multiple bonds occur in alternation with single bonds throughout all or part of a molecule (Mehta and Mehta, 2005). This bond arrangement allows π electrons to move freely throughout the conjugated system and become delocalized. As a result, the potential energy of the molecule is reduced, increasing its stability (Mehta and Mehta, 2005). Hooke's law (Smith and Geoffrey, 2005) can approximately predict the frequencies corresponding to bond vibrations in small molecules. However, the application of Hooke's law becomes limited for large molecules since vibrational coupling is more common (Freeman, 1974; Smith and Geoffrey, 2005). Vibrational coupling in a large molecule results from the interaction of bond vibrations involving atoms of similar mass and can cause a shift in Raman peaks (Freeman, 1974). In many cases, vibrational coupling can even occur between bond vibrations that are not directly connected in a molecule (Freeman, 1974).

The intensity of Raman peaks is inversely proportional to the fourth power of the excitation wavelength (Smith and Geoffrey, 2005). Even though a shorter excitation wavelength may generate Raman peaks of higher intensity, it has a greater possibility of causing thermal decomposition and photodissociation in samples (Som and Foing, 2012; Montillaud et al., 2013). Therefore, the intensity and signal integration times for a sample must be limited in order to prevent excessive irradiation. The strength of the signal that reaches the detector of a Raman spectrometer is highly dependent on the physical properties of a target, such as grain size, internal boundaries, surface texture, and the presence of coatings (Wang et al., 1998; Wang, 1999).

Differences in π electron distribution affect the direction in which peak positions shift when comparing the Raman spectra of similar PAHs, since vibrational frequency increases with bond order (Socrates, 2001). However, vibrational frequencies can also be influenced by many other phenomena such as steric repulsion (when two neutral atoms close to each other their outer (valence) electrons interact to repel, and this repulsion energy increases very rapidly as the distance between the atoms decreases. This increase in energy as atoms are crowded together is called steric repulsion or steric hindrance), hydrogen bonding, and inductive effects which are often caused by the presence of functional groups. Also, the degree of influence that such phenomena have is difficult to measure using Raman spectroscopy alone (Socrates, 2001). Therefore, the use of additional techniques such as ultraviolet–visible or nuclear magnetic resonance (NMR) spectroscopy might be necessary to determine the molecular structures of PAHs, which contain multiple functional groups or various hydrocarbon side chains. An understanding of how ring system variation affects Raman shifts in similar PAHs was achieved by first noting the position of certain peaks in the Raman spectrum of naphthalene that corresponded to specific molecular vibrations proposed by Srivastava and Singh (2007). The recurrence of these peaks in the Raman spectra of similar PAHs allowed shifts in their positioning to be determined as a function of changes in the size and arrangement of the ring system. When the Raman spectra of naphthalene and anthracene were compared, anthracene displayed vibrational frequency decreases for C—C and in-phase C=C stretching, as well as ring breathing. However, frequency increases were observed for C—C stretching coupled with C—H bending and ring deformation. C–C–C bending, C=C stretching and out-of-phase C—H stretching did not appear to greatly change in frequency. The Raman peaks resulting from molecular vibrations in functional groups are greatly influenced by the immediate chemical environment. For example, the Raman shift corresponding to the symmetric stretching vibration of a nitrite is observed between 1220 cm^{-1} and 1285 cm^{-1} when the nitrogen atom is directly bound to an oxygen atom that also shares a bond with a carbon atom (Larki, 2011). However, if the nitrogen atom is directly bound to an aromatic entity, the same Raman shift is observed between 1318 cm^{-1} and 1357 cm^{-1} (Larki, 2011). Therefore, it is not accurate to assume that a molecular vibration associated with a certain functional group will always correspond to a Raman peak that occurs within a specific narrow wavenumber range.

As expected, differences in π electron distribution affect the direction in which peak positions shift when comparing the Raman spectra of similar PAHs. Vibrational frequencies are influenced by other properties such as steric repulsion, hydrogen bonding, and the types and locations of functional groups. These affect Raman peak positions associated with C—C and in-phase C=C stretching, ring breathing, C—C stretching coupled with C—H bending, ring deformation, C–C–C bending, C=C stretching, and out-of-phase C—H stretching. As a result, no robust systematic variations were found between PAH structures and positions of these major Raman peaks.

7.10.6 Raman spectroscopy—Available information sources

Burke (2001) has provided a review on the applications of fluorescence spectroscopy to fluid inclusion studies. McCreery (2000) presents a source of fundamental information on Raman spectroscopy that details a thorough, consistent discussion on the theory and practice of Raman spectroscopy for chemical analysis. A review by Frezzotti et al. (2012) summarizes the recent developments of Raman spectroscopy in fluid inclusion research to provide support for laboratory analyses. Chen et al. (2016) details UV Raman spectroscopy of hydrocarbon-bearing inclusions in rock salt that discusses most of the Raman peaks in HCFIs and non-HCFIs. Burruss (2003a,b) discussed the Raman spectroscopy of fluid inclusions and Loppnow et al. (2004) detailed Raman spectroscopy as a popular technique for investigating complex organic matter. Dubessy et al. (2012) published a book on the applications of Raman spectroscopy on earth sciences with detailed application and instrumentation for different aspects in earth sciences. Vratislav et al. (2015) also provided a detailed discussion on Raman and IR spectroscopic analysis of hydrocarbon and no-hydrocarbon compounds. The books, review papers, and other communications can be used for further reference. Several websites are available with Raman spectroscopic data on minerals and chemical compounds that can be used as valuable resources. Some examples are given here: Raman Spectra Database, Siena Geofluids Lab (http://www.dst.unisi.it/geofluids/raman/spectrum_frame.htm); Raman Spectra Database Lyon (http://www.ens-lyon.fr/LST/Raman); the RRUFF database is creating a complete set of high-quality spectral data from well-characterized minerals (https://rruff.info/); SpectraOnline mineral and chemical database (Spectra.galactic.com/SpectraOnline), etc.

7.10.7 Raman spectroscopy-Other applications

Raman spectroscopy is a remarkable analytical tool for fluid inclusion studies. The Raman spectrum depends heavily on the instrumental conditions and factors such as the spectral resolution and laser wavelength as well as the presence of different components in a mixture and their relative concentrations that can modify the appearance of the spectral response obtained. Raman spectroscopy can act as a good analytical tool for the analysis of biomarkers (unique chemical species or patterns that are responsible for that particular material), where Raman spectroscopy can differentiate the key molecular species based on their spectral signatures. The Raman spectrum of graphitic carbons can provide some information of geological relevance by avoiding or minimizing the analytical biases and can act as a versatile tool for the detection and characterization of graphitic carbons in geological, environmental, and extraterrestrial samples.

7.11 Conclusions

HCFIs contain entrapped petroleum fluids that have migrated through the rocks in the geological past and are of special interest to the petroleum industry. The characterization of HCFIs using noncontact, nondestructive, optical Raman spectrometry presents considerable challenges. Fluorescence (background and from the aromatic compounds in HCFIs) and the presence of overlapping bands from the host mineral are the possible undesirable effects during Raman analysis, and can generally mask the

Raman signals. To overcome these hurdles, special wafer preparation techniques, along with the use of fluorescence quenchers, were employed to obtain Raman signals from natural HCFIs. Further, proper selection of excitation wavelength, and appropriate choice of laser power and optics, yielded the best possible Raman signals in HCFIs from the Mumbai Offshore Basin, India. A laser wavelength of 785 nm used for the present study identified Raman peaks corresponding to cyclohexane, benzene, ethylene, carbonyl sulfide, sulfur dioxide, nitrogen, and carbon monoxide in HCFIs, along with the presence of calcium carbonate and water. The present chapter elucidates the potential of Raman spectroscopic methods using a 785 nm laser excitation for detecting the chemical constituents of HCFIs.

References

Ahmadjian, M., Brown, C.W., 1976. Petroleum identification by laser Raman spectroscopy. Anal. Chem. 48, 1257–1259.

Asher, S.A., 1984. Ultraviolet resonance Raman spectrometry for detection and speciation of trace polycyclic aromatic hydrocarbons. Anal. Chem. 56, 720–724.

Atamas, N.A., Yaremko, A.M., Seeger, T., Leipertz, A., 2004. A study of the Raman spectra of alkanes in the Fermi-resonance region. J. Mol. Struct. 708 (1–3), 189–195. https://doi.org/10.1016/j.molstruc.2004.05.036.

Bakker, R.J., 2004. Raman spectra of fluid and crystal mixtures in the systems H_2O, H_2O–NaCl and H_2O–$MgCl_2$ at low temperatures: applications to fluid inclusion research. Can. Mineral. 42, 1283–1314.

Bény, C., Guilhaumou, N., Touray, J.C., 1982. Native-sulphur-bearing fluid inclusions in the CO_2–H_2S–H_2O–S system—microthermometry and Raman microprobe (MOLE) analysis—thermochemical interpretations. Chem. Geol. 37, 113–127.

Berlman, B., 1965. Handbook of Fluorescence Spectra of Aromatic Molecules. Academic Press, New York.

Bodnar, R.J., 1990. Petroleum migration in the Miocene Monterey formation, California, USA: constraints from fluid inclusion studies. Mineral. Mag. 54, 295–304.

Bourdet, J., Burruss, R.C., Bodnar, R.J., Eadington, P.J., 2011. Assessment of UV-Raman for analysis of petroleum inclusions. In: European Current Research on Fluid Inclusions (ECROFI–XXI), Montanuniversität Leoben, Austria, 9–11 August 2011, Abstract Volume, pp. 50–51.

Bruno, J.Z., Wilhoit, R.C., Dickson, C.H., 1978. Selected Raman spectral data. Supplementary, Vol. no. D-23 of the TRC-API44 hydrocarbon project publication.

Burke, E.A.J., 1994. Raman microspectrometry of fluid inclusions. In: De Vivo, B., Frezzotti, M.L. (Eds.), Fluid Inclusions in Minerals: Method and Applications. Short Course of Working Group (IMA) "Inclusions in Minerals". Polytechnic Institute and State University, Virginia, pp. 25–44.

Burke, E.A.J., 2001. Raman microspectrometry of fluid inclusions. Lithos 55, 139–158.

Burke, E.A.J., Lustenhouwer, W.J., 1987. The application of a multichannel laser Raman microprobe (Microdril-28) to the analysis of fluid inclusions. Chem. Geol. 61, 11–17.

Burruss, R.C., 2003a. Petroleum fluid inclusions, an introduction. In: Samson, I., Anderson, A., Marshall, D. (Eds.), Fluid Inclusions: Analysis and Interpretation. Mineralogical Association of Canada, Short Course Series, vol. 32, pp. 159–174.

Burruss, R.C., 2003b. Raman spectroscopy of fluid inclusions. In: Samson, I., Anderson, A., Marshall, D. (Eds.), Fluid Inclusions: Analysis and Interpretation. Mineralogical Association of Canada, Short Course Series, vol. 32, pp. 279–289.

Carey, P.R., 1999. Raman spectroscopy, the sleeping giant in structural biology, awakes. J. Biol. Chem. 274, 26625–26628.

Carey, F.A., Sundberg, R.J., 2007. Advanced organic chemistry. Part A: structure and mechanisms. Springer.

Chen, Y., Wang, X., Bodnar, R.J., 2016. UV Raman spectroscopy of hydrocarbon-bearing inclusions in rock salt from the Dongying sag, eastern China. Org. Geochem. 101, 63–71.

Cloutis, E., Paul, S.P., Applin, D., Goltz, D.M., 2016. Identification and discrimination of polycyclic aromatic hydrocarbons using Raman spectroscopy. Icarus 274. https://doi.org/10.1016/j.icarus.2016.03.023.

Court, R.W., Sephton, M.A., Parnell, J., Gilmour, I., 2007. Raman spectroscopy of irradiated organic matter. Geochim. Cosmochim. Acta 71, 2547–2568.

Dearing, T.I., Thompson, W.J., Rechsteiner, C.E., Marquardt, B.J., 2011. Characterization of crude oil products using data fusion of process Raman, infrared, and nuclear magnetic resonance (NMR) spectra. Appl. Spectrosc. 65, 181–186.

Dhamelincourt, J.M., Beny, C., Dubessy, J., Poty, B., 1979. Analyse d'inclusions fluides à la microsonde MOLE à effet Raman. Bull. Minéral. 102, 600–610.

Dubessy, J., Audeoud, D., Wilkins, R., Kosztolanyi, C., 1982. The use of the Raman microprobe mole in the determination of the electrolytes dissolved in the aqueous phase of fluid inclusions. Chem. Geol. 37, 137–150.

Dubessy, J., Poty, B., Ramboz, C., 1989. Advances in C–O–H–N–S fluid geochemistry based on micro-Raman spectrometric analysis of fluid inclusions. Eur. J. Mineral. 1, 517–534.

Dubessy, J., Caumon, M.C., Rull, F., Sharma, S., 2012. Instrumentation in Raman spectroscopy: elementary theory and practice. In: European Mineralogical Union (EMU) Notes in Mineralogy, Vol. 12, pp. 83–172. Chapter 3.

Freeman, S.K., 1974. Applications of laser Raman spectroscopy. J. Chem. Educ. 51 (10), A504.

Frezzotti, M.L., Tecce, F., Casagli, A., 2012. Raman spectroscopy for fluid inclusion analysis. J. Geochem. Explor. 112, 1–20.

Gardiner, D.J., Graves, P.R., 1989. Practical Raman Spectroscopy. Springer-Verlag, Berlin, Heideberg. ISBN-13: 978-3-540-50254-8.

Goldstein, R.H., 2001. Fluid inclusions in sedimentary and diagenetic systems. Lithos 55, 159–193.

Guilhaumou, N., 1982. Accurate analysis of fluid inclusions by the laser molecular microprobe (MOLE) and by microthermometry. Travaux. Lab. Geol. Ecole Norm. Sup. Paris 14, 68 (in French).

Guilhaumou, N., Jouaffred, D., Velde, B., Beny, C., 1988. Raman microprobe analysis on gaseous inclusions from diagenetically altered Terres Noures (S-E, France). Bull. Mineral. 111, 577–585.

Hansen, S.B., Berg, R.W., Stenby, E.H., 2002. How to determine the pressure of a methane containing gas mixture by the means of two weak Raman bands, υ_3 and $2\upsilon_2^+$. J. Raman Spectrosc. 33, 160–164.

Horsfield, B., McLimans, R.K., 1984. Geothermometry and geochemistry of aqueous and oil bearing fluid inclusions from Fateh field, Dubai. Org. Geochem. 6, 733–740.

Jehlička, J., Beny, C., 1999. First and second order Raman spectra of natural highly carbonified organic compounds from metamorphic rocks. J. Mol. Struct. 480–481, 541–545.

Jehlička, J., Urban, O., Pokorný, J., 2003. Raman spectroscopy of carbon and solid bitumens in sedimentary and metamorphic rocks. Spectrochim. Acta A Mol. Biomol. Spectrosc. 59, 2341–2352.

Karlsen, D.A., Skeie, J.E., 2006. Petroleum migration, faults and overpressure, b part I: calibrating basin modeling using petroleum in traps—a review. J. Pet. Geol. 29, 227–256.
Karlsen, D.A., Nedkvitne, T., Larter, S.R., Bjørlykke, K., 1993. Hydrocarbon composition of authigenic inclusions, application to elucidation of petroleum reservoir filling history. Geochem. Cosmochim. Acta 57, 3641–3659.
Karlsen, D.A., Skeie, J.E., et al., 2004. Petroleum migration, faults and overpressure. Part II. Case history: the Haltenbanken Petroleum Province, offshore Norway. In: Cubitt, J.M., England, W.A., Larter, S. (Eds.), Understanding petroleum reservoirs: towards an integrated reservoir engineering and geochemical approach. vol. 237. Geological Society, London, Special Publications, pp. 305–372.
Kettanah, Y.A., 2013. Hydrocarbon fluid inclusions in the Argo salt, offshore Canadian Atlantic margin. Can. J. Earth Sci. 50, 607–635.
Kihle, J., Hurum, J.H., Liebe, L., 2012. Preliminary results on liquid petroleum occurring as fluid inclusions in intracellular mineral precipitates in the vertebrae of Pliosaurus funkei. Nor. J. Geol. 92, 341–352.
Kneipp, K., Kneipp, H., Itzkan, I., Dasari, R.R., Feld, M.S., 1999. Ultrasensitive chemical analysis by Raman spectroscopy. Chem. Rev. 99, 2957–2975.
Larki, P.J., 2011. Infrared and Raman Spectroscopy-Principles and Spectral Interpretation. Elsevier, ISBN: 978-0-12-386984-5.
Larouche, N., Stansfield, B.L., 2010. Classifying nanostructured carbons using graphitic indices derived from Raman spectra. Carbon 48 (3), 620–629. https://doi.org/10.1016/j.carbon.2009.10.002.
Lisk, M., O'Brien, G.W., Eadington, P.J., 2002. Quantitative evaluation of the oil-leg potential in the Oliver gas field, Timer Sea, Australia. Am. Assoc. Petrol. Geol. Bull. 86, 1531–1542.
Lopponow, G.R., Shoute, L., Schmidt, K.J., Savage, A., Hall, R.H., Bulmer, J.T., 2012. UV Raman spectroscopy of hydrocarbons. Philos. Trans. R. Soc. Lond. A362, 2461–2476.
Magnotti, G., KC, U., Varghese, P.L., Barlow, R.S., 2015. Raman spectra of methane, ethylene, ethane, dimethyl ether, formaldehyde and propane for combustion applications. J. Quant. Spectrosc. Radiat. Transf. 163, 80–101.
Marshall, C.P., Edwards, H.G.M., Jehlicka, J., 2010. Understanding the application of Raman spectroscopy to the detection of traces of life. Astrobiology 10, 229–243.
McCreery, R.L., 2000. Raman Spectroscopy for Chemical Analysis. Wiley-Interscience, New York.
McMillan, P.F., 1985. Vibrational spectroscopy in mineral sciences. In: Kieffer, S.W., Navrotsky, A. (Eds.), Microscopic to Macroscopic: Atomic Environments to Thermodynamic Properties. Reviews in Mineralogy, Mineral. Soc. Am., Washington, DC, pp. 9–63.
McMillan, P.F., 1989. Raman spectroscopy in minerology and geochemistry. Annu. Rev. Earth Planet. Sci. 1989 (17), 255–283.
McMillan, P., Hofmeister, A., 1998. Infrared and Raman spectroscopy. In: Hawthrone, F. (Ed.), Spectroscopic methods in mineralogy and geology. Reviews in Mineralogy, vol. 18, pp. 99–159.
McMillan, P.F., Dubessy, J., Hemley, R., 1996. Applications in earth, planetary and environmental sciences. In: Turrell, G., Corset, G. (Eds.), Raman Microscopy. Academic Press, pp. 289–351.
Mehta, B., Mehta, M., 2005. Organic Chemistry, first ed. Prentice-Hall of India Pvt. Ltd.
Mernagh, T.P., Wilde, A.R., 1989. The use of the laser Raman microprobe for the determination of salinity in fluid inclusions. Geochim. Cosmochim. Acta 53, 765–771.
Mitsui, M., Ando, N., Nakajima, A., 2007. Mass spectrometry and photoelectron spectroscopy of Tetracene cluster anions, (Tetracene) ($n = 1 - 100$): evidence for the highly localized nature of polarization in a cluster analogue of Oligoacene crystals. J. Phys. Chem. A 111 (39), 9644–9648. https://doi.org/10.1021/jp076134h.

Montillaud, J., Joblin, C., Toublanc, D., 2013. Evolution of polycyclic aromatic hydrocarbons in photodissociation regions. Hydrogenation and charge states. Astron. Astrophys. 552, 15.
Munz, I.A., 2001. Petroleum inclusions in sedimentary basins: systematics, analytical methods and applications. Lithos 55, 195–212.
Munz, I.A., Iden, K., Johansen, H., Vagle, K., 1998. The fluid regime during fracturing of the Embla field, central trough, North Sea. Mar. Pet. Geol. 15 (8), 751–768.
Nai, Z., ZuoJi, T., YingYang, L., HuiTong, W., FuQing, A., JianHua, M., 2007. Raman characteristics of hydrocarbon and hydrocarbon inclusions. Sci. Chin Ser. D Earth Sci. 50, 1171–1178.
Nasdala, L., Smith, D.C., Kaindl, R., Zieman, M.A., 2004. Raman spectroscopy: analytical perspectives in mineralogical research. In: Beran, A., Libowitzky, E. (Eds.), Spectroscopic methods in mineralogy: EMU notes in mineralogy, vol. 6, pp. 281–343.
Orange, D., Knitile, E., Farber, D., Williams, Q., 1996. Raman spectroscopy of crude oils and hydrocarbon fluid inclusions: a feasibility study. In: Dyar, M.D., McCammon, C., Scaefer, M.W. (Eds.), Mineral spectroscopy: a tribute to Roger G. Burns. vol. 5. Geochem. Soc. Spec. Publ, pp. 65–81.
Panczer, G., De Ligny, D., Mendoza, C., Gaft, M., Seydoux-Guillaume, A.-M., Wang, X., 2012. Raman and fluorescence. EMU Notes Mineral. 12, 61–82. Chapter 2.
Pasteris, J.D., Kuehn, C.A., Bodnar, R.J., 1986. Applications of the laser Raman microprobe RAMANOR U-1000 to hydrothermal ore deposits: Carlin as an example. Econ. Geol. 81, 915–930.
Pasteris, J.D., Wopenka, B., Seitz, J.C., 1988. Practical aspects of quantitative laser Raman microprobe spectroscopy for the study of fluid inclusions. Geochim. Cosmochim. Acta 52, 979–988.
Piranon, J., Pradier, B., 1992. Ultraviolet-fluorescence alteration of hydrocarbon fluid inclusions. Org. Geochem. 18, 501–509.
Pironon, J., Bourdet, J., 2008. Petroleum and aqueous inclusions from deeply buried reservoirs: experimental simulations and consequences for overpressure estimates. Geochim. Cosmochim. Acta 72, 4916–4928.
Pironon, J., Thiéry, R., Teinturier, S., Walgenwitz, F., 2000. Water in petroleum inclusions: evidence from Raman and FT-IR measurements, PVT consequences. J. Geochem. Explor. 69–70, 663–668.
Raman, C.V., 1928. A new radiation. Ind. J. Phys. 387.
Raman, C.V., Krishnan, K.S., 1928. The optical analog of the Compton effect. Nature 121, 711.
Roberts, S., Beattie, I., 1995. Micro-Raman spectroscopy in the Earth Sciences. In: Microprobe Techniques in the Earth Sciences. Minerological Association Series. Chapman & Hall, London, pp. 387–408.
Roedder, E., 1984. Fluid inclusions. Rev. Mineral. Geochem. 12, 1–644.
Rosasco, G.J., Roedder, E., Simmons, J.H., 1975. Laser-excited Raman spectroscopy for nondestructive partial analysis of individual phases in fluid inclusions in minerals. Science 190, 557–560.
Rull, F., 2012. The Raman effect and the vibrational dynamics of molecules and crystalline solids. EMU Notes Mineral. 12, 1–60. Chapter 1.
Ryder, A.G., Przyjalgowski, M.A., Feely, M., Szczupak, B., Glynn, T.J., 2004. Time-resolved fluorescence micro spectroscopy for characterizing crude oils in bulk and hydrocarbon-bearing fluid inclusions. Appl. Spectrosc. 58, 1106–1115.
Schoenherr, J., Littke, R., Urai, J.L., Kukla, P.A., Rawahi, Z., 2007a. Polyphase thermal evolution in the infra-Cambrian Ara Group (South Oman Salt Basin) as deduced by maturity of solid reservoir bitumen. Org. Geochem. 38, 1293–1318.

Schoenherr, J., Urai, J.L., Kukla, P.A., Littke, R., Schléder, Z., Larroque, J.M., Newall, M.J., Nadia, A.A., Hisham, A.S., Rawahi, Z., 2007b. Limits to the sealing capacity of rock salt: a case study of the infra-Cambrian Ara salt from the South Oman salt basin. Am. Assoc. Pet. Geol. Bull. 91, 1541–1557.
Schopf, J.W., Kudryavtsev, A.B., Agresti, D.G., Czaja, A.D., Wdowiak, T.J., 2005. Raman imagery: a new approach to assess the geochemical maturity and biogenicity of permineralized Precambrian fossils. Astrobiology 5, 333–371.
Seitz, J.C., Pasteris, J.D., Wopenka, B., 1987. Characterization of CO_2-CH_4-H_2O fluid inclusions by microthermometry and laser Raman microprobe spectroscopy: inferences for clathrate and fluid equilibria. Geochim. Cosmochim. Acta 51, 1651–1664.
Shoute, L.C.T., Schmidt, K.J., Hall, R.H., Webb, M.A., Rifai, S., Abel, P., Arboleda, P.H., Savage, A., Bulmer, J.T., Loppnow, G.R., 2002. UV Raman spectroscopy of oilsands-derived bitumen and commercial petroleum products. Appl. Spectrosc. 56, 1308–1313.
Smekal, A., 1923. The quantum theory of dispersion. Naturwissenschaften 11, 873–875.
Smith, E., Geoffrey, D.G., 2005. Modern Raman Spectroscopy—A Practical Approach. John Wiley & Sons Ltd, England.
Socrates, G., 2001. Infrared and Raman Characteristic Group Frequencies: Tables and Charts, third ed. J. Wiley and Sons, Chichester.
Som, S.M., Foing, B.H., 2012. Thermal degradation of organic material by portable laser Raman spectrometry. Int. J. Astrobiol. 11, 177–186.
Srivastava, A., Singh, V.B., 2007. Theoretical and experimental studies of vibrational spectra of naphthalene and its cation. Ind. J. Pure Appl. Phys. 45, 714–720.
Tarcea, N., Popp, J., 2012. Chapter 5. Raman data analysis. In: Dubessy, J., Caumon, M.-C., Rull, F. (Eds.), Raman Spectroscopy Applied to Earth Sciences and Cultural Heritage. EMU Notes Mineral, 12, pp. 193–226.
Teinturier, S., Pironon, J., 2004. Experimental growth of quartz in petroleum environment. Part I: procedures and fluid trapping. Geochim. Cosmochim. Acta 68, 2495–2507.
Touray, J.C., Beny-Bassez, C., Dubessy, J., Guilhaumou, N., 1985. Microcharacterization of fluid inclusions in minerals by Raman microprobe. Scan. Electron Microsc. (I), 103–118.
Verma, D., Jadav, G.N., Biswal, T.K., Jena, S.K., Sharma, N., 2012. Characterisation of hydrocarbon-bearing fluid inclusion in sandstones of Jaisalmer Basin, Rajasthan: a preliminary approach. J. Geol. Soc. Ind. 80, 505–514.
Vratislav, H.V., Huraiová, M., Slobodník, M., Thomas, R., 2015. Geofluids. Developments in Microthermometry, Spectroscopy, Thermodynamics and Stable Isotopes, first ed. Elsevier, ISBN: 978-0-12-803241-1.
Wang, A., 1999. Some grain size effects on Raman scattering intensity for in situ measurements on rocks and soils—experimental tests and modeling. Lunar Planet. Sci. 00. abstract 1644.
Wang, A., Haskin, L.A., Cortez, E., 1998. Prototype Raman Spectroscopic Sensor for in situ mineral chacrterisation in planetary surfaces. Appl. Spectrosc. 52 (4), 477–487.
Wood, R.W., 1930. Improved technique for the Raman effect. Phys. Rev. 36, 1421.
Wopenka, B., Pateris, J.D., Freeman, J.J., 1990. Analyis of individual fluid inclusions by Fourier transform infrared and Raman microspectrometry. Geochim. Cosmochim. Acta 54, 519–533.
Zhou, Q., Xiao, X.M., Tian, H., Wilkins, R.W.T., 2013. Oil charge history of bitumens of differing maturities in exhumed Palaeozoic reservoir rocks at Tianjingshan, NW Sichuan Basin, southern China. J. Pet. Geol. 36, 363–382.
Zhou, Q., Xiao, X.M., Pan, L., Tian, H., 2014. The relationship between micro-Raman spectral parameters and reflectance of solid bitumen. Int. J. Coal Geol. 121, 19–25.
Zhu, Z.Y., Gu, R.A., Lu, T.H., 1998. Applications of Raman Spectroscopy in Chemistry. Northeast University Press, China, p. 171.

CHAPTER 8

Conclusions and future perspectives in HCFI analysis

V. Nandakumar[a] and J.L. Jayanthi[b]
[a]Scientist - G, National Facility for Geofluids Research and Raman Analysis, National Centre for Earth Science Studies (NCESS), Thiruvananthapuram, Kerala, India
[b]Project Scientist - C, National Facility for Geofluids Research and Raman Analysis, National Centre for Earth Science Studies (NCESS), Thiruvananthapuram, Kerala, India

The study of fluid inclusions has played a pivotal role in greatly improving our understanding of the earth processes in which the fluid phase participate. It includes ore deposit genesis, metamorphism, igneous processes, diagenesis, catagenesis, petroleum migration, etc. Recently, earth scientists have shown interest in understanding the role of fluid phase in various geological processes, and the most direct approach to the study of ancient crustal fluids is through the analysis of fluid inclusions. The interaction of paleo fluids at different stages of basin development is significant in the formation and migration of hydrocarbon fluids as a process subsequent to compaction, diagenesis, and catagenesis and is also responsible for the world's significant petroleum reserves. Fluid inclusion analysis in the sedimentary basin is significant in petroleum exploration activities and our efforts are to bring out the relevant information based on the analysis of HCFIs: ways and methods to process the sedimentary rock samples, preparation of wafers to clearly decipher the HCFIs from the associated aqueous fluid inclusions, and the various methods to analyze the HCFIs using optical techniques.

The study of fluid inclusions can help in understanding the diagenetic controls on porosity evolution and thermal history. Further, fluid inclusion work has been tried for understanding the pathways of hydrocarbon migration and is used to determine the hydrocarbon recharge/expulsion history with the help of basin modeling techniques. Petroleum fluoresces with UV illumination, and the color of fluorescence is linked to maturity and API gravity (Lumb, 1978; McLimans, 1987; Tsui, 1990), thereby providing an important clue to the composition of oil in the HCFIs. The fluorescence emission ratio (F620/F560) was also found to be useful to determine

Hydrocarbon Fluid Inclusions in Petroliferous Basins
https://doi.org/10.1016/B978-0-12-817416-6.00007-1

the unknown API gravity of oils in micron-sized HCFIs and the correlation with fluorescence emission and API gravity was also established (Nandakumar and Jayanthi, 2016).

Petrographic analysis of fluid inclusions in the context of understanding the diagenetic history has also proven to be useful. Individual events of petroleum migration can be recognized by cement superpositional or crosscutting relationships (Goldstein, 2001). A detailed study of crosscutting relationships, together with homogenization temperature measurements, is quite useful in understanding reservoir systematics. Furthermore, the extraction and analysis of hydrocarbon fluid inclusions provide useful information for determining the hydrocarbon migration history (Hall et al., 1996; Goldstein & Reynolds, 1994) and for evaluating the composition of hydrocarbons. Recently, the utility of a 785 nm diode laser for the determination of Raman shifts in natural HCFIs was also reported (Jayanthi et al., 2017). A thorough understanding of the PVTX properties of fluids is essential for determining the phase behavior of fluid inclusions and estimating the conditions of entrapment. In recent years, attempts have been made to interpret the PVT relationships of oil in fluid inclusions to integrate with petroleum system modeling for the better evaluation of temperature and pressure of entrapment.

8.1 Futuristic applications of HCFI analysis

Dry wells in terms of oil findings, drilling of near missed oil pools, and an assessment on the efficient production/hydrocarbon potential in established reserves are the major challenges associated with petroleum exploration. The development in fluid inclusion studies can address some of these challenges such as petroleum migration, hydrocarbon quality, timing of migration and migration mechanism, effective sealing of reservoirs over geologic time, etc. and petroleum geoscientists can continuously focus on developing fluid inclusion techniques. The petroleum industry has to emphasize the utility of such a versatile technique for which the main bottleneck is because of the limited understanding of it and the lack of awareness on the fluid phase relationships in petroleum fluid inclusions.

The analysis of hydrocarbon fluid inclusions and its application to exploration and reservoir appraisal has experienced rapid development during the last decade. The traditional correlation studies can also include fluid inclusion data. Analytical developments enable the characterization of maturity and source facies of petroleum, which can be used together with data on fluid samples and core extracts. This has led to the recognition of early charge histories with both lower maturity and different sources involved (Karlsen et al., 1993; Lisk et al., 1996; George et al., 1996, 1997, 1998). The developments have also been directed toward an improved understanding of the physical properties of petroleum inclusion fluids by including microthermometry, volumetric measurements by confocal laser scanning microscopy and chemical data in PVT modeling (Aplin et al., 1999; Munz et al., 1999a,b). These methods are in their early stages concerning precision and accuracy. Nevertheless, data necessary for fluid flow modeling, such as fluid densities, viscosities, pressure, and temperature, can be determined. Numerical modeling of reservoir filling histories, such as the models of England (1989), England and Mackenzie (1989) and England et al. (1987, 1995), can therefore be constrained by analytical data from fluid inclusions.

8.1.1 Petroleum migration

Documentation of petroleum inclusions in basins with limited well control, reservoir-equivalent outcrop samples, or dry holes in adjoining structures can very well identify carrier beds to understand whether hydrocarbons have moved through targeted stratigraphic sections. Detection can be made possible using UV-based thin section studies or bulk, destructive screening techniques like GC, MS, etc. These observations can define the stratigraphic horizons through which migration has occurred, thereby allowing focusing of exploration efforts toward appropriate lithologies and toward assessment of the probable reservoir quality in these lithologies. Alternative sources of petroleum, such as the inclusions inherited in clastic rocks intercalated with shales and the beds that can act as source and reservoir horizons containing fluid inclusions, can also be studied.

8.1.2 Biomarker analysis of liquid petroleum inclusions

Biomarker analysis of liquid petroleum inclusions correlates HCFIs to source lithologies or oil in production platforms, thereby placing useful hints on migration pathways and information on the distribution of source rock kitchen and maturity at expulsion. These analyses may help to identify new play concepts by documenting unexpected source rocks and migration histories. Recent advances like Raman spectroscopic analysis have allowed meaningful organic geochemical data to be extracted from increasingly diverse sample types with low inclusion abundance. Carbon isotopic ratios on dry gas inclusions can aid in distinguishing between biogenic and thermogenic sources. The presence of biogenic gas does not necessarily affect an oil exploration focus, whereas identification of thermogenic gas may indicate an overmature source rock, and this may influence exploration strategies.

8.1.3 Hydrocarbon quality

Once petroleum inclusions are identified, phase relationships observed during microthermometric analyses, coupled with estimated PVT properties of petroleum fluids, can be used to evaluate risk associated with hydrocarbon type. The patented technique developed by NCESS using nondestructive spectroscopic analyses can be employed in identifying the quality of hydrocarbons in terms of its API gravity (Patent No. 315456 dated 03/07/2019). Evidence for earlier oil migration could be observed in reservoirs that currently produce gas and scenarios of in situ cracking of oil to gas vs. displacement and deasphaltening of early oil by later gas can be addressed using fluid inclusion data.

8.1.4 Timing relationships

Relationships between hydrocarbon charging and porosity evolution are important for addressing the risks associated with reservoir quality. Early hydrocarbon charging can sustain porosity in tight lithologies, whereas fracture-controlled migration of petroleum can occur both on a bulk scale and on a mineralogical scale, which can be well understood using secondary fluid inclusion trails. Combining microthermometric data with basin modeling or other geologic information can place constraints on the timing of petroleum charging or the age of migration events.

8.1.5 Reservoir sealing

Evaluation of inclusion abundance and distribution in reservoir horizons can aid in identifying regional or local sealing intervals, as well as their effectiveness over geologic time. This is to be distinguished from instantaneous, present-day sealing capacities that are defined, for instance, by capillary entry pressures of core samples. Evidence for petroleum migration to shallow levels may indicate a lack of an effective seal and suggest loss or dispersal of migrating hydrocarbons.

8.1.6 Migration mechanisms

The distribution of petroleum inclusions is often inconsistent with classical conceptual models of petroleum migration and suggests that thick stratigraphic intervals of rock can be exposed to laterally migrating hydrocarbons. This may indicate that buoyancy forces can be subordinate in controlling petroleum transport, as compared, for instance, to transient total pore pressure fluctuations, subsurface water potential gradients, and/or inhomogeneously distributed capillary forces, all of which can impede or counter the density driven flow.

8.1.7 Fluid inclusions and diagenetic systems

Most diagenetic systems in sedimentary rocks are defined by the temperature, pressure, and salinity of the fluids active in them. Fluid inclusions are samples of the fluids involved during diagenesis and are a record of diagenetic systems. Therefore, fluid inclusions are one of the best tools providing temperature, pressure, and fluid chemistry data on ancient diagenetic systems that cannot be obtained by any other means.

8.1.8 Oil exploration and development

Fluid inclusion techniques can address significant variables, even in microscale that affect petroleum exploration such as the evidence for migration, product type and quality, timing of migration with respect to porosity evolution, and effectiveness of petroleum seals over geologic time, etc. (Hall et al., 1996). However, fluid inclusion techniques remain effectively underutilized to date in exploration activities.

Once hydrocarbon fluid inclusions are identified, with microthermometric observations we are able to determine the conditions of entrapment of the fluids, which gives an understanding of the oil/gas window and source rock maturity that can lead to the nature of the oil. An assessment of API gravity of oils in HCFIs using fluorescence emission techniques also gives a direct understanding of the quality of oil in a reservoir.

The thermal history of sedimentary basins is an important theme for oil companies and there are many techniques that are used, including %Ro, apatite fission track analysis, and numerical simulation (Basinmod 2 D and Petromod 3D). Fluid inclusion data need to be integrated with other techniques used to investigate the thermal history of sedimentary basins. Fluid inclusion data and various results obtained can act as a proxy to routine parameters inputting into PSM.

8.1.9 Fluid correlation in space and time

Fluid transport directions are determined using the tool of mapping of petroleum compositional characteristics. Biomarkers and maturity parameters in the range C_{19}–C_{35} have been vital in fluid correlation studies. The origin of these compounds, in terms of the biological link and thermal behavior, has been the focus of extensive research (Mackenzie, 1984). They are therefore useful for correlation of oils with each other and with their proposed source rocks and for assessment of their thermal maturity (Mackenzie, 1984). Since the biomarkers and maturity parameters occur only as trace components in oils, this technique is not suited when fluids of condensate or wet gas composition are involved. The samples used for such studies include drill stem test (DST) samples and core extracts. Core extracts are obtained by extraction with organic solvents and may include immobile phases, such as tar mats and adsorbed reservoir surface coatings, extracts of indigenous organic matter, in addition to the remains of the reservoir fluids (Larter & Aplin, 1995). The content of nitrogen-sulfur-oxygen bearing (NSO) compounds is commonly higher in core extracts than in sandstone oils (Leythaeuser and Rückheim, 1989; Horstad et al., 1990; Stoddart et al., 1995; Larter & Aplin, 1995; Baylis, 1998). Direct comparison between data from the two types of samples is therefore problematic and needs specific consideration. The literature shows that the gasoline range (C_4–C_{12}) compounds can also be used for fluid correlation (Ten Haven, 1996; Odden, 1999; Odden and van Graas, 1998). The gasoline compounds have the advantage of being abundant in all types of petroleum fluids. The method is especially useful for condensate compositions or even wet gas. Only sandstone samples can be used for this type of study, since core extracts show a loss of light hydrocarbons ($<C_{15}$). Until now, fluid correlation has been limited to the spatial distribution of present-day petroleum.

The expansion of analytical methods for studying petroleum in fluid inclusions opens a new window for fluid correlation with time. Off-line crushing methods can provide biomarkers and maturity parameters. Comparison between data from present-day fluids and fluid inclusions may establish compositional differences in both time and space. This can lead to the recognition of early charge histories with both lower maturity and different sources involved (Karlsen et al., 1993; Lisk et al., 1996; George et al., 1996, 1997, 1998). Online GC analysis of petroleum inclusion fluid is a suitable technique for determination of gasoline range compounds but has so far not been used in correlation studies. With the increasing interest in gasoline range compounds, this technique will be a valuable tool in the future.

8.1.10 PVT characterization

Most fluid inclusion studies are calculating fluid densities and isochores, in order to estimate the pressure and temperature conditions of trapping, either by combining information from solids or from coexisting fluid phases. For petroleum inclusions, it is important to estimate the fluid properties at pressure and temperature conditions other than the trapping conditions. This may include saturation pressure at a specified temperature or the gas-to-oil ratio (GOR) at any specified pressure and temperature, and these calculations can be done by means of an equation of state. The most accurate results for complex petroleum fluids can be obtained by two- or three-constant cubic equations (Firoozabadi, 1988). For a more extensive review of PVT modeling, the textbooks of Pedersen et al. (1989) and Danesh (1998) are recommended.

The total fluid composition in mole percent is the input for calculation of two-phase inclusion using an equation of state. In addition, the critical properties of each component must be characterized, and binary interaction coefficients between the different components must be known. Homogenization temperature is necessary for isochore calculation. The characterization of the fluid composition is the most delicate aspect for PVT modeling. Reservoir petroleum fluids normally contain hundreds of different components. To treat all components separately is impossible, and therefore components with similar physical characteristics are instead grouped together as pseudo-components. Critical properties for the pseudo-components (i.e., single-carbon-number groups and the residue) can be derived from the molecular weights and densities by various correlation schemes (Kesler & Lee, 1976; Riazi & Daubert, 1987).

Although analysis of petroleum inclusions has made enormous progress in the last decade, sufficient information on the composition of petroleum inclusion fluids for PVT modeling appears to be problematic to achieve. Isochores have been calculated by combining microthermometry data with compositional data of present-day hydrocarbon fluids (Narr & Burruss, 1984; McLimans, 1987; Bodnar, 1990; Munz et al., 1999a). This is, however, only a valid approach for cases with recent reservoir filling where the composition of the inclusion fluid is approximately the same as the present-day reservoir fluid. Problems arise with this method when the composition of the inclusion fluid is not identical to the present-day reservoir fluid. This can partly be solved by an iterative approach (Aplin et al., 1999) where it is assumed that the inclusion fluid is most accurately modeled if the fluid, which is used as input for the calculations, is genetically related.

A model composition of the inclusions is calculated by using the present-day reservoir fluid composition as a regression parameter. The necessary data from fluid inclusion analyses are microthermometry and the volumetric ratio between vapor and liquid at room temperature, determined by confocal laser scanning microscopy.

A procedure for compositional characterization of inclusion petroleum, online GC based, was suggested by Munz et al. (1999b). It is like the procedure used for present-day fluids in reservoir engineering. Compounds in the range from C_1 to C_5 were treated as individual components. The range from C_6 to C_{30+} was divided into pseudo-components by integrating ranges instead of peaks. Pseudo-components were defined as single-carbon-number groups, and the average values from the literature for the critical properties of these pseudo-components were used. The GC analyses show a high precision for the gas components, but are more problematic for the heavy end. A more accurate result may therefore be obtained by using properties of the heavy components as a tuning parameter, rather than the gas content.

Petroleum always coexists with an aqueous fluid phase in sedimentary basins. Even in reservoirs with a high petroleum saturation, the formation of water cannot be entirely displaced. Fluid immiscibility relationships must therefore be considered. Solubility of water in petroleum has been disregarded in the methods described above, based on two arguments (Munz et al., 1999b): The solubility of water in petroleum is minimum and the properties of water would not be significantly different from the more polar compounds of petroleum. For the heavy components of petroleum, omission of the very small amounts of dissolved water, therefore, will not significantly affect the results. Even though the solubility of petroleum compounds in water is low, it is not to be neglected. The highest solubilities are found for methane and light aromatics. The properties of methane are significantly different from those of water. Dissolved methane in water may therefore change the PVT properties of the aqueous fluids (Hanor, 1980), and need to be considered in the estimation of trapping pressure and temperature conditions. The pressure and temperature conditions for the trapping of two coexisting fluid phases with no mutual solubility can be estimated by the intersecting isochore principle (Roedder & Bodnar, 1980). For two fluids with a mutual solubility, however, the criterion for immiscibility is equal to the homogenization temperature, where the low density fluid homogenizes to the vapor phase and the high-density fluid to the liquid phase (Roedder, 1984; p. 256). Ideally, coexisting petroleum and aqueous inclusions should therefore show the same homogenization temperature. Numerous examples show that petroleum inclusions typically homogenize at lower temperature than the coexisting aqueous inclusions (Nedkvitne et al., 1993; Munz et al., 1998, 1999a). The reason for this is that the mutual solubility is too small to be observed by microthermometry. The analyzed homogenization of aqueous inclusions represents the final homogenization. Phase transitions (e.g., partial homogenization, clathrate melting) related to dissolved methane should take place at low temperature, but are commonly difficult to observe. The homogenization of petroleum inclusions, however, represents only a partial homogenization. The final homogenization would have been dissolution of the aqueous phase into the then single-phase petroleum. Again, due to the extremely small amounts of water, this phase transition has not been reported. Observations of an aqueous fluid within the petroleum inclusions generally relate to heterogeneous trapping. The final homogenization temperature does not have any relevance and cannot be measured without decrepitation. Immiscibility is therefore difficult to show from microthermometry, and recognition of equilibrated, coexisting petroleum and aqueous inclusions must rely on textural criteria only. If textural criteria indicate the contemporaneous trapping of petroleum and aqueous inclusions, methane saturation of the aqueous inclusions can be assumed, and the aqueous homogenization temperature represents the trapping temperature for both types of inclusions (Nedkvitne et al., 1993). The trapping pressure may then be found by intersection between the petroleum inclusion isochore and the trapping temperature. Petroleum fluids may also experience phase separation at geological conditions. A situation with three immiscible fluid phases (gas, oil, water) will be the consequence of phase separation, and, again, this should be recognized by equal homogenization temperatures. Munz et al. (1999a) showed that the difference in homogenization temperature between cogenetic aqueous and petroleum inclusions is an important qualitative indicator for the degree of undersaturation of the petroleum inclusions (oil or gas). In a situation where an undersaturated oil gets saturated with gas, the homogenization temperature will be that of the coexisting aqueous inclusions (Munz, 2001).

8.1.11 Geochemical analytical techniques applicable to fluid inclusions

Petroleum is a mixture of many organic compounds and is commonly analyzed from the organic and inorganic points of view through a variety of techniques. In particular, the optical parameter of Vitrinite Reflectance is widely utilized to obtain indications on the maturity of potential source rocks. Organic compounds are usually analyzed by Gas Chromatography and Pyrolysis. A more detailed analysis of organic components may be carried out by means of gas chromatographic mass spectrometry. Atomic absorption spectrophotometry and inductively coupled plasma spectrometry are commonly utilized to identify inorganic components at trace level concentrations. Mass Spectrometry is also widely utilized for isotopic investigations of crudes and natural gases. Devices, such as sniffers, etc., are also utilized during surface geochemical prospecting (Martinelli, 2003).

8.2 Integration of fluid inclusion data with petroleum system modeling

Demaison and Huizinga (1994) defined the petroleum system as a 'dynamic petroleum generating and concentrating physicochemical system functioning in geologic space and time.' Integrating the concept of petroleum system with sedimentary basins is a recently evolving research topic. It may be useful to first consider the distribution of petroleum in the various types of basins and then the vertical distribution of petroleum within basins.

Distribution of hydrocarbons within basins follows a regular pattern. Considered vertically, oil gravity decreases with depth. Heavy oils tend to be shallow and, with increasing depth, pass down into light oils, condensate, and finally gas, until the point at which hydrocarbons and porosity are absent. Oils tend to become lighter not only downward but also laterally toward a basin center. Typically, heavy oils occur around basin margins, and condensate and gas in the center. The causes of these vertical and lateral variations of oil and gas are of considerable importance.

The physical and chemical processes involved in the generation, migration, and accumulation of petroleum are well understood to such an extent that their potentiality can be quantified as a function of the geological history of a sedimentary basin. Basin modeling by numerical simulations has become a routine rather than a valuable tool for petroleum exploration activities. Sophisticated software packages have been developed for deterministic forward models which allow the temporal reconstruction of the history of a sedimentary basin and the concomitant evolution of most of the processes relevant to the formation of petroleum accumulations (Welte et al., 1997). In order to undertake a basin modeling study, a number of input parameters are to be obtained from well data. Invariably, for certain parameters, assumptions based on geological reasoning can be made. The required input parameters (Poelchau et al., 1997) for petroleum system modeling can include:- rate of subsidence, uplift and deformation of sedimentary basin fill; depositional conditions and types of sediments (sedimentation rates, environments, organic matter accumulation, etc.); age and thickness of all sedimentary strata; rock properties such as porosity, permeability, density, thermal conductivity, heat capacity, and compressibility; fluid properties (in water, oil, and gas) such as composition, density, viscosity, etc.; thermal history of the basin, i.e., heat input from the basement and conditions for conductive and convective heat transport, etc.; kinetics of organic matter transformation in source rocks; hydrodynamics and conditions for fluid redistribution; trap formation, e.g., by tectonic processes, and their timing.

Basin modeling can be performed in one dimension (for a single well), in two dimensions (for a geologic profile combining several wells), or even in three dimensions. The main advantages of basin modeling, if integrated with seismic information, are twofold, i.e., quantitative predictions as to where petroleum accumulations can be expected in a sedimentary basin. Most importantly, the temporal evolution of all relevant processes and their effects can be reconstructed. The maps for individual geological time periods can be obtained which show regional variations in the maturity of the source rocks (areas where they are immature, mature, or overmature). The superposition of these maps, which show different aspects (like type, size, and distribution of potential traps) of structural evolution for the same time period, allow to rank petroleum prospects.

8.2.1 Thermal modeling-thermal calibration

In petroleum system modeling software (PSM) like Petromod (PetroMod 2018), after the first thermal calibration of the model, the results are further improved by using the Heat Flow Calibration tool (either use the model with the assigned heat flow maps from the McKenzie Crustal Model (3D Maps) or the model with the Trend Assignment Maps). The present borehole temperature is used for present-day thermal calibration and VRo and Tmax are used for paleothermal history calibration in such software. The paleo-temperature estimated from the fluid-inclusion study provides extra support for thermal history modeling. Fluid inclusion data help in calculating the paleo-temperature (the Th data obtained during microthermometric analysis) that can be used in PSM to build the thermal Model. In geologically older Proterozoic basins, where VRo data could never be obtained (e.g., Vindhyan basin, Kadappa basin, etc. in India), fluid inclusion data can help to predict the maturity that can further be used in thermal maturity modeling. The fluid inclusion study also can derive the maturity of oil even in case of dry wells or wells with limited oil presence, which can be an aid to predict the depocenter and can be corroborated with PSM to make a robust model.

The Th data obtained through fluid inclusion studies can be used to calibrate the thermal models generated using PSM software (Fig. 8.1). Further, a thermal history plot of a dry well we have been discussing throughout in the book (RV-1, a dry well from the Mumbai offshore basin) has also been generated using fluid inclusion Th data (Fig. 8.2) in Petromod. The plot explains the Th variations in the basin throughout the period and in different formations. The Th falls within the oil window in the Panna Formation where we got good oil presence as hydrocarbon fluid inclusions. This brings out the wide range utility of fluid inclusion data in PSM for thermal calibration, and our studies are ongoing with this aspect.

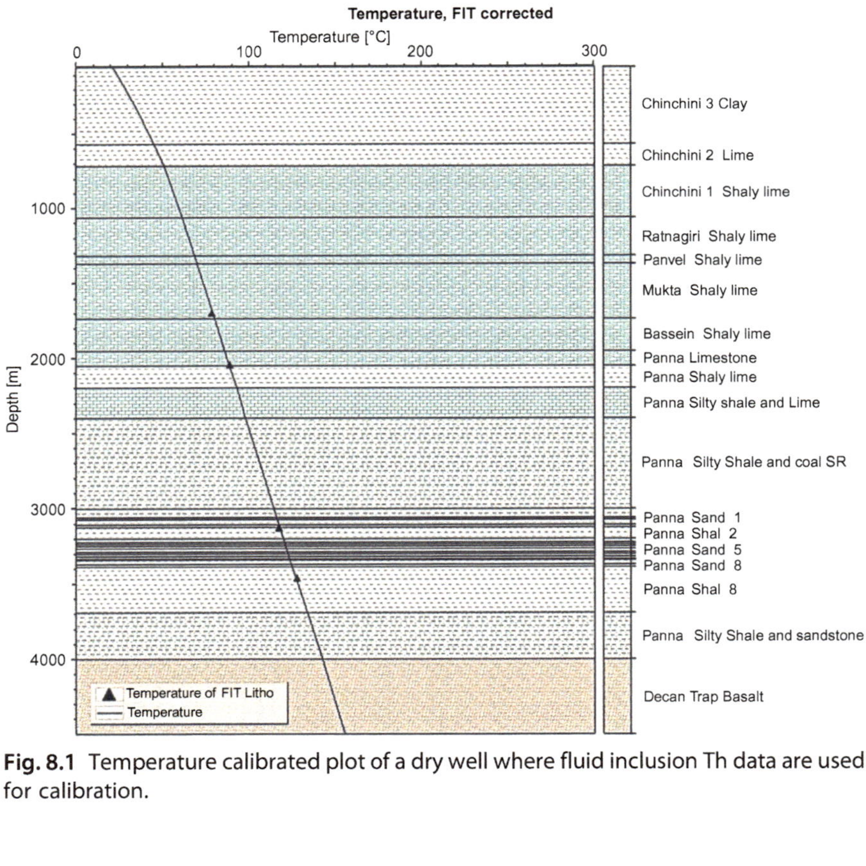

Fig. 8.1 Temperature calibrated plot of a dry well where fluid inclusion Th data are used for calibration.

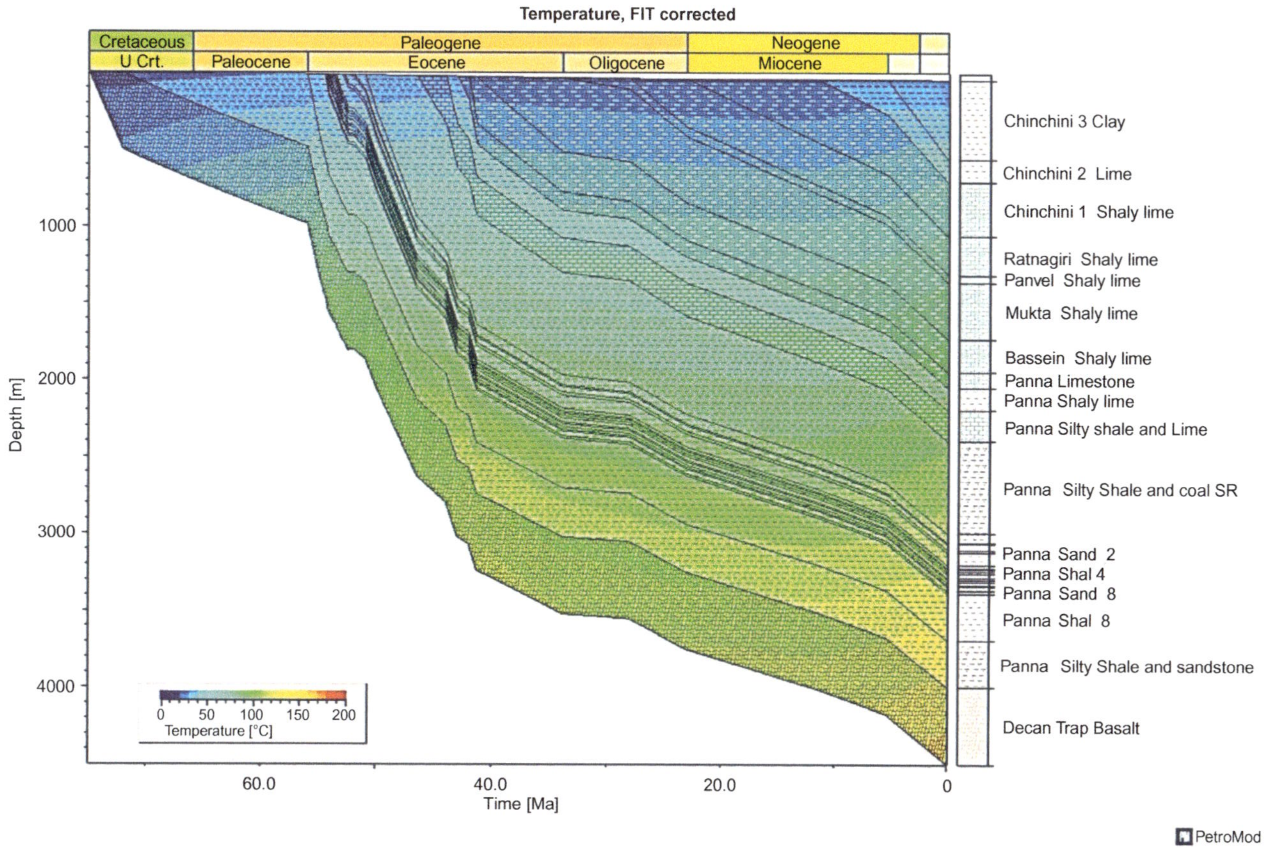

Fig. 8.2 Thermal history plot of the RV-1 well of the Mumbai offshore basin with Petromod software with the calibrated fluid inclusion Th data.

8.2.2 Compositional modeling-calibration of API/GOR

Compositional calibration in PSM is possible using the derived API gravity data obtained through fluorescence studies as discussed in Chapter 6. API gravity is an arbitrary scale adopted by the American Petroleum Institute for expressing the density of oils. Crude oils can be classified according to API wherein the higher the API gravity, the lighter the oil. Fig. 8.3 shows an example of compositional modeling using API/GOI data in PSM. The giant accumulation in the four-component master model in Fig. 8.3 has an API of about 55 and a GOR of 31.95. The field data obtained for this heavier oil are an API of 30 and a GOR of 100. So calibration is needed for the fluids of this accumulation to its measured properties. API/GOR calculation from the fluid Inclusion study can be used for the calibration of petroleum system modeling. This calibrated compositional model would help in the prediction of hydrocarbon composition.

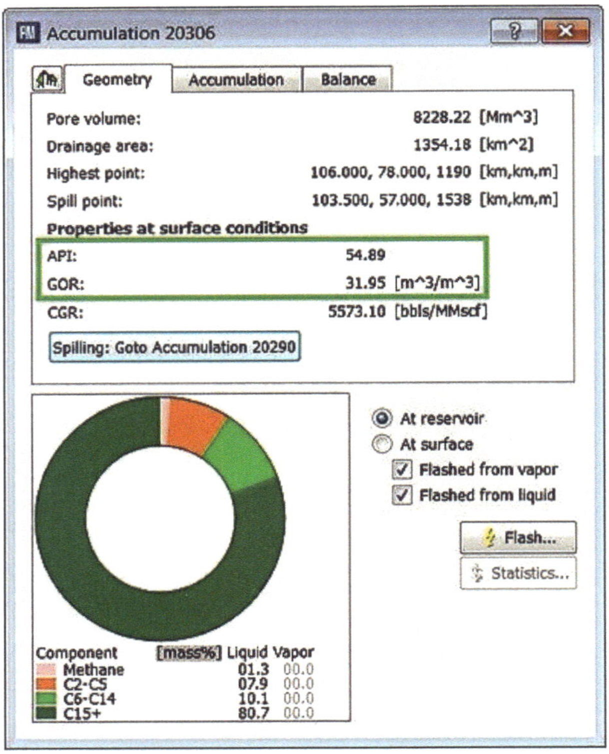

Fig. 8.3 Example of a compositional calibration model using PSM where the fluid inclusion API/GOR data can predict the hydrocarbon composition.

In case of no/little discovery in exploration blocks (e.g., Kutch-Saurashtra, Kerala-Konkan in India or in a less explored part of any basins in the world), the fluid inclusion study gives a detailed insight of API/GOR information of the expected hydrocarbon. The derived data of API/GOR through fluid inclusion studies also can be used in the PSM to calibrate the PVT model. The reservoir engineers can make use of the fluid inclusion data to estimate the predicted production profile for value assessment, and the cash flow model of prospects can further use the calibrated PVT model. The production profile and reservoir flow simulation model, created based on the fluid properties from integrated fluid inclusion and the PSM study, can help in placing and optimizing the exploration wells. The composition prediction at the exploration stage also helps in calculating the economics of the prospect for investment purposes.

8.2.3 Migration modeling

Migration modeling is one of the high-end tools to predict the hydrocarbon presence in the exploration and development stage (High Resolution LGR Modeling). PSM offers a different migration mechanism in the exploration stage for the prediction—hybrid, flowpath, and invasion percolation. The fluid inclusion study can help track the migration pathway by identifying the carrier beds (migrating secondary hydrocarbon trails) even where oil discovery has not happened. In less explored areas, the fluid inclusion study can help in establishing the possibility of having an active petroleum system. However, the fluid inclusion study can bring about a revolution to calibrate the migration pathways of the model. Hence, the fluid inclusion study can provide a new avenue to the Oil and Gas industry for finding new plays for future exploration.

8.3 Insight—A dream to reality

We emphasize the need for utilizing fluid inclusion studies as a tool for oil exploration activities. The types of information that can be obtained through the study are vital inputs for finding out the composition and P–T estimation aiding to unravel the petroleum system and even calibrating modeling software. Fluid inclusion techniques remain underutilized in the petroleum industry because of the unknown significance and limited understanding of fluid phase relationships within petroleum fluid inclusions. The proliferation of fluid inclusion studies aimed at the exploration of petroleum can be helpful to locate oil pools with more certainty and aid in the competent production of the wells.

Sedimentation in basins spanning millions of years will experience several subterranean fluid processes like migration of oils and charging of reservoirs leaving true samples of the prevailing fluids as inclusions in authigenic growths, diagenetic cements and as secondary trails in minerals forming the rocks. These inclusions are authentic representatives of the pore fluids that moved through different formations that are archived in the geological formations, even though the original fluids must have migrated away in the past geologic times. The petroleum industry can be supported by fluid inclusion studies as they can yield information on the evolution, migration, and entrapment of petroleum. Hydrocarbon fluid inclusions (HCFIs) are true representatives of the fluids without much alteration and evade any further infiltration.

Application of fluid inclusion techniques in the sedimentary realm got a fillip in recent times and the pioneers in the field are Prof. RH Goldstein (University of Kansas), Dr. J. Parnell (University of Aberdeen, UK), Prof. Dag Karlsen (University of Oslo), Prof. Eadington (CSIRO), Dr. Jan Kihle (IFE, Norway) Prof. Robert C. Burrus (USGS), Dr. Jan Kihle (IFE, Norway), Prof. Alan Ryder and Prof. Martin Feely (NUI, Ireland), etc. Unsuccessful attempts to develop fluid inclusion techniques were made in India until 2010. The repeated failure of ONGC in striking oil in the Kerala Offshore Basin, where all the 15 wells drilled did not yield oil, was the impetus to develop a fluid inclusion technique by the Center for Earth Science Studies, Thiruvananthapuram, India. Conducive lithology, depocenter thickness, %Ro, Rock-eval pyrolysis data giving indications as to the type of organic matter and thermal maturity existing in the Kerala basin compared to the Mumbai offshore basin led to the belief that oil should and may exist in trace quantities. Hence, the technique of fluid inclusions was thought of to detect oil and to assess its quality through spectroscopic techniques, and a project in this regard was submitted to the Ministry of Earth Sciences, Government of India by the lead author.

Though trained in the application of fluid inclusion techniques in hard metamorphic rocks, its application in sedimentary rocks posed a big challenge. The Oil and Natural Gas Corporation, Government of India supplied samples at 5 m intervals from two offshore exploratory wells, viz., RV-1 well from the Mumbai Basin and KK-4C-A-1 well from the Kerala-Konkan Basin. Wafer preparation techniques were developed and several rounds of discussions made with Dr. Jan Kihle (IFE, Norway) and Prof. Alan Ryder

(NUI, Ireland) in assembling the hardware. Special wafer preparation techniques were developed to suppress the background fluorescence and to gather specifically the fluorescence from oil inclusions. Serious petrographic work ensued, and we were able to decipher late-stage fluid inclusions of sedimentary origin (smaller size of the order 5–10 μm, lots of Brownian motion, presence in the authigenic zones and in secondary trails crosscutting cements, grain boundaries, and specifically at a lower Th max of 140°C) compared to the higher Th max of greater than 200°C, in situ ones from the provenance lithology of the sediments. Hydrocarbon fluid inclusions and the coeval aqueous inclusion assemblages, mostly with high salinity, were targets. The Microlaser Raman Spectroscope was put to use in two specific areas: (1) Fluorescence analysis using emission spectra (2) Species identification using Raman analysis.

All the objectives envisaged were achieved. The quest for oil offshore to Kerala was the driving force behind the success of the research program. When the lead author started to consider sedimentary basins, he read widely about sedimentary basins, its evolution, and the properties of hydrocarbon fluid inclusions. Some of the publications that helped were Goldstein and Reynolds (1994), Kihle (1995), North (1985), Munz (2001), Goldstein (2001), Burruss (2003), Blamey and Ryder (2009), McLimans (1987), etc. We convey our experience in HCFIs through this book and hope this could be useful for the global scientific community being introduced into the area of HCFI research and make the readers understand about the utility of HCFIs in future petroleum exploration activities. An insight into the various tools that are recently used and developed for deriving vital information from fluid inclusions is detailed in this book that could be used for furtherance of the research activities in this area.

References

Aplin, A.C., Macleod, G., Larter, S.R., Pedersen, K.S., Sørensen, H., Booth, T., 1999. Combined use of confocal laser scanning microscopy and PVT simulation for estimating the composition and physical properties of petroleum in fluid inclusions. Mar. Pet. Geol. 16, 97–110.

Baylis, S.A., 1998. Geochemical comparison of core extracts and oil samples in reservoir. Org. Geochem. 29, 463–484.

Blamey, N.J., Ryder, A.G., 2009. Hydrocarbon fluid inclusion fluorescence: a review. In: Reviews in Fluorescence 2007. Reviews in Fluorescence, vol. 2007. Springer, New York, NY, p. 2009, https://doi.org/10.1007/978-0-387-88722-7_13.

Bodnar, R.J., 1990. Petroleum migration in the Miocene Monterey formation, California, USA: constraints from fluid-inclusion studies. Min. Mag. 54, 295–304.

Burruss, R.C., 2003. Petroleum fluid inclusions, an introduction. In: Samson, I., Anderson, A., Marshall, D. (Eds.), Fluid Inclusions. Analysis and Interpretation. Mineralogical Association of Canada, Short Course Series, vol. 32, pp. 159–174.

Danesh, A., 1998. PVT and phase behaviour of petroleum fluids. Develop. Petrol. Sci. 47. Elsevier 387 pp.

Demaison, G., Huizinga, B.J., 1994. Genetic classification of petroleum systems using three factors: charge, migration, and entrapment. In: Magoon, L.B., Dow, W.G. (Eds.), The Petroleum System—From Source to Trap. AAPG Memoir, vol. 60, pp. 73–89.

England, W.A., 1989. The organic geochemistry of petroleum reservoirs. Org. Geochem. 16, 415–425.

England, W.A., Mackenzie, A.S., 1989. Some aspects of the organic geochemistry of petroleum fluids. Geol. Rundsch. 78, 291–303.

England, W.A., Mackenzie, A.S., Mann, D.M., Quigley, T.M., 1987. The movement and entrapment of petroleum fluids in the subsurface. J. Geol. Soc. Lond. 144, 327–347.

England, W.A., Muggeridge, A.H., Clifford, P.J., Tang, Z., 1995. Modelling density-driven mixing rates in petroleum reservoirs on geological time-scales, with application to the detection of barriers in the Forties Field (UKCS). In: Cubitt, J.M., England, W.A. (Eds.), The Geochemistry of Reservoirs, vol. 86. Geological Society London Special Publication, pp. 185–201.

Firoozabadi, A., 1988. Reservoir-fluid phase behavior and volumetric prediction with equation of state. J. Petrol. Technol., 397–406.

George, S.C., Lisk, M., Eadington, P.J., Quezada, R.A., Krieger, F.W., Greenwood, P.F., 1996. Comparison of palaeo oil charges with currently reservoired hydrocarbons using the geochemistry of oil-bearing fluid inclusions. In: Paper SPE 36980 Presented at the 1996 SPE Asia Pacific Oil and Gas Conference in Adelaide, Australia 28–31, pp. 159–171.

George, S.C., Krieger, F.W., Eadington, P.J., Quezada, R.A., Greenwood, P.F., Eisenberg, L.I., Hamilton, P.J., Wilson, M., 1997. Geochemical comparison of oil-bearing fluid inclusions and produced oil from the Toro sandstone, Papua New Guinea. Org. Geochem. 26, 155–173.

George, S.C., Lisk, M., Summons, R.E., Quezada, R.A., 1998. Constraining the oil charge history of the south pepper oilfield from the analysis of oil-bearing fluid inclusions. Org. Geochem. 29, 631–648.

Goldstein, R.H., 2001. Fluid inclusions in sedimentary and diagenetic systems. Lithos 55, 159–193.

Goldstein, R.H., Reynold, T.J., 1994. Systematics of Fluid Inclusions in Diagenetic Minerals, SEPM (Society for Sedimentary Geology) Short Course 31, USA ISBN #1–56576-00805.

Hall, D.L., Sterner, S.M., Shentwu, W., Bigge, M.A., 1996. Applying Fluid Inclusions to Petroleum Exploration and Production. Fluid Inclusion Technologies, Inc. USA.

Hanor, J.S., 1980. Dissolved methane in sedimentary brines: potential effect on the PVT properties of fluid inclusions. Econ. Geol. 75, 603–609.

Horstad, I., Larter, S.R., Dypvik, H., Aagaard, P., Bjørnvik, A.M., Johansen, P.E., Eriksen, S., 1990. Degradation and maturity controls on oil field petroleum column heterogeneity in the Gullfaks field, Norwegian North Sea. Org. Geochem. 16, 497–510.

Jayanthi, J.L., Nandakumar, V., Anoop, S.S., 2017. Feasibility of a 785 nm diode laser in Raman spectroscopy for characterizing hydrocarbon-bearing fluid inclusions in Mumbai Offshore Basin, India. Petrol. Geosci. 23 (3), 369–375.

Karlsen, D.A., Nedkvitne, T., Larter, S.R., Bjørlykke, K., 1993. Hydrocarbon composition of authigenic inclusions: application to elucidation of petroleum reservoir filling history. Geochim. Cosmochim. Acta 57, 3641–3659.

Kesler, M.G., Lee, B.I., 1976. Improve prediction of enthalpy of fractions. Hydrocarbon Process 55, 153–158.
Kihle, J., 1995. Adaptation of excitation-emission micro-spectroscopy for characterization of single hydrocarbon fluid inclusions. Org. Geochem. 23, 1029–1042.
Larter, S.R., Aplin, A.C., 1995. Reservoir geochemistry: methods, applications and opportunities. In: Cubitt, J.M., England, W.A. (Eds.), The Geochemistry of Reservoirs, vol. 86. Geol. Soc. London Spec. Publ, pp. 5–32.
Leythaeuser, D., Rückheim, J., 1989. Heterogeneity if oil composition within a reservoir as a reflection of an accumulation history. Geochim. Cosmochim. Acta 53, 2119–2123.
Lisk, M., George, S.C., Summons, R.E., Qezada, R.A., O'Brien, G.W., 1996. Mapping hydrocarbon charge histories: detailed characterisation of the south pepper oil field, Carnavon Basin. APPEA J. 36, 445–464.
Lumb, D.M., 1978. Organic luminescence. In: Lumb, D.M. (Ed.), Luminescence Spectroscopy. Academic Press, New York, pp. 93–148.
Mackenzie, A.S., 1984. Applications of Biological Markers in Petroleum Geochemistry Advances in Petroleum Geochemistry, 1. Academic Press, London, pp. 115–214.
Martinelli, G., 2003. Petroleum geochemistry. In: Petroleum Engineering—Upstream. EOLSS, Reggio Emilia.
McLimans, R.K., 1987. The application of fluid inclusions to migration of oil and diagenesis in petroleum reservoirs. Appl. Geochem. 2, 585–603.
Munz, I.A., 2001. Petroleum inclusions in sedimentary basins: systematics, analytical methods and applications. Lithos 55, 195–212.
Munz, I.A., Iden, K., Johansen, H., Vagle, K., 1998. The fluid regime during fracturing of the Embla field, Central Trough, North Sea. Mar. Pet. Geol. 15, 751–768.
Munz, I.A., Johansen, H., Holm, K., Lacharpagne, J.-C., 1999a. The petroleum characteristics of the Frøy field and the Rind discovery, Norwegian North Sea. Mar. Pet. Geol. 16, 633–651.
Munz, I.A., Johansen, H., Johansen, I., 1999b. Characterisation of composition and PVT properties of petroleum inclusions: Implications of reservoir filling and compartmentalisation. In: Paper SPE 56519 presented at the 1999 SPE Annual Technical Conference and Exhibition in Houston, 3–6 October 1999.
Nandakumar, V., Jayanthi, J.L., 2016. Hydrocarbon fluid inclusions, API gravity of oil, signature fluorescence emissions and emission ratios: an example from Mumbai offshore, India. Energy Fuel 2016 (30), 3776–3782.
Narr, W., Burruss, R.C., 1984. Origin of reservoir fractures in little knife field, North Dakota. AAPG Bull. 68, 1087–1100.
Nedkvitne, T., Karlsen, D.A., Bjørlykke, K., Larter, S., 1993. Relationship between reservoir diagenetic evolution and petroleum emplacement in the Ula field, North Sea. Mar. Pet. Geol. 10, 255–270.
North, F.K., 1985. Petroleum Geology. Allen & Unwin Inc. USA. ISBN: 0-04-553003-3.
Odden, W., 1999. A study of natural and artificially generated light hydrocarbons (C4–C13) in source rocks and petroleum fluids from offshore Mid-Norway and the southermost Norwegian and Danish sectors. Mar. Pet. Geol. 16, 747–770.
Odden, W., van Graas, G.W., 1998. Application of light hydrocarbons (C 4–C 13) to oil/source rock correlations. Org. Geochem. 28 (12), 823–847.
Pedersen, K.S., Fredenslund, A., Thomassen, P., 1989. Properties of Oils and Natural Gases. Gulf Publishing.
Poelchau, H.S., Baker, D.R., Hantschel, T., Horsfield, B., Wygrala, B., 1997. Basin simulation and the design of the conceptual basin model. In: Welte, D.H., Horsfield, B., Baker, D.R. (Eds.), Petroleum and basin evolution. Springer, Berlin/Heidelberg/New York, pp. 5–70.

Riazi, M.R., Daubert, T.E., 1987. Characterization parameters for petroleum fractions. Ind. Eng. Chem. Res. 26, 755–759.
Roedder, E., 1984. Fluid inclusions mineral. Soc. Am. Rev. Min. 12. p. 256.
Roedder, E., Bodnar, R.J., 1980. Geologic pressure determination from fluid inclusion studies Annu. Rev. Earth Planet. Sci. 8, 263–301.
Stoddart, D.P., Hall, P.B., Larter, S.R., Brasher, J., Li, M., Bjorøy, M., 1995. The reservoir geochemistry of the Eldfisk field, Norwegian North Sea. In: Cubitt, J.M., England, W.A. (Eds.), The Geochemistry of Reservoirs, vol. 86. Geological Society London Special Publications, pp. 257–279.
Ten Haven, H.L., 1996. Applications and limitations of Mango's light hydrocarbon parameters in petroleum correlation studies. Org. Geochem. 24, 957–976.
Tsui, T.F., 1990. Characterizing fluid inclusion oils via UV fluorescence microspectrophotometry—a method for projecting oil quality and constraining oil migration history (abstract). AAPG Bull. 74, 781.
Welte, D.H., Horsfield, B., Baker, D.R. (Eds.), 1997. Petroleum and Basin Evolution. Springer-Verlag, Berlin. 535 pp.

Index

Note: Page numbers followed by *f* indicate figures, and *t* indicate tables.

F

I

K

Kolibri 360 GmbH
Ferdinand-Jühlke-Straße 7
99095 Erfurt, Deutschland
produktsicherheit@kolibri360.de